COMMUNICATIONS SATELLITE DEVELOPMENTS: SYSTEMS

Edited by
Gilbert E. LaVean
Defense Communications Agency
Reston, Virginia

William G. Schmidt
CML Satellite Corporation
Washington, D. C.

Volume 41
PROGRESS IN
ASTRONAUTICS AND AERONAUTICS

Martin Summerfield, Series Editor-in-Chief
Princeton University, Princeton, New Jersey

Technical papers selected from the AIAA 5th Communications Satellite Systems Conference, April 1974, subsequently revised for this volume.

Published by the American Institute of Aeronautics and Astronautics in cooperation with The MIT Press

American Institute of Aeronautics and Astronautics
New York, New York

The MIT Press
Cambridge, Massachusetts and London, England

Library of Congress Cataloging in Publication Data
Main entry under title:

AIAA Communications Satellite Systems Conference, 5th,
Los Angeles, 1974.
Communications satellite developments, systems.

(Progress in astronautics and aeronautics; v. 41)
"Technical papers selected from the AIAA 5th Communications Satellite Systems Conference, April 1974, subsequently revised for this volume."
Includes bibliographies.
1. Artificial satellites in telecommunication——Congresses.
I. LaVean, Gilbert E. II. Schmidt, William G. III. Title. IV. Series.
TL507.P75 vol. 41 [TK5104] 629.1'08s [621.38'0422] 75-45244
ISBN: 0-915928-05-1

Table of Contents

Progress in Astronautics and Aeronautics	Martin Summerfield, Series Editor PRINCETON UNIVERSITY
VOLUMES	EDITORS
1. Solid Propellant Rocket Research. 1960	Martin Summerfield PRINCETON UNIVERSITY
2. Liquid Rockets and Propellants. 1960	Loren E. Bollinger THE OHIO STATE UNIVERSITY Martin Goldsmith THE RAND CORPORATION Alexis W. Lemmon Jr. BATTELLE MEMORIAL INSTITUTE
3. Energy Conversion for Space Power. 1961	Nathan W. Snyder INSTITUTE FOR DEFENSE ANALYSES
4. Space Power Systems. 1961	Nathan W. Snyder INSTITUTE FOR DEFENSE ANALYSES
5. Electrostatic Propulsion. 1961	David B. Langmuir SPACE TECHNOLOGY LABORATORIES, INC. Ernst Stuhlinger NASA GEORGE C. MARSHALL SPACE FLIGHT CENTER J. M. Sellen Jr. SPACE TECHNOLOGY LABORATORIES
6. Detonation and Two-Phase Flow. 1962	S. S. Penner CALIFORNIA INSTITUTE OF TECHNOLOGY F. A. Williams HARVARD UNIVERSITY
7. Hypersonic Flow Research. 1962	Frederick R. Riddell AVCO CORPORATION
8. Guidance and Control. 1962	Robert E. Roberson CONSULTANT James S. Farrior LOCKHEED MISSILES AND SPACE COMPANY
9. Electric Propulsion Development. 1963	Ernst Stuhlinger NASA GEORGE C. MARSHALL SPACE FLIGHT CENTER

10. Technology of Lunar Exploration. 1963	Clifford I. Cummings and Harold R. Lawrence JET PROPULSION LABORATORY
11. Power Systems for Space Flight. 1963	Morris A. Zipkin and Russell N. Edwards GENERAL ELECTRIC COMPANY
12. Ionization in High-Temperature Gases. 1963	Kurt E. Shuler, Editor NATIONAL BUREAU OF STANDARDS John B. Fenn, Associate Editor PRINCETON UNIVERSITY
13. Guidance and Control – II. 1964	Robert C. Langford GENERAL PRECISION INC. Charles J. Mundo INSTITUTE OF NAVAL STUDIES
14. Celestial Mechanics and Astrodynamics. 1964	Victor G. Szebehely YALE UNIVERSITY OBSERVATORY
15. Heterogeneous Combustion. 1964	Hans G. Wolfhard INSTITUTE FOR DEFENSE ANALYSES Irvin Glassman PRINCETON UNIVERSITY Leon Green Jr. AIR FORCE SYSTEMS COMMAND
16. Space Power Systems Engineering. 1966	George C. Szego INSTITUTE FOR DEFENSE ANALYSES J. Edward Taylor TRW INC.
17. Methods in Astrodynamics and Celestial Mechanics. 1966	Raynor L. Duncombe U.S. NAVAL OBSERVATORY Victor G. Szebehely YALE UNIVERSITY OBSERVATORY
18. Thermophysics and Temperature Control of Spacecraft and Entry Vehicles. 1966	Gerhard B. Heller NASA GEORGE C. MARSHALL SPACE FLIGHT CENTER
19. Communication Satellite Systems Technology. 1966	Richard B. Marsten RADIO CORPORATION OF AMERICA

20. Thermophysics of Spacecraft and Planetary Bodies Radiation Properties of Solids and the Electromagnetic Radiation Environment in Space. 1967 — Gerhard B. Heller, NASA GEORGE C. MARSHALL SPACE FLIGHT CENTER

21. Thermal Design Principles of Spacecraft and Entry Bodies. 1969 — Jerry T. Bevans, TRW SYSTEMS

22. Stratospheric Circulation. 1969 — Willis L. Webb, ATMOSPHERIC SCIENCES LABORATORY, WHITE SANDS, AND UNIVERSITY OF TEXAS AT EL PASO

23. Thermophysics: Applications to Thermal Design of Spacecraft. 1970 — Jerry T. Bevans, TRW SYSTEMS

24. Heat Transfer and Spacecraft Thermal Control. 1971 — John W. Lucas, JET PROPULSION LABORATORY

25. Communication Satellites for the 70's: Technology. 1971 — Nathaniel E. Feldman, THE RAND CORPORATION; Charles M. Kelly, THE AEROSPACE CORPORATION

26. Communication Satellites for the 70's: Systems. 1971 — Nathaniel E. Feldman, THE RAND CORPORATION; Charles M. Kelly, THE AEROSPACE CORPORATION

27. Thermospheric Circulation. 1972 — Willis L. Webb, ATMOSPHERIC SCIENCES LABORATORY, WHITE SANDS, AND UNIVERSITY OF TEXAS AT EL PASO

28. Thermal Characteristics of the Moon. 1972 — John W. Lucas, JET PROPULSION LABORATORY

29. Fundamentals of Spacecraft Thermal Design. 1972 — John W. Lucas, JET PROPULSION LABORATORY

30. Solar Activity Observations and Predictions. 1972	Patrick S. McIntosh and Murray Dryer ENVIRONMENTAL RESEARCH LABORATORIES, NATIONAL OCEANIC AND ATMOSPHERIC ADMINISTRATION
31. Thermal Control and Radiation. 1973	Chang-Lin Tien UNIVERSITY OF CALIFORNIA, BERKELEY
32. Communications Satellite Systems. 1974	P. L. Bargellini COMSAT LABORATORIES
33. Communications Satellite Technology. 1974	P. L. Bargellini COMSAT LABORATORIES
34. Instrumentation for Airbreathing Propulsion. 1974	Allen E. Fuhs NAVAL POSTGRADUATE SCHOOL Marshall Kingery ARNOLD ENGINEERING DEVELOPMENT CENTER
35. Thermophysics and Spacecraft Thermal Control. 1974	Robert G. Hering UNIVERSITY OF IOWA
36. Thermal Pollution Analysis. 1975	Joseph A. Schetz VIRGINIA POLYTECHNIC INSTITUTE
37. Aeroacoustics: Jet and Combustion Noise; Duct Acoustics. 1975	Henry T. Nagamatsu, Editor GENERAL ELECTRIC RESEARCH AND DEVELOPMENT CENTER Jack V. O'Keefe, Associate Editor THE BOEING COMPANY Ira R. Schwartz, Associate Editor NASA AMES RESEARCH CENTER
38. Aeroacoustics: Fan, STOL, and Boundary Layer Noise; Sonic Boom; Aeroacoustic Instrumentation. 1975	Henry T. Nagamatsu, Editor GENERAL ELECTRIC RESEARCH AND DEVELOPMENT CENTER Jack V. O'Keefe, Associate Editor THE BOEING COMPANY Ira R. Schwartz, Associate Editor NASA AMES RESEARCH CENTER

39.	**Heat Transfer with Thermal Control** Applications. 1975	M. Michael Yovanovich UNIVERSITY OF WATERLOO
40.	Aerodynamics of Base **Combustion.** 1975	**S. N.** B. Murthy, Editor PURDUE UNIVERSITY **J. R.** Osborn, Associate Editor PURDUE UNIVERSITY **A. W.** Barrows and J. R. Ward, **Associate** Editors BALLISTICS RESEARCH LABORATORIES
41.	Communications Satellite Developments: Systems. 1976	**Gilbert E. LaVean** **DEFENSE COMMUNICATIONS AGENCY** **William G. Schmidt** **CML SATELLITE CORPORATION**

(Other volumes are planned.)

PREFACE

On April 6, 1965, just a decade ago, the first commercial operational satellite was placed in synchronous orbit. By June 1965 this satellite was providing commercial satellite service to the Atlantic region of the world. It was not until July 1969 that worldwide service was made available. In a single decade, communication satellites have increased the transoceanic communications capability by more than an order of magnitude. In addition to reduced operating costs and improved performance, users have been provided with completely new capabilities such as transoceanic television and high-speed digital service.

Only recently has this revolution in international communications capability been able to affect the United States domestic scene. The Federal Communications Commission (FCC) encountered divergent points of view that slowed down the development of a policy for the operation of U.S. domestic satellite systems such as existed in Canada (TELESAT) and the U.S.S.R. (MOLNIYA). In 1972, after five years of investigation, the Office of Telecommunications Policy (OTP) and the FCC established the "Open Skies" policy for domestic satellite systems. This policy permits the development of many U.S. domestic communication satellite systems, limited only by the economic base that is generated by the demand for this class of service. Therefore, U.S. policy and existing international agreements now recognize two categories of commercial communications satellite systems–International Systems (INTELSAT and Special Service Satellites) and Domestic Systems (U.S. and foreign).

The Communications Satellite Systems Conferences have spanned the past decade and have contributed significantly to the current successes in this field. Communications satellite technology has now attained a high level of maturity and is having an ever increasing impact on our society. Therefore, the AIAA 5th Communications Satellite Systems Conference was organized with a greater emphasis on the overall system aspects of communication satellites. This emphasis resulted in introducing sessions on policy, spectrum utilization, and geopolitical/economic/national requirements, in addition to the usual sessions on technology and system applications. This was considered essential because, as the communications satellite industry continues to mature during the next decade, it must assume an even more productive and responsible role in the world community. Therefore, the professional systems engineer must develop an ever increasing awareness of the world environment, the most likely needs to be satisfied by communication satellites, and the geopolitical constraints that will determine the acceptance of this capability and the ultimate success of the technology. The papers from the Conference are organized into two volumes of the AIAA Progress in Astronautics and Aeronautics series; the first book (Volume 41) emphasizes the system aspects, and the second book (Volume 42) highlights recent technological innovations.

Communications satellite technology has developed at a remarkable rate over the past decade. This technology explosion has created an apparently paradoxical situation in which the communications satellite systems engineer has produced excellent solutions looking for problems to solve. This dilemma has resulted from the inability of the international sociological and political machinery to respond in an effective manner to reap the benefits of this new capability. The Policy section (Chapter I, Volume 41) reviews the development of communications satellite policy as it pertains to international and domestic systems. Future policy is strongly driven by technology, but in the short term the systems engineer must develop systems that conform to the constraints of existing policy or expect long delays before new concepts are incorporated into operational systems.

The tremendous capability of communications satellites, the relatively large initial investment, and the almost immediate international implications require extensive negotiation among interested parties (and there are usually many) whenever a proposed system is presented for ratification by the affected parties. As a result of this process and the ensuing strict regulations and agreements, it has been difficult to offer new or unusual services or to extend established services in a flexible manner. The difficulty in establishing interntional organizations to cope with the problems associated with the financing and operating of these systems is discussed in Chapter II, Volume 41. The role of communications satellites in future societies and in future military strategies are also major topics of this chapter.

The communication requirements stated by the user and the geopolitical constraints together define the general problem to be solved by the systems engineer. From this information it should be possible to develop a set of system performance criteria that can be used as the basis to evaluate a range of system alterntives. Frequently, however, it is difficult to define specific values for the system criteria without first performing a sensitivity analysis in relation to the specific parameter in question. This sensitivity analysis usually is accomplished by assuming a gross system representation and allowing the parameter in question to vary across a reasonable range of values while assessing the impact on cost and performance. The resulting parametric study is used to determine the desired operating point for the parameters that drive the system design. Chapter III of Volume 41 (System Analysis and Tradeoffs) includes some excellent examples of this kind of system tradeoff analysis. These analyses apply to communications satellite systems and navigational satellite systems, and to the complex problems encountered when these functions are combined.

Communications satellite technology has matured rapidly over the past decade, and it is now possible to provide new services to an ever-increasing community of users. The new services include wideband digital service, color television, secure voice, and rapid and flexible extension of service. The new users served include mobile platforms (ships and aircraft), remote terminals, and small direct-access users. This is a greatly expanded role over the large fixed-terminal point-to-point deployment that has been the mainstay of INTELSAT. Improvements in component reliability have made it feasible to increase satellite complexity and at the same time reduce the size of the crew that operates and maintains the Earth terminals. To allow a large number of users to obtain direct access to these new services, the terminals must be economical from both the initial investment and operation and maintenance (O&M) standpoints. Terminals can be made simpler and more economical by placing an increased burden on the spacecraft (i.e., increased satellite radiated power and onboard processing). This seems to be the logical trend for the future, and so the new generations of satellites now in the planning stages are being provided with narrower beam antennas and onboard processing. Chapter IV of Volume 41 (Advanced Concepts) presents new concepts that employ processing satellites to achieve antenna beam switching and navigational positioning. The impact of introducing a private line service for a large community of users is discussed therein also.

Communications satellite technology now can provide operational systems that have demonstrated high availability and superior performance. The INTELSAT system has become the principal means of international voice communication and, of course, the only means for transoceanic television broadcast and wideband digital service. The increased demand has required INTELSAT to develop improved satellites with expanded capability. The current generation of INTELSAT IV's is being supplemented by INTELSAT IV-A's, which promise greater capacity, principally through frequency reuse. At present, however, INTELSAT is no longer the only operational system. Several U.S. domestic systems (Western Union, RCA GLOBCOM, COMSAT/AT&T-GTE, American Satellite) are operational or about to become operational, and the U.S. will soon deploy fully operational military systems. Foreign domestic satellite service is developing rapidly also. Most of these systems, however, are (or will be) nationally owned and operated. The TELESAT system is already available for service in the Canadian region. Other countries, such as Algeria and Brazil, are availing themselves of the tranponder leasing service offered by INTELSAT. The interruptible service that Algeria is obtaining will give it an excellent system at a very reasonable annual cost. Also, MARISAT will soon be supplying service for maritime ships. The DoD and NASA have supported most of the technological developments that have established the basis for this rapidly expanding industry. Except for the U.S.S.R., the DoD and NASA still have the only major space launching capabilities in the world. The major operational DoD satellite communications system is the Defense Satellite Communications System (DSCS). In 1974, the new "Phase II" DSCS satellites were declared operational in the Pacific and Atlantic regions of the world. The tactical operational service of DoD is to be established by leasing a portion of the MARISAT spacecraft in early 1976. By 1978 the FLEETSAT/AFSATCOM system will provide increased capability for this class of user. For the more distant future, DoD is looking to survivable satellite techniques that will enhance the command and control capability of the strategic forces. Chapter V of Volume 41 (Communications Satellite Systems) describes several of these systems.

Chapter I of Volume 42 presents an excellent cross section of papers covering the technologies available for future satellites. The key areas of mass versus communications power and of benefit versus risk are emphasized in the evolution of the satellite concept. In the course of these discussions, most of the technologies being considered for use in the 1980 generations of satellites are exposed. There appears to be a general trend in satcom spacecraft toward multiple antenna beams because of frequency-reuse, higher EIRP, or user discrimination requirements. Such satellites impose stringent requirements upon the positioning and orientation subsystem of the satellite, as well as upon the attitude control subsystem. A three-axis stabilization concept is usually proposed, rather than the spin-stabilization typical of the current generation systems. Nuclear-powered spacecraft have long been a topic of considerable controversy within the satellite community. In general, the nuclear power advocates were at a disadvantage since the solar cell approach was clearly more attractive on the weight-limited vehicles of today. However, the advent of high performance radioisotope power systems may change all that.

There is an increasing number of applications for satellite communications to support networks involving hundreds of potential terminals. A premium must be placed by operators of these networks on the need to keep the initial cost of each terminal as low as possible and, at the same time, to have inherent in the basic terminal design such features as will minimize the recurring costs of operation and maintenance. Tradeoffs of minimizing the initial cost of the terminal against the requisite terminal availability, together with the operational experience of Telesat Canada in utilizing small unattended receive-only Earth stations, are presented in Chapter II of volume 42 (Terminal Technology). The impact of elliptical satellite orbits (i.e., Molniya) on Earth terminal design also is discussed.

The third major element of a satcom system is the control subsystem, and this is the most all-pervasive of the three major elements because it encompasses every user, every terminal, and every satellite. As such, it must be integral to the total system design and operation and be responsive in time to the demands placed upon it. From the overall point of view, there are various types of control. Systems, satellites, terminals, operational modes, links, and channels are all subject to control procedures. The section on Control Technology (Chapter III, Volume 42) deals with the complex control terminal operations and the difficulty of integrating a mix of terminal sizes and types. These papers emphasize the need to automate these operations to reduce the response time and improve system performance.

The area of technology called baseband or communications processing encompasses modulation, multiplexing, multiple access, and coding for both the transmission path and the data source itself. A form of multiple access that has become popular of late is "single channel per carrier" (SCPC) and may be defined best as a frequency division multiple-access (FDMA) method in which each carrier is limited to one voice channel, or its data equivalent, and assigned on a dedicated basis. Because of its marked similarity to FDMA demand assignment methods, SCPC is sometimes used as a stepping stone into an FDMA demand assignment (DA) system or as a dedicated subset of such a system. SCPC is normally applied to networks of small, lightly loaded terminals in relatively isolated circumstances or as an auxiliary capability of a large Earth station. These are relatively new applications of communication technology; the impacts of the propagation media and the motion of the platforms require reassessment. These problem areas and the ability of the modems and coding equipment to combat them are addressed in Chapter IV of Volume 42 (Multiple Access, Modulation, and Coding).

One of the trends of satellite communciations systems has been the greater use of portions of the frequency spectrum other than the SHF region that spawned most of the technical and operational growth of the satcom technology. The UHF and EHF segments are of particular interest, the former for its promise of low-cost mobile terminals and the latter for very wide spectral allocations that minimize interference with terrestrial communications. Each of these regions also has unique transmission path characteristics. The EHF band and the upper portion of the SHF band are impacted to a significant degree, proportional to the frequency of interest, by rain attenuation. The UHF band and the lower part of the SHF band are influenced by ionospheric scintillations. Because of the importance of information on these phenomena to satellite communciations systems engineers, five papers on these phenomena (four of them from the AIAA 12th Aerospace Sciences Meeting) have been selected for inclusion in Chapter V of Volume 42 (Transmission Path Effects).

The systematic coverage provided by this two-volume set will serve on the one hand to expose the reader new to the field to a comprehensive coverage of communications satellite systems and technology, and on the other hand to provide also a valuable reference source for the professional satellite communication systems engineer.

We wish to thank our wives, Barbara LaVean and Ruth Schmidt, for their patience and understanding.

Gilbert E. LaVean
William G. Schmidt

November 1975

CHAPTER I—POLICY

POLITICAL CONSTRAINTS INFLUENCING COMMUNICATION SATELLITE SYSTEMS

Gilbert E. LaVean*

Defense Communications Agency, Reston, Va.

The inherent capability of a synchronous communication satellite to provide direct communications to a third of Earth's surface also causes the most perplexing problems associated with communication satellites. First, the satellite is a potential interference source to the same one-third of Earth's surface and therefore presents a difficult frequency management problem. Radio communications planning always has required an extensive frequency management effort, but the expanded region of possible impact and the publicity usually associated with the development of communication satellite systems has increased severely the magnitude of this consideration. Second, to develop an effective operating system, the corporate participants must establish basic agreements on cost sharing, system operation, and revenue sharing. As the number of participants increases, the difficulty in obtaining these necessary agreements is aggravated.

In recognition of these problems, the U.S. took a leadership role in establishing the interim INTELSAT system. As a result of this effort, the world was able to benefit from the technology that the U.S. had developed in the areas of boosters, advanced communications, and electronics. The U.S. Congress fully supported these efforts and passed the Communications Act of 1962 to provide the national initiative to get INTELSAT under way. This Act formulated national policy "to establish, in conjunction and in cooperation with other countries, as expeditiously as practicable, a commercial communications satellite system, as part of an improved global communications network, which will be responsive to public needs and national objectives, which will serve the communication needs of the United States and other countries, and which will contribute to

---

*Deputy Chief, Satellite Communications Engineering Division, Defense Communications Engineering Center.

world peace and understanding." In 1964 the United States and thirteen other nations signed the interim agreement that resulted in the formation of the International Telecommunication Satellite Organization (INTELSAT). In February 1973, the Definitive Agreements went into effect, and over eighty nations ratified it. More than one hundred countries, territories, or possessions were leasing satellite services (approximately 10,000 half circuits) on a full-time basis in 1974. The Communications Satellite Corporation (COMSAT) was established by Congress in 1963 as the U.S. representative to INTELSAT and has been the primary reason for the tremendous success of INTELSAT.

The 1962 Communications Act, however, addressed a broader scope of requirements and the long-term application of all communication satellite systems. It provided for the possibility that a broad spectrum of requirements (such as governmental needs, international carrier, domestic carrier, and mobile users) might not be satisfied by a single system. The Declaration of Policy and Purpose of the Act states: "...(d) It is not the intent of Congress by this act to preclude the creation of additional communication satellite systems, if required to meet unique governmental needs or if otherwise required in the national interest." Attention is drawn to the last phrase of the sentence in the quote from the 1962 Act "... or if otherwise required in the national interest." This phrase was the object of lengthy discussion during the ensuing hearings[1] and was the result of the Church-Lausche amendment. From the foregoing discussion, it is clear that Congress always has intended to support the existence of commercial and government systems and that the Government should utilize these combined resources to the best interest of the United States Government.

Thus, INTELSAT was developed for a single-purpose, international carrier service, and the system is designed in the most effective manner to achieve that purpose. It is not surprising, then, if it does not efficiently perform the other divergent missions that could be required of a satellite communications system. Service for small mobile users requires a different system design, and systems designed for military and domestic applications are also likely to emphasize different features than the INTELSAT system. The U.S. policy therefore has permitted the development of both military and commercial communication satellite systems.

The evolution of the commercial application of communication satellite has resulted in specialized systems optimized to meet the needs of particular groups of users. The policy governing these systems also has evolved, and the Office of Telecommunications Policy (OTP) and Federal Communications Commission (FCC) have generally taken the initiative in formulating the new policy. The FCC has limited the participation of COMSAT in systems other than INTELSAT as shown in this 1972 statement: "When the Satellite Act was enacted, Congress, in order to assure that the fledgling corporation (COMSAT) received needed communications expertise and guidance, made provision for communication carriers to own up to 50% of COMSAT's Board of Directors... COMSAT was created by Congress primarily for the important and immediate purpose of representing and promoting this nation's interests in the establishment and operation, in conjunction with other nations, of a global international communication satellite system. That mission, with the aid and support of A.T. & T. and other carriers, has been achieved with a high degree of success. COMSAT is now seeking entry into the domestic communications field to compete with A.T. & T. and other carriers in supplying new and improved domestic communications services. However, in this field, the underlying considerations which motivated Congress to permit and encourage A.T. & T.'s ownership in COMSAT are no longer controlling. On the contrary, the competitive roles which COMSAT and A.T. & T. are assuming in the domestic communications field dictate the need for maximum independence from each other and an arms-length relationship."[2] Thus, currently (in 1974) 99.7% of the outstanding COMSAT stock is held by public shareholders.[3] COMSAT was permitted to form a subsidiary corporation, COMSAT GENERAL, to enter into competitive commercial ventures.

The polarization between international and domestic systems is a natural outgrowth of the regulatory authority that each nation has over the conduct of business within its national boundaries and is recognized in the INTELSAT operational agreement.[4]

It is clear that INTELSAT recognizes a responsibility (and a potential market) to satisfy domestic needs if in so doing it does not interfere with the primary mission of INTELSAT, i.e., global international telecommunications. The FCC[5] considers it is consistent for the U.S. to participate in the INTELSAT consortium for international communications and at the same time to allow competition among systems, which might include INTELSAT facilities, to provide domestic service.

The FCC established the authority to regulate a separate Domestic Communications Satellite System in FCC 72-1198.[6]

This order states: "The Second Report adopted a policy of affording a reasonable opportunity for entry into the domestic communications satellite field by qualified applicants subject to certain showings and conditions." This generally has entailed showing financial and technical viability, but strict regulation is placed on spacecraft characteristics and position, and on ground-station locations. This competitive policy was in contrast to a controlled monopoly, such as exists with COMSAT in the international arena. There was a heated controversy for several years over the advantages of the different policy options. The decision in favor of open competition reflects the current political environment of increased competition, less regulation, and stricter antitrust law enforcement that prevails at the national level (both Congressional and Executive branches).

This general philosophy of increased competition has brought the communications industry into a dynamic new era which has culminated in the recent entry of specialized carriers into the common carrier field and the current (1975) antitrust suit against A.T. & T.

These actions will have a long-lasting effect on the industry, and the outcome of the series of interconnect hearings in process by the FCC also will impact the cost and responsiveness of service from carriers other than A.T. & T. The FCC has ordered[7] "...assuring that all carriers providing retail interstate satellite services (whether or not affiliated with Bell System Companies) have access, at nondiscriminatory terms and conditions, to local loop and inter-exchange facilities as necessary for the purpose of originating and terminating such interstate service to their customers." Several specialized carriers, including domestic satellite companies, are claiming that A.T. & T. is discriminating against them. The FCC is proceeding with rulings on these cases. It can be seen that, if the specialized carriers are forced to construct their own local terrestrial distribution system, they will not be competitive with A.T. & T. Thus, resolution of these important issues will have a serious impact on the desirability of the new domestic systems to satisfy user requirements.

As a result of the FCC ruling on domestic communication satellite policy,[8] applications from eight companies were filed; two applications have since been withdrawn, and five system operators were authorized to proceed with their plans. These five were: COMSAT GENERAL/A.T. & T., Western Union Telegraph Company, American Satellite Corporation, General

Telephone and Electronics (G.T. & E.), and RCA Global Communications. CML, a company which was jointly owned by COMSAT General, MCI, and Lockheed, was approved by the FCC, but no definitive operating plans have been submitted to the FCC. Since that time, G.T. & E. has merged its efforts with A.T.& T, and the CML group has undergone several restructurings. In 1974 IBM purchased the major portion of the stock (55%) with COMSAT GENERAL retaining 45%. The FCC has ruled that IBM and COMSAT can each own a maximum of 49%, which means that a third partner must be admitted into the corporation. AETNA Life Insurance has been selected as the third partner, and the final apportionment of the stock is under negotiation. This arrangement has been contested by several entities and again must be resolved by the FCC. The shakeout of this infant industry is likely to continue for several years, at least until the market is well defined and all side issues and policy matters have been resolved.

Another important area of international regulation and control involves the International Telecommunications Union (ITU), which establishes frequency planning and spectrum utilization policy. This is the instrument that the world states have organized to determine which frequency bands should be set aside for providing specified services (e.g., satellite fixed, satellite mobile, television broadcast, etc.) on an international basis. The current primary operational commercial satellite communications band, established by the ITU, is in the 4- and 6-GHz band, and the primary operational military bands are 225-400 MHz (FLEETSATCOM/AFSATCOM) and 7 and 8 GHz (DSCS). The ITU generally meets every ten years or so on an irregular basis and reassesses the previous allocations to determine the necessity for change. Therefore, the regulations established at these meetings will have a long-term impact. The general meetings to be held in 1979 will develop regulations that are expected to last through the year 2000.

The diverse and far-reaching impacts of communication satellites are apparent or should be evident by observing the great difficulty in arriving at the useful operating agreements alluded to in previous paragraphs. The tremendous capability of communications satellites, the relatively large initial investment, and the almost immediate international implications require extensive negotiation among interested parties (and there are usually many) whenever a proposed system is presented for ratification by the affected parties. As a result of this process and the resulting strict regulations and agreements, it has been difficult to offer new or unusual services, or extend established services, in a flexible manner.

Very little advantage has been taken of the inherent ability of satellites to extend service instantly between terminals located anywhere within the coverage area. Some progress is being made in providing maritime service, but it has been very difficult to establish a flexible international service for ships or aircraft operating in oceanic regions. Other flexible international services for small terminals, such as demand access, require extensive review by the participating countries. Therefore, technology has provided a capability today that is seriously constrained by the organizational structure that has evolved to regulate it.

References

[1]U.S. Congress, House of Representatives, Committee on Government Operations, "Satellite Communications (Military-Civil Roles and Relationships)," House Rept. 178, Eighty-ninth Congress, first session, Washington, D.C., GPO 1965, p. 46.

[2]Federal Communications Commission, Docket 16495, "Establishment of Domestic Communication-Satellite Facilities by Non-Governmental Entities," FCC Record 72-1198, 89897, Dec.22,1972, p. 19.

[3]COMSAT, "New Ventures for a New Decade," COMSAT Ann. Rept. to the President and Congress, 1974, p. 57.

[4]International Telecommunications Satellite Organization, "Agreement and Operating Agreement," Treaties and Other International Acts Series, entered into force, Feb. 12, 1973.

[5]Federal Communications Commission, Docket 16495, "Establishment of Domestic Communication-Satellite Facilities by Non-Governmental Entities," FCC Record FCC 70-306, 46001, March 24, 1970.

[6]Ref. 2, p. 1.

[7]Federal Communications Commission, Docket 16495, "Establishment of Domestic Communication-Satellite Facilities by Non-Governmental Entities," FCC Record FCC 72-531, 79456, June 16, 1972, p. 16.

[8]Ref. 7, p. 1.

CHAPTER II—GEOPOLITICAL ENVIRONMENT AND SYSTEM REQUIREMENTS

Communication satellite technology has developed at a remarkable rate over the past decade. This technology explosion has created an apparently paradoxical situation in which the communication satellite systems engineer has produced excellent solutions looking for problems to solve. This dichotomy has resulted from the inability of the international sociological and political machinery to respond in an effective manner to reap the benefits from this new capability. Therefore, when operating within the current organizational structure, the tendency is to replace existing services with a more economical substitute. This has been accomplished successfully by replacing long-haul point-to-point submarine cables with communication satellite services (i.e., the primary mission of INTELSAT). Regulation and payment for the service and distribution of the profits are accomplished by essentially the same process that has evolved over many years to support the International Record Carriers. This organizational framework is undergoing extensive changes in order to cope with the new problems introduced by wide utilization of communication satellites. The impact of communication satellites on possible future societies is discussed in the paper by Burgess.

To resolve the issues pertaining to the operation of the system and to obtain the unusually high initial investment requires an effective organization. The procedure for developing the organizational structure to implement operational systems is described by Gosewinckel. If there is a lack of an established need, or no awareness of that need, for a new service, negotiations have proceeded slowly. This trend generally has continued until some situation causes a sense of urgency to enter into the normal process. The difficulties in establishing a flexible commerical service for aircraft and ships at sea are good examples of this time-consuming process. Certainly there has always been a real need for better maritime communication service, and the required technology has been available for many years, but the implementation of a viable operational system has caused a reassessment of certain international and national interests.

The political obstacles to establishing an operational communication satellite system to satisfy domestic needs are not as insurmountable, and many countries are proceeding with their own domestic systems. Communication satellites offer an excellent solution for providing good communications service to countries that have large unpopulated regions, such as Canda and the U.S.S.R. The Philippines is another country that is well suited for the application of domestic satellites, and the paper by Cheadle discusses the attractiveness of communication satellites for developing nations. Indonesia, Brazil, and Algeria are among those planning domestic systems. Several nations are planning to avail themselves of the special transponder rates offered by INTELSAT, and others are planning to purchase satellites from U.S. industry, while still other countries like Japan and France, plan to develop their own space industry.

The flexibility that communication satellites can provide to the military environment is extremely important. Communication satellites can furnish reliable communications to mobile platforms, such as aircraft and ships at sea. The need for ground mobile forces to move often and with short notice requires the usage of rapidly deployable terminals. These restrictions, communication in motion and rapidly deployable terminals, normally impose the severe constraint of many terminals with small antennas attempting to operate simultaneously in the transponder. The paper by Bond discusses the difficult problems that have to be solved by the Department of Defense (DoD).

To field an operational military system that has a significant improvement in capability normally requires 8 to 10 years. Therefore it is necessary to expend money at a rather high rate for many years before an operational capability is demonstrated. Most other communications systems (i.e., LOS, HF) can be purchased incrementally with a corresponding incremental improvement in communications capability. Thus, for other systems, the system cost is easily spread out over many years without critical components (i.e., spacecraft) having a large impact on total system performance. This independence of network elements is not true in satellite communications, because a satellite failure can render hundreds of millions of dollars of other equipment inoperable. This long-term commitment (8 to 10 years) is difficult to maintain in a governmental organization that is likely to incur significant philosophical changes at least every 4 years. Therefore, a recognition of this problem and a vehicle to provide long-term commitment are essential to developing effective military communication satellite systems.

ALTERNATIVE FUTURES FOR MANKIND AND THE ROLE OF COMMUNICATIONS SATELLITES

Eric Burgess*

Science and Technology News Service, Northridge, Calif.

Abstract

Several possible world futures are discussed, and the role to be played by communications satellites in these several futures is postulated. In a continued fragmented world nationalism, communications satellites are expected to continue to serve special interest groups without realizing their full potential. Maximum use of communications satellites is most likely to take place in a people-inhomogenous world federation rather than a world state.

I. Introduction

After a brief period of imaginative expansion associated with the exploration of space and the other planets of the solar system, many people now are accepting limitations of thought and action as doomsday prophets foresee inevitable catastrophes for the human race and possibly a headlong plunge into fossildom. Space-age consciousness, newly born on planet Earth, is being eroded by an eruption of negative thinking of either/or alternatives, rather than acceptance of a wide range of optional futures. Many advanced thinkers in different disciplines recently have expressed strong opinions that the human race is, indeed, at a turning point and that it is not possible for the species to continue as it has in the past.[1] They arrive at such a conclusion from several unprecedented conditions being experienced today.

Presented as Paper 74-492 at the AIAA 5th Communications Satellite Systems Conference, Los Angeles, Calif., April 22-24, 1974.

*Communications Consultant.

First, a major reduction in the global death rate has not been accompanied by a global reduction in the birth rate. As a consequence, there is a rapidly increasing global population exceeding the increase in food production, especially in the underdeveloped countries, with a swing from population concentration in technologically advanced nations to a population concentrated in much less technologically advanced nations. Depending upon which demographic model is chosen, the world population is expected to double within 30 years and to continue to increase until it hopefully stabilizes somewhere between 7 and 15 billion people within 50 to 100 years. Today half the people who have ever lived on Earth are alive, and these are predominently children of the space age, since nearly half of the world's population is less than 15 years old.

Without applications of the highest technology, the Earth cannot support the expected large population, and large proportions of the population must be thrust below starvation levels, with grave possibilities of a major Malthusian collapse, coupled with severe repercussions into the better-fed sections of humanity because of global interdependences.

Second, the burgeoning population is rapidly using up readily available and easily accessible raw materials: fossil fuels, minerals, metals, food, paper. Supplies of these "essentials" to meet the demands of the increasing population can be provided only through application of high technology: deep mining, recycling, ocean floor mining, synthesis of transportable fuels, synthesis of proteins, enzyme chemistry, greater use of electronic instead of written communications, and so on. All of these processes are energy-intensive, and yet, so far, mankind has not embarked seriously on a major effort to develop alternative energy sources to replace the dwindling oil reserves of the world and still refers to energy in terms of barrels of oil.

Best estimates are that, with a concerted effort yet to be started, most of the alternative energy sources cannot be generally available on a large scale before the 21st century. Large sources of energy also are needed to harvest food crops to feed the burgeoning populations, since mechanized western agriculture consumes about 20 times as many calories as are available in the food it produces. In terms of energy in and energy out, paddies are more efficient than western agriculture.

Third, industrialized society is somewhat of an anachronism superimposed upon the biosphere and geosphere of planet Earth. It is geared to maximize yield by continued growth in the rate of pushing chemicals and calories through the system. This is all very well for a developing society as for a developing fetus, but its continuation in a mature society is analogous to cancerous growths in a biological organism.

By contrast to the industrial system, the biosphere is geared to survival. Maximizing yield has never been a survival factor. Thus, the aims of the biosphere and of the industrial human sphere are incompatible. All systems convert or use materials and energy. But the industrialized system breaks with natural order. It erroneously assumes that the biosystem on which it feeds can provide an infinite source of materials (chemicals) and of energy (calories), and that the physical world provides an infinite dump for wastes. It ignores the fact that the biosphere has taken many millions of years to concentrate special materials from the geosphere. And when the industrial system consumes these materials and dumps them as wastes into the ocean, they can be replaced only by an artificial biosphere. Thus, survival of the human species, which has become essentially a dependent species, depends upon radical changes to the present industrial exploitation of the biosphere.

Fourth, over the next hundred years or so, radical changes to life styles worldwide are mandatory if mankind is to house and feed the anticipated 7 to 15 billion people before a stable population is reached. Moreover, it is unlikely that this population boom can tolerate 10% of all human activities being wasted in weapon development, as is the case today.

Many of the new people will be born in villages but must migrate to cities, since the agricultural communities cannot support them. Cities are expected to proliferate and grow enormously until all coastal regions are first dotted with cities and then merged into megalopoli. Urban villages as part of agro-industrial complexes can feed billions of people; but it is probable that meat substitutes will be essential, and the world will have to abandon paper as a means of communication.

Ultimately, people concentrated in cities must lose individual freedom and accept conformity, dehumanized and strictly controlled. This process already has started, with new bureaucracies springing up to meet each new brushfire crisis and promulgation of restrictions and controls of many kinds. Even in the supposedly enlightened "free" world, the process of control and conformity is apparent, with more and more restrictions on individuals and with police chiefs heading the pecking order and budget priority lists among departments of cities. It seems inevitable that the world must become increasingly artificial and its population homogenized to sustain its total population at any reasonable level of affluence. Both a natural environment and a diversified people cannot survive in an overpopulated world.

Fifth, classical control systems for the industrial system superimposed on the bio- and geosystems are not working. Worldwide inflation, violent economic fluctuations, and a general inability to supply basic needs of population (food, housing, transportation, clothing) are symptomatic of the archaic and ineffective nature of this control system. The system consists of social, economic, legal, political, religious, educational, and institutional factors established mainly in the 16th and 17th centuries. In the high-technology society of today, it hunts wildly when perturbed. And the effects will be even more disturbing in the worldwide high technology needed to sustain the burgeoning world population of the near future.

For mankind to survive, some economists concede now that many traditional basics of human society must change, just as those of a developing bio-organism must change when it, too, reaches maturity. These may include personal property rights, goals of growth, economies of exploitation and waste, excessive consumption to maximize profits for special interest groups by throughput of materials, and use of money as a medium of exchange. Such changes may be thrust upon mankind by crises of catastrophic nature in the coming decades. Against these unprecedented factors in human society being experienced today, several possible world futures can be examined.

II. Possible World Futures

The future may not be predetermined but rests solely with mankind's actions and choices within the next few decades. And the high technology of today certainly offers a number of choices to mankind. A first major decision is whether or not to limit growth. As mentioned earlier, most biological species experience two periods as living systems: a period of rapid growth from conception to maturity, followed by a period of zero growth during reproductive years and afterwards. Perhaps societies or planetary intelligences have to do likewise. Perhaps the reproductive years of a planetary consciousness have to be devoted to disseminating its progeny among the stars while it remains in a zero-growth state on its own planet.

The current concept being practiced by most world societies on Earth is to try to stimulate growth to overcome the economic problems by pumping chemicals and calories faster and faster through the system by use of advanced technology. Carnivorously, society creates machines to consume Earth itself. This accelerated throughput eases the situation, but it is a palliative and not a remedy. Ultimately, the human species has to face the fact that it is on a physically limited planet, and it has to accept maturing: it can no longer grow physically in numbers; it can no longer increase its appetite for Earth's resources; instead it must diet.

Recently, several writers have suggested some possible futures for mankind which can be summarized as competitive nationalism continued, emergence of a superstate, and development of a world confederation.[2] Each has, of course, several variants, and each leads to a very different type of human being, and a different need for communications, especially communications satellites.

III. Competitive Nationalism

If the world continues very much as it is today with no central world governing body and tithing its activities to protecting itself against itself, the logical outcome is for nations to scramble competitively for diminishing resources and to develop increased ability to destroy each other. Although increased warfare and world terrorism are possible futures, destruction also can be by economic pressure or withholding of vital raw materials.

The consequence must be deterioration of industrial nations while the world generally becomes increasingly polluted and its people reduced to bare subsistence levels. This is the future postulated in H.G. Wells' Things To Come. But his solution through technocracy and a rise back from barbarism does not seem likely, since during mankind's decline most of the easily accessible fuels and raw materials will have been exploited.

IV. Superstate

A world superstate could arise in at least two ways, stemming from current bureaucratic trends worldwide. There might be agreement between two major world powers, or imperialistic conquest by one power. The latter, however, carries with it the risk of an all-out nuclear holocaust and seems less likely. Superpowers may be thrust into a superstate because of their inability to support the large populations of the underdeveloped world as they aggressively demand their share of the world's wealth. Such a superstate would aim at preserving its own people and letting the rest of the world fend for itself, literally starving the world into submission and hopefully decimating the population of the underdeveloped nations. This superstate is the Brave New World of Huxley. The "have" countries might merge through economic cartels, monetary and political agreements, economic and technical agreements. Already the trend is evidenced between the U.S. and the Soviet Union, between the Soviet Union and Europe, and to a lesser extent among other national groupings. Special deals to fill gas tanks of affluent nations while denying India petrochemicals for fertilizers are symptomatic of the superstate trend.

As "have-not" nations struggle for survival, their environments must deteriorate rapidly so that the world environment must suffer even though the superstate can control its environment. The end result is an artificial planet as the superstate gains control of the whole world and attempts to clean up the mess. Such a superstate will have to restrict personal freedom and might even push to homogenous populations through genetic engineering. Every attempt would be made to seek conformity and uniformity and through this to produce a certain measure of happiness among the homogenous population.

V. World Confederation

An inhomogenous human population might be preserved on a global scale by a condederation of nations under some as-yet-to-be-evolved world governing body. In such a confederation, the have nations would share their resources, riches, and know-how with the have-nots. To support all of the billions of people, many of today's affluent societies would have to accept a reduced standard of living, and the environment of the planet would have to be controlled strictly and probably become a totally artificial environment. But by such a world system, the diversity of human cultures and people might be preserved.

Technology applications also might reach a stage of sophistication, especially when throughput for throughput's sake has been eliminated, at which all production of human essentials is by computer-controlled machines. Then both the simplistic capitalistic system of amassing personal profits and wealth and the simplistic communist system of rewarding only for production might have to be abandoned in favor of a moneyless society where all of the necessities of life are provided free. In such a society, each person would "borrow" for the time being those capital goods he needs to use infrequently. Such might include means of transportation, houses or apartments, vacation accommodations and pleasure crafts, tools, library facilities, and so on. Trends in this direction already are observable in the world today. A major change ultimately might be to educate children to use leisure in fulfilling their lives instead of training them in vocational skills.

In each of these possible futures and their innumerable variations, communications must play an increasingly important role, since, as populations increase, communications must replace commuting, and society must concentrate on moving ideas and information instead of people and things. Even manufacturing processes might be coded for producing goods where they are needed in fully automated factories controlled from central produce dispatchers. Movement of goods across the country or across the world as necessary in the present economic system to take advantage of cheap labor may be eliminated. And, as the world needs more and more communications, it must have less expensive ways of providing such communications. Communication satellites appear to supply the answer.

VI. Role of Communications Satellites

Most of anticipated roles of communications satellites in the possible futures of mankind have been discussed at length in many papers presented over the past 25 years. Unfortunately, and in some cases fortunately, many of the roles have not been realized. Only a few can be discussed in this short paper.

VII. Things-To-Come Scenario

So far, communications satellites have been predominantly the tools of special interests, and, except for overseas communications and television relays form overseas, they have been used little by the majority of mankind. Wide-band inexpensive links for the diverse communications needs of humanity have remained largely unsatisfied after over a decade of practical experience with satellites and over 25 years since such satellites were first proposed in technical papers and popular articles by Clarke[3] and Burgess[4,5] soon after World War II. Some critics have stated that the development of communications satellites has been analogous to the canal companies developing the railroads and the railroads then developing the airlines and that this is why there has not been full application of communication satellite potential.

Neither industry nor government nor, indeed, the great majority of the public, the taxpayer who has invested money in the space program so that communications satellites could be in orbit instead of on paper, seems to recognize the immense communications cost savings and versatility offered by the concept of communications via relays in space extending far beyond the limited concepts of replacing Earth-based landlines and microwave links in a people-oriented communications network. The immense potential for assisting education has virtually been untapped, and yet, at the time of the Congressional hearings for the Public Television Act of 1967, the president of the Interuniversity Communications Council stated: "A broad educational telecommunications network is essential for the continued advancement of education within current and future economic restraints."[6]

More recently, a leading educator on the West Coast wrote to the Federal Communications Commission: "The current financial and social crisis in American schools cannot be resolved without more rapid and massive application of modern technology ... (in which) ... wide-band communication may be technically and economically the most critical single factor."

On the whole, however, United States educators have been disorganized and fragmented and have not made clear what their needs for educational communications, let alone communications satellites, might be in the years ahead. Although the current usage of communications satellites concentrates on large, expensive antennas for ground stations coupled to existing landline systems, many scientists have shown that educational satellites could be incorporated into much less expensive systems. There have been some experimental setups that have demonstrated such systems on a pilot scale using NASA satellites.[7]

But education is only one application of communications satellites which remains virually untapped. High rates of communications between computers and terminals of all kinds offer many diverse applications throughout the whole of human society. And cost and efficiency over the present dial-up telephone systems seem to offer great potential. The telephone system was designed generally for people, not for machine communications. It is ironical that the cost of processing information by computer has been considerably reduced by applications of the highest technology, whereas the cost of communicating among computers and between computers and people still rests on the expensive ground-based telephone system and has derived little or no benefits from innovative applications of the high technology of satellite systems.

Under the first optional future in which the world continues to muddle through with elected representatives merely acting as buffers between special interest groups and the professionals who run cities, states, and countries, and most of human activity consists of reacting to brush-fire crises and trying to defend itself from itself on a local and international scale, the future for communications satellites looks bleak. Past experience suggests that it will be unlikely for satellite communications to find full applications in meeting the needs of mankind for rapid and inexpensive communications. Rather, satellites will continue to serve only relatively small sections of humanity and will continue to be controlled by the military, overseas communications companies, and specialized utility companies. There is no shortage of ideas for uses of communications satellites, but these ideas never seem to get into orbit because of procrastination, institutional roadblocks, and other human failings.

The advent of the space shuttle, if this vehicle is not shot down before it is launched, may, however, provide opportunities to push wider applications of communications satellites. And private enterprise operations to carry payloads into space may push us even further along the road to using the potential of satellites more fully. For example, opportunities to put large antennas in space inexpensively could negate many of the current arguments against wider use of satellites because of frequency spectrum limitations and mutual interference at ground stations when using small antennas there. However, the nation and the world seem to be locked into a fashionable type of thinking concerning the use of these satellites, and this may prove difficult to break once the habit has been around for several decades. A satellite hierarchy geared to the thinking of the public utilities is about as useful as an energy hierarchy geared to the thinking of the oil companies. Neither can serve mankind effectively in the long haul, since large companies have been notoriously noninnovative, just as large bureaucracies have been. But all must inevitably collapse in the fall to new barbarism if mankind fails to organize into a global society with some common and generally acceptable aim for the multifaceted peoples of the world.

VIII. Imperialistic-Superstate Scenario

Inability of the current control system to maintain economic stability and societal fluctuations of increasing amplitude may push world societies into an imperial superstate scenario. In such a world future, the population will be homogenized and controlled in large cities, and communications satellites may quickly assume a new role, that of providing a communications control system to preserve the superstate and to monitor the activities of all of its people. Since the superstate also will control global communications and many military frequencies would be relinquished with global disarmament following establishment of the superstate, additional frequency allocations can be made for snooping and civilian spying of all kinds. Have-not nations would be largely out of the system, left to themselves to perish, and monitored and surveyed only to prevent them from becoming a threat to the superstate, e.g., the breeding ground for new pathogens.

This possible future has been dealt with at length by many authors, and the extension of satellites communications to a superstate need not be belabored. Suffice it to say that, in such a people-homogenous world state, communications satellites will be used to control and to police large human populations concentrated in megalopoli. Such control and continuous surveillance may have to be quite absolute, that is, until genetic engineering can produce reliable copies of "desirable" citizens. One suggestion for genetic engineering, for example, is to extract ova and sperm from maturing individuals and then sterilize the individuals. After an individual's death, a tribunal would go through the records of its life and decide if the individual should be reproduced. If in favor, the tribunal would authorize use of that individual's genetic material; otherwise the material would be destroyed.

Since the lifetime activities must be thoroughly recorded and documented, this task alone, spread over all humans, could occupy large communications systems via satellite. It is but one example of the need for information in the superstate scenario. Generally such a state would rely heavily upon satellite communications for surveillance and transfer of information to computers, data banks, and for human/machine interfaces. Perhaps these links also would be employed for direct control of people through implanted electrodes, even direct relaying of programs to individual pleasure centers of the brain.

Communications satellites thus might play a role in such a world scenario toward replacing man by a machine, which ultimately continues to evolve as living systems have in the past, so that the biosphere of Earth becomes entirely an electromechanical sphere in which there is no major role for the human system once it has started the process into operation.

IX. People: Inhomogenous World Federation

The third major alternative appears to be the only option that retains some hope for the preservation of what we have come to regard as human qualities in man. It may be idealistic and utopian, but it also may be essential to the survival of the species. The strength of mankind is in the diversity of its genes on an individual basis. This diversity provides an ability to resist new pathogens and surmount major environmental changes. Inhomogeneity thus must be an important survival factor that humanity must preserve at all costs.

In a world confederation with some kind of world governing body that permits individual cultural groups to preserve their individuality and live peacefully in a global society, communications satellites will achieve their greater utility to mankind. As general information relays, they can become accelerators of cultural evolution, which is almost as important as biological evolution for the survival of the species. They can act as resource conservers by moving ideas and information to relieve pressure from transportation systems for goods and people. They can provide experience in education and recreation to replace the need of society earlier to occupy people in work. In a workless world, such informational activities will be vital to maintaining people's well being.

Many of the utopian ideas for uses of communications satellites have been suggested elsewhere. It is symptomatic of our times that, despite much talk, there has been little action. Perhaps when mankind has been pushed to the brink, and providing he does not fall over, a new species of man will develop on Earth which will get things done rather than talk about why it cannot do things.

In conclusion, homosapiens may not survive as a species without radical but evolutionary changes to global operations as a living system integrated with the biosystem and geosystem of the planet. Possible alternative futures for mankind offer diverse new communications satellite roles, with such satellites playing more active roles in the superstate or the world confederation than they are likely to play in continued competitive nationalism of today. Satellites can assist the needed integration of the geo, bio, and human spheres of global activities through observations, monitoring, communications in the widest sense, and command functions.

In any continuation of fragmented world nationalism, satellites may not develop rapidly or fully beyond serving narrow special-interest groups and are most unlikely to realize their full potential. In part this may be caused by frequency spectrum wastage arising from fragmented nationalism and also from the economic and political systems under which such satellites communications systems must be designed, developed, and implemented either to make a profit or to serve a military purpose. In a people-homogenous world state, communications satellites most probably would perform entirely different roles in controlling and policing large human populations concentrated in megalopoli but with little military applications. In a people-inhomogenous world federation, communications satellites might realize their full potential to maximize human fulfillment in providing general information relays, cultural evolution accelerators, and resources conservers, and enable mankind to change from moving people and things to moving ideas and information.

References

[1] Burgess, E., ed., The Next Billion Years, NASA Ames Research Center, Calif., 1974.

[2] Murray, B.C., Future Reality, Harper and Row, New York, 1975.

[3] Clarke, A.C., "Extra-terrestrial Relays," Wireless World, London, Oct. 1945.

[4] Burgess, E., "Into Space," Aeronautics, London, Nov. 1946.

[5] Burgess, E., "The Establishment and Use of Artificial Satellites," Aeronautics, London, Sept. 1949.

[6] Burgess, E., "Will US Education Lose in Satellite Race?," The Christian Science Monitor, March 3, 1971.

[7] Burgess, E., "Hawaiian Educators Use Satellites," The Christian Science Monitor, Aug. 3, 1971.

LONG-RANGE PLANNING FOR TELECOMMUNICATIONS IN A MULTINATIONAL ENVIRONMENT

W. G. Gosewinckel*

Overseas Telecommunications Commission,
Sydney, Australia

Abstract

The magnitude and character of the problems which face the coordinated planning and development of the world's telecommunications networks arise from the need for multinational negotiations and agreements in an environment where particular objectives and methods of achieving them are complicated by basic differences of approach, often reflecting national policies. The importance of these differences, which arise from social and ethnic relationships, geographical factors, government regulatory processes, and the availability of financial and technical resources, is discussed together with a description of how several worldwide telecommunications organizations have tackled the problem of long-range planning in such an environment.

Introduction

When one looks at international telecommunications, one first is struck by the very high growth characteristic. Taking international telephony as an example, current growth on a global basis is of the order of 25%/annum, or doubling every three years. Individual traffic streams have a very much higher growth characteristic: typically those streams between developing countries or where high-quality service recently has been introduced.

Presented as Paper 74-490 at the AIAA 5th Communications Satellite Systems Conference, Los Angeles, Calif., April 22-24, 1974.

* Assistant General Manager.

Other international telecommunications services such as telex and higher-speed forms of digital transmission have exhibited even higher growth rates for some years. For example, growth rates for the international telex service have been over 30%/annum for several years and are expected to continue at that rate for at least the next five years.

The next 10 years will see the progressive development of the international telephone service with the introduction of international subscriber dialing, already in use between some major European countries and the United States, whereby subscribers directly dial almost any subscriber in the world. In addition to these traditional services, new services are emerging, some of which can be expected to become major services in the future. These include several forms of visual telecommunications, including international conferences by television and visual telephone systems. It also is expected that services tailored to meet the need for very high-speed transmission of data between computer installations throughout the world will be developed further.

These developments will bring with them the need not only to increase the capacity of international links substantially but also to increase their availability and reliability. The concept of service continuity in international telecommunications already is moving away from the situation where emergency systems are brought into service only during major system failures. Instead, the trend is to provide service via facilities that are available continuously and that employ diverse transmission media and routings and have sufficient redundancy that service can be maintained, even in the presence of a relatively major system failure.

Another interesting characteristic of international telecommunications and one of the important factors that has stimulated demand is that, in almost every country of the world, the rates charged to the subscriber for telephone, telex, data services, etc., are decreasing. This, coupled with instant, high-quality service, means that a greater percentage of the public is using international services. This progressive reduction in the price charged to the subscriber is a reflection of the progressively reducing cost of the intercontinental link. Figure 1 indicates how the per-circuit costs for submarine cables have declined progressively and quite dramatically as a result of new technology and economies of scale. The graph indicates how the annual cost per circuit per kilometer has dropped from the early British 80-circuit submarine cable design as used in the existing Australia-Canada cable

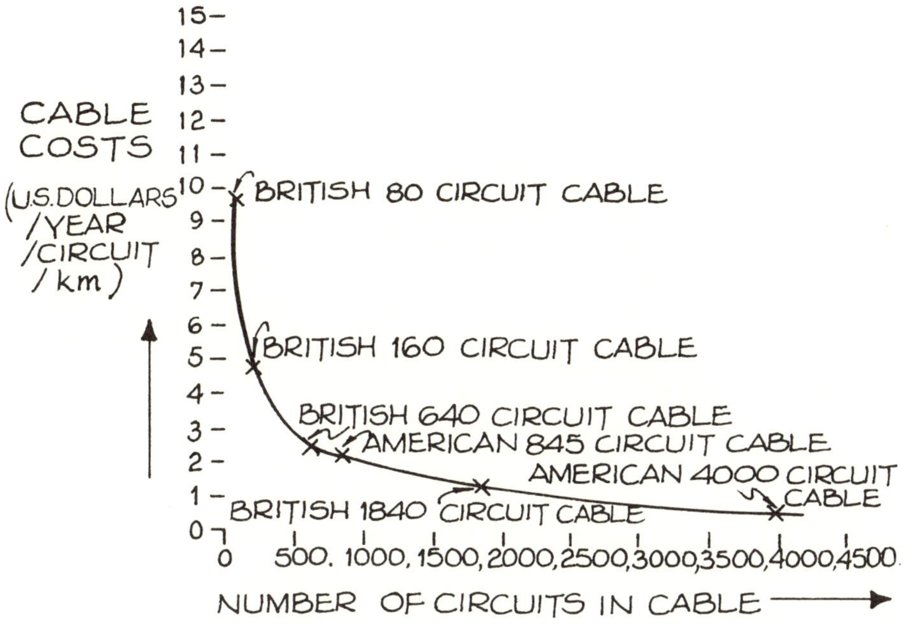

Fig. 1 Decline of cable costs.

(COMPAC) to 0.1 of the cost of the latest British design, which is planned for use between Britain and Canada, and even lower again for the latest U.S.-designed 4000-circuit cable system.

Similarly, the costs associated with the use of communications satellites have declined progressively since their first use for international telecommunications. Figure 2 shows how the charge made for the use of Intelsat satellites has fallen from $32,000 (U.S.) per half circuit per year in 1964 to the current charge of $11,160 (U.S.), with an expectation by the Intelsat organization of a decline to approximately $2500 (U.S.) by the end of the 1970's. Although some costs other than those associated with transmission plant are increasing, the total costs and thus the charge to the customer can be expected to continue the already evident downward trend.

In summary, the future international scene is seen to be characterized by rapidly increasing demand for existing

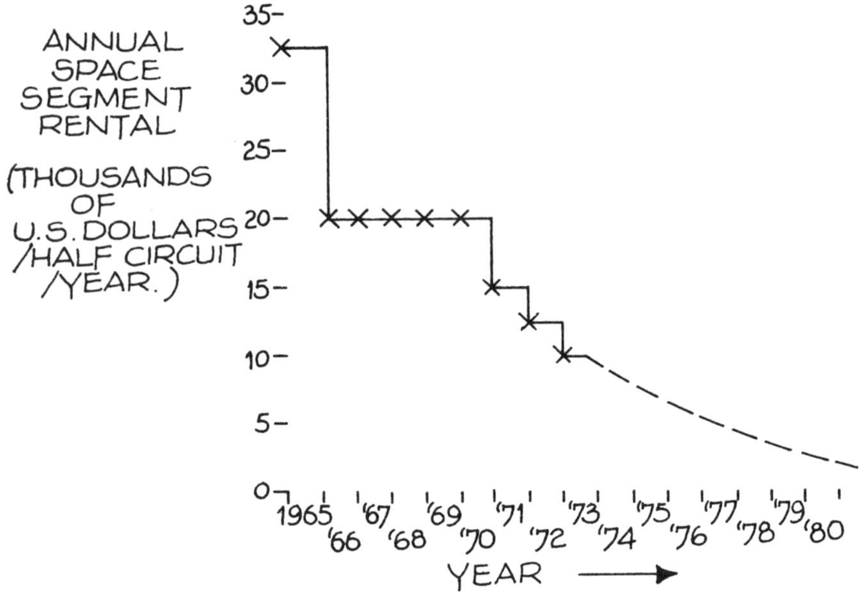

Fig. 2 Annual space segment rental vs year.

services, the emergence of many new services, and a need to increase the availability and reliability of all services progressively.

Technological Change and the Challenge of the Future

In the past, new technology and innovation have met and, in many cases, actually stimulated demand. In the field of intercontinental submarine telephone cables, the capacity has risen from the 48 telephone circuits of the first trans-Atlantic cable in 1956 to the latest system design of over 4000 telephone circuits for transoceanic use. Separately, the spectacular development of commercial communications satellites has produced successive generations of satellites since 1965 with capacities of some 250, 1200, and 5000 telephone circuits, whereas the capacity of the next generation of Intelsat satellites is forecast as being in the neighborhood of 12,500 circuits.

Submarine cable technology was responsible for elevating intercontinental telephony from the vagaries associated with high-frequency radio to a quality similar to that encountered on local telephone connection. Similarly, it was satellite technology that made available for the first time sufficiently large increments of transmission capacity to make intercontinental television a reliable reality. Over the next 20 years, it is expected that these two basic forms of telecommunications technology will continue to provide the means of interconnecting the continents of the world. When one considers the future requirement for greater reliability, the availability of these two very different types of systems assumes great significance. Although different in almost every respect, the two technologies provide systems with many complementary characteristics. For example, each operates in a totally different environment (one in space, the other on the ocean floor) and, as one might expect, situations affecting the operation of one are unlikely to affect the other simultaneously. Secondly, the access of each system is different, the cable providing essentially point-to-point communications, whereas a satellite can be accessed from any location within its coverage zone.

The transmission delay, of great importance for telephony, associated with satellites is comparatively high and independent of distance whereas that of cables is lower and almost directly proportional to distance. Similarly, the costs of a satellite link are the same irrespective of the distance traversed, whereas circuits derived from cable systems are dependent almost directly upon the cable length.

These differing and often complementary characteristics provide the designers of the international networks of the future with almost ideal ingredients. For example, the different media used by an Earth system can, with balanced use, lead to increased reliability, whereas the differing cost vs distance characteristic can lead, by selection, to reduced overall costs.

The rapid technological advances of the past 10 years in both of these fields of communication can be expected confidently to continue. Thus, the major challenges for the future are not seen to arise so much in the area of technical development but more from the need to develop strategies on a global basis to insure that modern technology is utilized in a coordinated and effective manner.

Facilities and Network Planning

As with domestic telecommunications systems, the planning of international networks has two distinct phases: first, planning for the establishment of major facilities such as Earth stations, satellites, submarine cables, microwave systems, switching centers, etc.; and secondly, a plan for their coordinated interconnection to form a network adequate in terms of capacity and reliability.

In the international field, decisions on both of these matters require bilateral and in many cases multilateral agreement. These planning processes take, at least in theory, the same form as planning for national telecommunications networks but are complicated by many factors that derive from their multinational character:

1) Bearing in mind the varying stages of development of different countries and the state of development of their domestic networks, it is not surprising that different views are held on the weight to be placed on such aspects as the desirable reliability and need for diversification of international circuits and the technical performance of the international system as a whole.

2) Furthermore, the stage of development of a country has a substantial bearing on the importance attached to individual services; to a developing country, the telegraph service may be the most important of all, whereas to a highly developed country, the wide-band data service could assume a great degree of significance, particularly for forward-planning purposes.

3) The imposition of government views takes a different form from country to country, depending on whether the international service is the responsibility of a department of state, a public corporation, or private companies, such as in the U.S., where government control is applied by regulation through the Federal Communications Commission (FCC).

4) The position in some countries is complicated further by the fact that the provision and operation of cable and satellite services are operated by separate organizations which produces a competitive rather than complementary situation.

These many differences found between countries in the telecommunications world make their active participation in the planning of global networks, involved as it must be in relatively large financial commitments, a very complex task. As in other fields, it is difficult (and sometimes impossible) to get individual countries to agree on any particular matter, even more so where financial gain or loss is implicit and where the best solution financially for one may not be the best solution financially for all.

Against this background of rapidly expanding demand and multinational involvement, it is interesting to examine two worldwide organizations that have recognized the problems associated with the planning, implementation, and utilization of telecommunications systems on an international basis and have, over the past two years, established planning forums for their resolution.

The two organizations are Intelsat, an international joint venture consisting of 85 countries who collectively own the world's commercial satellite system, and the British Commonwealth Telecommunications Organization, which plans, designs, provides, and operates international networks linking Commonwealth countries with each other and with non-Commonwealth countries.

It is interesting to note that both of these organizations share one important characteristic, and one that has meant the difference between a theoretical exercise in planning and a real and effective one. Each organization operates under a formal international agreement involving financial commitment by the participants. In the case of Intelsat, the costs associated with the in-orbit satellite system are shared by each participating country in accordance with that country's use of the system. Similarly, in the Commonwealth organization, each partner to the scheme may make use of the total submarine cable and satellite facility network and meets his share of the total cost in accordance with the extent of his use. Thus, participating countries to planning studies conducted by these two organizations are aware that the results of the collective studies will impact on them financially as well as operationally.

Intelsat

The Intelsat organization was established in 1964, by agreement between the governments of several countries, to provide for the establishment and operation of a global system of commercial communications satellites. The in-orbit portable system, i.e., the satellites, are owned by participating countries in accordance with a quota determined by each country's use of the facilities, whereas the Earth stations located in each participating country are owned and controlled by that country's telecommunications authority.

The management of the Intelsat system at the policy level is vested in a Board of Governors, comprised of representatives of those countries that have a quota in excess of a certain minimum figure. The Board of Governors of Intelsat has established a number of specialist committees to advise them on several matters relevant to the management of the global satellite system. The most recently established committee is that for the long-range planning of satellite facilities. Its function is to study the global requirements expected to be met by the satellite system between the years 1978 and 1988; formulate a long-term implementation plan for satellites in each ocean area of the world to meet projected requirements, again over the period 1978-1988; and devise investment plans for the satellite system which produce an optimum relationship between satellite costs and Earth stations costs.

British Commonwealth Telecommunications Organization

Another somewhat similar planning organization dealing with telecommunications on a worldwide basis is that established within the British Commonwealth Telecommunications Organization. The Commonwealth Telecommunications Organization is an association of 23 Commonwealth countries that collectively represent more than one-quarter of the world's international telecommunications. Its purpose is to provide the efficient and effective communications between countries of the Commonwealth as a collaborative, financially based exercise.

The financial agreement that makes possible the collaborative global networks is based upon the concept that all network costs, such as those for satellite Earth stations, submarine cable system, international gateway exchanges, etc., are aggregated, with each partner meeting these costs in proportion to its actual use of the network. This arrangement enables the following: the coordination of facilities, leading to more efficient use of them than would otherwise be the case; the carriage of non-Commonwealth traffic, leading to economies of scale and thus enhanced profitability; and the encouragement of development programs that in turn stimulate traffic to the benefit of individual partner countries and the network as a whole.

The body primarily responsible for the oversight of the development of the network and the administration of the Organization's financial arrangements is the Commonwealth Telecommunications Council. Assisting the Council are a number of specialist groups covering such disciplines as planning, operations, tariffs, and finance. One of these groups, dealing with system development, is concerned specifically with developing techniques for evaluating long-term network options in terms of economic and operational consequences. This group has been charged with the task of formulating principles, objectives, and criteria for the long-term development of services and facilities of the British Commonwealth and to produce, in accordance with those principles, a long-term plan for the future.

The Commonwealth international telephone network is a hierarchical network similar in type to that defined by the International Telegraph and Telephone Consultative Committee (CCITT) in their recommendations for a worldwide routing plan for the international telephone service. The only important difference is that the Commonwealth network does not specifically categorize transit switching centers into particular levels but rather determines the network design, taking into account such factors as the capacity of interconnecting facilities and the resultant overall network costs. Figure 3 shows a small segment of the Commonwealth network and demonstrates how satellite and cable facilities are combined to form a cohesive and reliable network.

Problems with Other Planning and Operating Methods

Although the CCITT world routing plan was developed some 10 years ago, it is interesting and perhaps significant that, apart from the Commonwealth network, much of the world's communications have continued on essentially a point-to-point basis, with most countries separately developing a radial network to those other countries of the world with which they have significant volumes of communications. Although the advent of multiple-access satellites certainly has impacted the situation greatly, this uncoordinated and inefficient method of operation derives itself from history, where, because of high transit charges made at intermediate transit points, direct operation from one country to another without involving a third country, although costly, keeps the revenue with the two countries directly involved. A change in approach can be seen only as a practical possibility if effective financial arrangements can be established on a worldwide basis, such

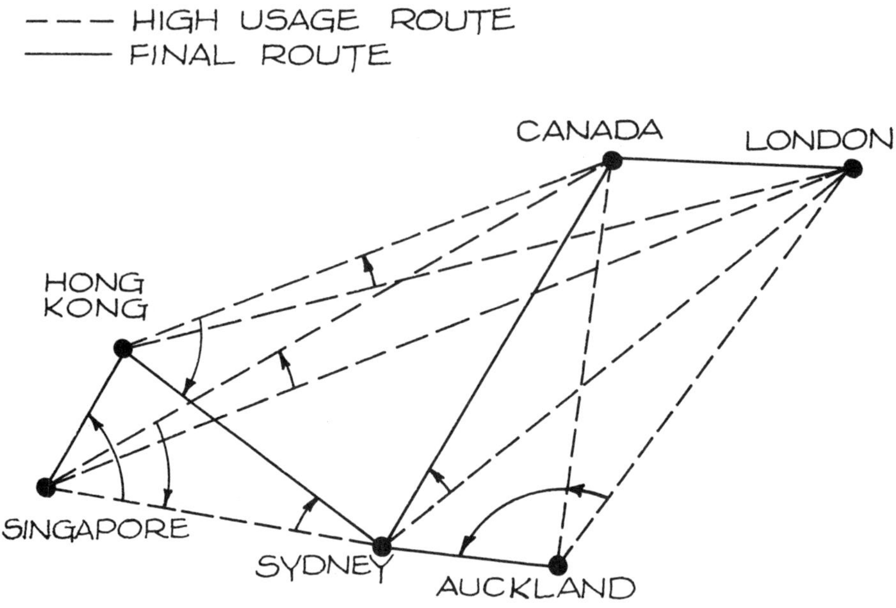

Fig. 3 Commonwealth telecommunications network: typical trunking arrangements.

that, as with Commonwealth countries, financial benefits of transit network operation are shared equitably between all participants in the scheme. Only after such arrangements are agreed to and endorsed by the countries of the world can we expect to have effective planning towards a unified, efficiently utilized worldwide network for the future.

A somewhat different forum set up with the aim of assisting the international planning process was established some years ago by the International Telecommunication Union (ITU). The ITU's World Plan Committee has been of considerable value in the past and is expected to play an important role in the future for the preparation of forecasts of international traffic and for the recording of each individual country's facilities and future plans. To date, the Plan Committee has not achieved the aim of some administrations of developing a fully detailed worldwide network model, primarily because of the difficulty of accurately reflecting into the work of the Committee the impact of commercial agreements between administrations, both for the provision of new facilities on the one hand and the routing of traffic on the other. However, the Plan Committee now is moving toward the provision of reliable data on which administrations may base their future plans.

It is becoming clear that the service objectives of most administrations will call for the parallel future development of cable and satellite systems, and that the advance of cable technology will create cables of such capacity as to require multilateral ownership in an increasing number of cases. Separately, it will be necessary to continue to develop traffic engineering methods in order to utilize facilities most efficiently. The challenge for the future is to develop viable means of planning and developing the total global system, taking account of available and still-to-be-developed technologies, using different technologies to complement each other, and providing the necessary grade of service at the minimum cost. This will require coordination and consultation between administrations at a level substantially higher than has been necessary previously, and will create planning organizational problems to which solutions will not emerge readily. A fundamental requisite to the generation of solutions is considered to be the recognition that effective facilities planning and facilities utilization on a worldwide scale require financial involvement in which the economic benefits of efficient planning flow to all participants.

Summary

1) The future is expected to bring with it continuing rapid growth in world telecommunications, with decreases in costs adding stimulus to the existing very high growth rates. New services will emerge, and existing services will become more sophisticated, thereby demanding a greater degree of reliability.

2) The main methods employed for intercontinental transmission will be submarine cables and communications satellites, which need to be seen by communicators not as competing but as complementary systems.

3) Technology almost certainly will be available when required to meet demand.

4) The practical means of achieving a properly coordinated effective and efficient world network of the future will rest on the ability of the users of technology (the

international telecommunications organizations of the world) to develop jointly the necessary financing and planning machinery on a global basis to insure that advanced technology is exploited effectively.

5) The Intelsat organization seems to provide a useful guide for the financing, planning, and implementing of major international facilities of the future, whereas the British Commonwealth Telecommunications Organization provides an example of how, given the right financial arrangements and incentives, these facilities can be interconnected effectively to form an efficient and effective integrated world network for the benefit of all.

6) Effective network development and utilization on an international basis can be performed only against a framework of financial involvement in both the planning and operational stages.

PHILIPPINE DOMESTIC SATELLITE SYSTEM

Geoffrey Cheadle*

Philippine Overseas Telecommunications
Corporation, Washington, D.C.

Abstract

The Philippines needs to expand its domestic telecommunications capabilities and augment the amount and type of services provided. In particular, educational television is needed by the many schools in remote areas. It appears that a domestic satellite system would be the quickest and most economical solution. Three main questions are discussed: 1) what types of service at which locations should be provided?; 2) what is the best overall technical/operational/ economic approach?; and 3) how can the project be financed? The Philippine Overseas Telecommunications Corporation has a plan. Services would include color television and duplex voice-grade channels. Fourteen stations are envisaged, beginning in the Mindanao-Visayan region in 1976, in addition to a base station at Tanay (the location of the existing international stations). The best technical/operational/economic approach involves the use of simplified nominal 10-m stations using narrow-band FM single-channel-per-carrier modulation.

Presented as Paper 74-491 at the AIAA 5th Communications Satellite Systems Conference, Los Angeles, Calif., April 22-24, 1974.

*Manager, Administration Department, International Telecommunications Satellite Organization (INTELSAT); also Brigadier General, U.S. Air Force (retired).

Initial implementation will involve lease of INTELSAT space segment capacity, with possible later implementation of a regional satellite program. Financing is expected to rely on commercial revenues in the absence (as yet) of subsidization.

Requirement

Geography

The Philippines is a country whose geography presents problems for domestic telecommunications networks. The country is large, scattered, rugged, and physically isolated. Being 1100 miles north-south by 700 miles wide and comprising about 116,000 square miles (which is about the area of Arizona), the Philippines is nevertheless scattered and fragmented. There are 7000 islands, of which only 2800 have names, 730 are inhabited, and two-thirds of the population lives on only two islands: Luzon and Mindanao.

The Philippines is rugged and mountainous with about a dozen active volcanoes. It is typhoon and earthquake-prone. All of these features, of course, make building terrestrial systems, such as microwave and troposcatter, extremely difficult. The situation also strongly calls for diversity of routing and availability of spare capacity for restoral of failed systems.

Natural Resources

The Philippines are blessed with a great amount and a great variety of natural resources, much of which are not yet fully developed. In agriculture there are rice, sugar cane, coconut, tobacco, and fiber plants. Gold, copper, lead, and manganese are mined in abundance. Fish abound. Forests provide large-scale lumber exports to Japan and the United States, and the extensive coastline with its many beautiful beaches is beginning, only now, to provide tourist revenues. There is a great need to extend international telecommunications into the hinterlands of the Philippines to stimulate and support a beginning large-scale growth of industry and commerce involving these many resources.

The People

The population of the Philippines has grown from 18½ million in 1946 to 40 million in 1973. The country enjoys 75% literacy, with 10 million people in school. English is spoken by 50% of the population, which also speaks some 70 native languages and dialects. There is a great need to upgrade the quality and scope of education in the remoter areas and to make up for the extreme shortage of teachers and teaching materials. The President of the Philippines has endorsed educational television as a means of bridging the gap. Agencies already exist to provide the materials and the programing.

The Requirement

The requirement for a domestic satellite telecommunications system thus rests on the twofold need for educational television in the remote areas and expanded commercial voice and data communications to all areas. Satellite communications are preferred to terrestrial because of ease and speed of implementation and compatibility with both television and voice-channel usage.

Technical/Operational/Economic Approach

Technical

The technical requirement is for a system using relatively inexpensive, easily maintainable Earth stations and the least expensive but capable space segment to provide both color television at a quality permitting rebroadcast plus a significant number (say 10-20 per station) of duplex voice-grade channels. Until about 1973, the biggest problem was how to achieve both television and communications channels without requiring excessive space segment capacity. It looked as if the capacity equivalent to an entire INTELSAT spot-beam transponder would be required. Fortunately, technology has progressed to permit the required service using only an INTELSAT global-beam transponder or equivalent. The lease charges of a global-beam transponder are half those of a spot-beam transponder. An alternate technical solution to the space segment requirement is the use of a "dedicated" or regional satellite. Such a satellite could be tailor-made for the Philippine requirement and would provide whatever capacity would be desired. The choice between use of INTELSAT and launch of a regional dedicated satellite is a matter of money, time, and risk. Current estimates as to what an entire global-beam transponder might lease for, under various conditions, range from $3 million in 1976 for a straight 1-yr lease of an ordinary transponder down to a possible $700,000/yr for use of spare capacity in a longer-term lease. On the other hand, estimates of implementations costs of a regional system using two satellites in orbit and a spare on the ground have been heard to run the gamut from $50 million to $100 million. Not only are there uncertainties in cost projections for both approaches, but there are a number of intangible considerations, such as risk factor, national prestige, and expansion capability. The current plan is to lease INTELSAT capacity for the first year or two and keep open the option of implementing a regional satellite program.

Operational

The operational hub of the domestic network will be a new central-control Earth station to be located at Tanay near Manila, where the two INTELSAT stations are located now. Four stations in the Mindanao-Visayan region (in the South) will open along with the base station in January 1976. Three additional stations will be operational by 1977 and seven more in the period 1977-1980, for a total of 15 Earth stations. If and when a regional satellite is launched, of course, a satellite control station will be required. It is probable that the effective use of communications channels will be augmented by some sort of centrally controlled demand assignment scheme. It is anticipated that use of the television capacity will include both educational and commercial television, whereas the voice-grade channels will carry ordinary long-distance telephone, leased private lines, telex and message service, and narrow-band data. A final operational scheme will be adopted which is simple, easy to operate, and easy to repair.

Economic

The economic approach, as touched on slightly in the foregoing, will be to keep the system simple and easy to maintain and operate. Fortunately, Earth stations filling this requirement (which also provide the necessary operational capability) now can be considered "off-the-shelf." Small stations with about 10-m sized antennas using simplified programed tracking schemes can be bought for somewhere around $200,000 each. The overall program poses an investment requirement of $6-8 million including nonrecurring costs such as site preparation, installation and test, etc. The largest cost will be the space segment. Whether this is provided by INTELSAT or regional dedicated satellite, it can be looked at as a recurring operational cost that will probably exceed 50% of the yearly outlay. No way of significantly reducing the space segment cost has been conceived.

Financing

It is considered no problem to obtain supplier-financing or similar help, say from the U.S. Export/Import Bank, for the purchase of equipment and associated costs. The problem is to identify the revenues necessary to make the system viable as time goes on. The Philippine Government has many other drains on its financial resources and cannot be expected to subsidize the domestic satellite program as a whole or the educational television service in particular. As yet, no foundation grants or U.S. Governmental assistance have been identified. There is a case for such assistance to stimulate the development of the country, but there are many bigger, perhaps more crucially urgent projects in competition.

It is expected, therefore, that the project will have to depend on its own operating revenues from commercial sources. There are two of these: commercial television and ordinary telecommunications services: voice and data. Commercial television is a part-time thing and, as INTELSAT has found, is good for the medium's "image" but not nearly good enough to support the system by itself.

The main source of revenues then boils down to commercial telecommunications. This would include long-distance telephone lines leased to the telephone companies, private leased lines (both voice and data), telex, and message service. Unfortunately, new rates cannot be established in isolation at the precise amounts needed to support the system because there are existing rates that would have to apply. Although the Philippines is a developing country and needs a domestic satellite system, there do exist two operational wide-band microwave systems generally linking north and south population centers. Although these existing systems are not perfect, they do provide quality service, so that satellite service is neither different in nature to the user nor of sufficiently greater quality to justify new and higher rates between the same pairs of stations. The current rates are low enough to pose a special problem to the satellite system planner. Suffice it to say at this point that aggressive marketing and careful implementation will be necessary to derive adequate revenue in the absence of significant gratuitous assistance.

Summary

In summary, the Philippines, by its geography and its developing status, needs a telecommunications expansion that is provided best by a domestic satellite system. The Philippine Overseas Telecommunications Corporation is proceeding with its plan to inaugurate such a system beginning in January 1976 with an initial five Earth stations, building to an eventual 15. The stations will be simple and will provide both rebroadcastable quality color television for education and commercial purposes and ordinary commercial telecommunications services. The space segment will be provided initially by lease from INTELSAT, with a possible later implementation of a regional dedicated satellite system. Financing probably will depend mostly on ordinary telecommunications revenues. It is expected that the domestic satellite system will satisfy ongoing needs as well as stimulate accelerated development of the Philippine "Hinterlands."

OVERVIEW: MILITARY SATELLITE COMMUNICATION SYSTEMS

Frederick E. Bond*
Defense Communications Agency, Washington, D.C.

Abstract

Since satellite communications systems exhibit a number of characteristics which are desirable for military application, the demand for use of this medium is increasing. Programs for system development and acquisition to date have been on a piecemeal basis. A need has evolved for a more unified approach. The Military Satellite Communications Systems Office (MSO) was established to meet this need. Initial activities of the MSO have included a requirements survey and analysis and a review of applicable technology for future systems. Issues pertinent to the selection of options for new systems include cost, satellite size and complexity, system control, frequency management, and survivability.

I. Introduction

The demand for Department of Defense (DOD) satellite systems to meet many diverse types of military communication requirements is growing steadily. Budget constraints and problems concerned with frequency allocation and assignments require that the systems meet these demands as efficiently and economically as possible. Satellite systems offer a number of distinct advantages that can be exploited for military uses. For example, they can provide

---

Presented as Paper 74-456 at the AIAA 5th Communications Satellite Systems Conference, Los Angeles, Calif., April 22-24, 1974.

* Deputy Director, MILSATCOM Systems Office.

practically global communication coverage and serve areas that are virtually isolated or lacking in significant terrestrial communication facilities. They are capable of supporting wide transmission bandwidths for applications involving very high data rates. For contingency operations, the use of transportable or mobile terminals permits rapid extension of major communication networks into new areas, or restoral of communications to locations whose facilities are destroyed or damaged.

For a generally different class of communication needs requiring service to moving platforms, such as aircraft, ships, and submarines, satellite circuits offer a more reliable alternative to the use of high-frequency ionospheric propagation for transmission beyond line of sight.

The use of satellite relay for critical command and control networks serving forces which are deployed in a combat area or amphibious landing zone offers a means for significant reduction in the logistics burden imposed by the use of conventional radio relay and troposcatter systems, and improvements in transmission performance for those circuits which rely on high-frequency propagation.

The characteristics of these applications, which are unique to military needs, include the following: 1) Security is required for both communications and system control. 2) The systems must be operable under conditions of jamming. 3) They must serve a wide variety of users with differing terminal sizes, data rates, and types of modulation. 4) They must be flexible and capable of rapid reconfiguration, with no time for political or diplomatic negotiations.

II. Current Military Satellite Communication Systems

The system presently serving the long-distance trunk requirement is known as the Defense Satellite Communications System (DSCS). Its primary mission is to support the World Wide Military Command and Control System (WWMCCS). The initial version, DSCS I, started in June 1966 with the launch of small and relatively simple shf satellites in near-synchronous orbit. The next generation (DSCS II) design is a larger, more powerful satellite including both narrow-beam and Earth coverage antennas. Two were launched first in November 1971. After considerable redesign effort, the next two were launched on December 13, 1973, and now are providing service to a variety of shf terminals with transmitter power ranging from 100 w to about 10 kw and antenna sizes ranging from 6 to 60 ft. Two more satellites were launched on May 20, 1975; the launch, however, was unsuccessful as the third stage failed to boost out of the parking orbit.

Activity in satellite applications in the military uhf band (240-400 MHz) began in 1967 with the launch of the small Lincoln Laboratory experimental satellite LES-5, demonstrating low-data-rate (75 BPS) transmission to aircraft, ships, land vehicles, etc. The more powerful LES-6 and TACSAT-1 satellites, launched in September 1968 and February 1969, respectively, permitted the technical and operational feasibility experiments to continue on a broader scale and eventually went into use as an interim operational system. TACSAT-1 ceased operation in December 1972. LES-6 is still performing with an Equivalent Radiated Power (ERP) reduction of about 8 db from its initial value of 27 dbw. Experience gained with this system has been incorporated into the concept of the Fleet Satellite Communications System (FLTSATCOM). FLTSATCOM will provide a global uhf relay as the primary transmission medium for beyond line of sight ranges for Navy ships and aircraft. For interim service, the Navy will be leasing uhf repeater channels on two COMSAT General Corporation MARISAT satellites. The first one is scheduled for launch in September 1975.

The FLTSATCOM satellite also will provide low-data-rate uhf communication channels serving high-priority Air Force needs. The Air Force portion of the system is known as the Air Force Satellite Communications System (AFSATCOM). In addition to the channels in the FLTSATCOM satellites, AFSATCOM terminals will use relays on other host satellites to obtain additional coverage.

Another program, which is still in the research and development stage,is the development of a survivable system to serve the most critical general war command and control functions -- SURV-SATCOM. Activity under this program includes the Lincoln Laboratory satellites LES-8 and 9 scheduled for lauch in late 1975.

In January 1974, a program memorandum was approved by the Defense Department covering the need for a family of small satellite terminals in the uhf and shf bands to support critical ground forces operations involving Army, Marine Corps, and Air Force units. Initially, the requirements will be served by channels on FLTSATCOM/AFSATCOM and DSCS satellites.

III. Need for a Unified Approach

The individual systems and programs described above are oriented toward specific user communities within the DOD. A need has evolved for a more unified technical approach for planning, development, acquiring,and operating future SATCOM systems. Some of the basic considerations driving this need are 1) satellite systems can potentially provide improved communication service for a wide spectrum of military missions and functions; 2) the degree to which all possible requirements could be satisfied is limited by a) the high cost of satellites and terminals, and b) allocations of radio frequency spectra and satellite orbit positions, and associated problems of interference; 3) priorities and needs of the user communities are sensitive to the effects of significant military and political events which occur throughout the world. Therefore, the systems should be as flexible as possible with the capability of providing efficient service through a wide range of scenarios, including contingencies, crises, limited wars, and global nuclear conflicts; 4) a number of the individual user requirements are intermittent with a low "duty cycle." This affords an opportunity for greatly improved efficiency using shared services within and among user communities; and 5) satellite system technology is dynamic. The entire community must be prepared to exploit new advances in planning future systems. In some cases this could impact operational doctrine, particularly as it relates to communication procedures.

To meet the need for the unified approach, the Director, Defense Communications Agency was given primary responsibility for performing the functions of System Architect for current and future DOD SATCOM system. In October 1973, the Military Satellite Communications Systems Office (MSO) was established within the Defense Communications Agency (DCA).

IV. MILSATCOM Office

The MSO Architectural role includes the following: 1) systems planning at DOD level for all future Military Satellite Communications systems, 2) total DOD requirements survey and analysis,

3) interprogram systems engineering, and 4) interoperation among DOD systems and with United Kingdom and NATO systems.

The primary focus of MSO activities to date has been in the areas of requirements analysis, generation of candidate options for future systems, and review of pertinent research and development activities.

V. Requirements Analysis

To begin the process of categorization and prioritization, a set of criteria is needed to evaluate each requirement. The criteria include operational factors (e.g., mission priority, circuit priority), technical factors, and cost considerations. As a first step in the process, the concept of "user community" was introduced. A loose definition of a "community" is a group of users sufficiently large and with enough common communication characteristics to potentially qualify for their own exclusive satellite communication system. Such characteristics might include user mission, terminal constraints, type, rate, and quality of service, survivability, message characteristics, and permissible type of control. Figure 1 lists the suggested user communities together with some primary and secondary interconnectivity requirements.

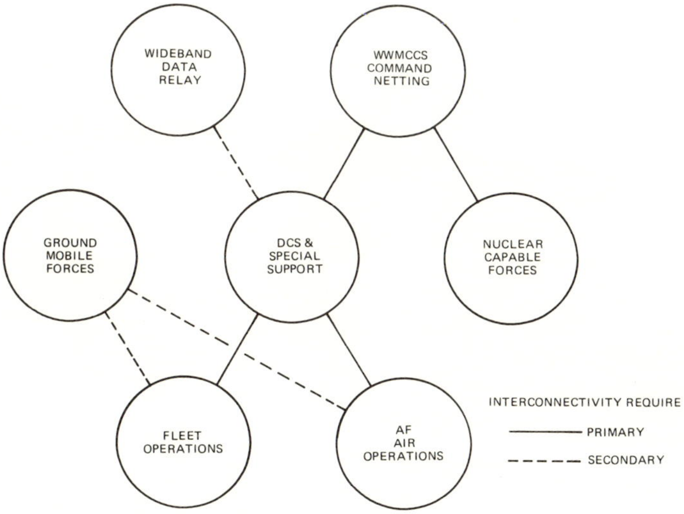

Fig. 1 Suggested user communities.

Figure 2 is a summary and comparison of the basic needs of the communities in terms of total capacity, relative number of terminals involved, and electronic and physical survivability. Generally speaking, airborne and shipborne mobile terminals are served by SATCOM systems in the military uhf band because of terminal cost constraints. This is especially important where a large number of terminals are involved. The Defense Communications System and wideband data communities require the largest capacity and the Nuclear Capable Forces community the smallest. With regard to anti-jamming and physical survivability, the Nuclear Capable Forces community and a portion of the WWMCCS Command Net require the most protection.

To complete the requirements analysis the totals for each user community have been categorized in three time periods: 1) Near Term -- 1975-80; 2) Mid Term -- 1980-84; and 3) Far Term -- 1985 and beyond. Furthermore, their sensitivity to scenario (ranging from day-to-day operations through general war) has been measured, as affected by sizing of force structures involved.

	SYSTEM CAPACITY	NO. OF TERMINALS	AJ	PHYSICAL SURVIVABILITY
● NUCLEAR CAPABLE FORCES (NCF) (MOBILE TERMINALS)	LOW	MODEST	SUBSTANTIAL	YES
● WWMCCS COMMAND NETTING	MODEST	FEW	SUBSTANTIAL	LIMITED
● HIGH–CAPACITY COMMS (FIXED AND TRANSPORTABLE)	VERY LARGE	MODEST	SELECTIVE (W/LOWER CAPACITY)	NO
DCS				
WIDEBAND DATA RELAY				
● NON–NCF (MOBILE TERMINALS)	LARGE	MANY	SELECTIVE	NO
GROUND MOBILE FORCES				
FLEET OPERATIONS				
AF AIR OPERATIONS				

Fig. 2 Comparison of needs of user communities.

	DSCS–II	AFSAT	FLTSAT
DCS & SPECIAL SUPPORT	X		
WWMCCS CMD NETTING	X		
INTELLIGENCE RELAY	X		
GMF	X	X	X
NCF (NON–SURVIVABLE)		X	
AF AIR OPERATIONS		X	
FLEET OPERATIONS	X		X

Fig. 3 Current commitments.

Figure 3 relates the user communities to the satellite communication systems which already exist or are presently in development and acquisition.

VI. Issues Pertinent to Selection of Options for Next Generation Systems

The pertinent issues include cost, satellite size and complexity, system control, frequency management,and survivability. System cost is a major consideration and must include research and development, production, launch, operation, and replenishment. Replenishment is principally related to satellite life, a factor which unfortunately is not well calculated. Experience to date strongly indicates that small, simple satellites last longer than large complex ones but there is no precise method for relating these factors, other than by statistical failure rates of components. One explanation advanced is that,if a satellite is too complex, it is too difficult to test exhaustively for all possible trouble sources prior to launch. More work needs to be done on failure mechanisms, design practices, etc., before the subject can be treated with more precision. In regard to system life, some have made a case for the very simple, small (perhaps one channel) spacecraft (DSCS-1 approach). With the trend toward more complex connectivity and the need for a more responsive satellite control, it may be extremely difficult to meet the burden imposed by such a large increase in the number of satellites. Thus, there are conflicting factors (cost, control, risk) relating to what might be an "optimum" satellite size and complexity to serve a given set of users.

The limitations in frequency allocation and potential interference problems are also an important aspect to be considered. The DSCS and potential follow-on systems occupy the shf band, 7250-8400 MHz. The military uhf band (240-400 MHz) is used for FLT-SATCOM, AFSATCOM, and potential follow-on systems. Congestion in the uhf band is large and growing. Clearly, frequency reuse is very difficult because most of the terminals use simple antennas with wide beams and the satellite antennas are for Earth coverage. Since the congestion problem is increasing, it is important that future system design must be reasonably flexible with respect to frequency assignment restrictions, and compatible with terrestrial systems.

In addition, channel utilization efficiency must be improved. The use of "dedicated" channels must be discouraged where possible. Development and application of demand assignment and sharing type techniques, with suitable control, must be accelerated.

Other candidate frequency bands for future military mobile use are under consideration, but no near-term solution is in sight.

Use of the shf band is also encountering problems. Obtaining assignments at specific locations in the world is becoming more difficult. Also, the potential high-data-rate transmission requirements discussed previously require large bandwidths. The use of higher frequencies (e.g., ehf) must be initiated to keep up with the increased demand. The higher bands introduce new problems, e.g., excessive attenuation through rain and fog. Propagation measurements are in process, but more data are needed.

VII. Impact of New Technology

In the antenna area the use of electronically controlled multibeam antennas will provide a significant increase in flexibility of coverage, uplink discrimination, and ability to reconfigure coverage areas rapidly. Also, the demonstrated ability to deploy a large spacecraft antenna (30 ft) opens the possibility for narrow beams at the military uhf band. This will permit more ERP for simple antennas and may pave the way for frequency reuse at uhf.

The advent of Time Division Multiple Access (TDMA) will allow more efficient use of satellite transmitter power. In the case of the DSCS, it will ease the power control problem now present with the large number of accesses. For user communities with short infrequent messages and a very large number of potential accesses (e.g., Fleet and Air operations), the use of TDMA imposes operational problems caused by the necessity to coordinate and synchronize prior to message transmission. A compromise system has been suggested using Frequency Division Multiple Access (FDMA) in the uplink, performing detection and processing in the satellite, and formatting TDMA data streams at the transmitter for downlink transmission.

The ability to perform analog functions (e.g., band filtering, limiting, detection, etc.) at Intermediate Frequency using digital circuits and the advances in Medium Scale Integration (MSI) and Large Scale Integration (LSI) techniques offer a potential for considerably more on-board satellite processing without sacrifice of reliability. Other technology improvements will offer lighter weight components, higher power amplifier efficiencies for solid state devices and tubes, and advanced systems for more efficient generation and storage of DC power. Collectively, these improvements offer higher capacity per unit weight (and, therefore, per unit cost) and greater survivability.

VIII. Summary and Conclusions

Military satellite communication requirements are increasing rapidly. Several practical limitations exist on achievable capacity and survivability. New technology will permit growth, but cost, frequency spectrum limitations, and other factors will always tend to limit the capability. The system architecture must necessarily include a continuing requirements analysis and tradeoffs with overall system costs. The evolution from present to follow-up systems will face a number of transition problems with respect to both the terminal and space segments. A significant challenge exists for both government and industry to plan and implement future systems which will assure the most effective use of this unique transmission medium.

CHAPTER III—SYSTEM ANALYSIS AND TRADEOFFS

The communication requirements stated by the user and the geopolitical constraints define the general problem for the system engineer. From this information it should be possible to develop a set of system performance criteria that can be used as the basis to evaluate a range of system alternatives. Developing performance criteria is an essential step in the system engineering process and will determine the ultimate success of the entire system design effort. This design step translates a number of very nebulous factors into quantifiable and measurable parameters that are agreed to by all the concerned parties.

Frequently it is difficult to define specific values for the system criteria without performing a sensitivity analysis in relation to the specific parameter in question. This sensitivity analysis is usually accomplished by assuming a gross system representation and then allowing the parameter in question to vary across a reasonable range of values while assessing the impact on cost and performance. The resulting parametric study is used to determine the desired operating point for the parameters that drive the system design. This process of system engineering is obviously an iterative one, and experience or the existence of an historical data base can save several iterations. The paper by Meyerhoff presents a technique for performing an iteration process to determine the operating point for one of the important system parameters.

It is also important for the system engineer to realize that the concerned and influential parties may not have an in-depth technical background to understand all the technical details, and therefore it is essential for the system engineer to be able to present his analysis in clear and concise terms. A tremendous amount of effort has been wasted on many programs by performing detailed system design before a consensus has been reached on the basic system objectives and performance criteria. Therefore, the system engineer must determine the important system issues and present sufficient tradeoff analyses to resolve them. The paper by Levit addresses generalized design and optimization techniques that will assist the system engineer in resolving critical issues. Upon resolution of the issues and the establishment of the peformance criteria, the system engineer selects a range of plausible system alternatives for extensive evaluation. These alternatives should be sufficiently different from a cost standpoint to provide a real choice to the evaluator.

The alternatives should be ranked according to their capability to satisfy the various evaluation criteria. This type of comparison is demonstrated in the paper by Stiglitz. The overall performance and cost of each alternative should be determined and a relative cost effectiveness assessment performed. The recommended alternative or alternatives should include detailed implementation plans that clearly indicate the date the system will become operational.

The inherent capabilities of communication satellites, the large coverage area, and wideband channels are also well suited for certain navigational needs. A relatively small number of navigational satellites can provide global coverage, and by receiving broadband signals from several of these spacecraft the user can accurately determine his location. This capability is demonstrated in the paper by Lee and Wade.

The same satellites could be used for both communication and navigation. It is technically possible to use the same signals for both purposes, but the frequency allocation agreements, referred to previously, would have to be modified to permit both services in the same frequency band. The spectrum utilization tradeoffs associated with combining different services on a single satellite are addressed in the paper by Long.

POWER BALANCING IN MULTIBEAM SATELLITES

H. J. Meyerhoff*

COMSAT Laboratories, Clarksburg, Md.

Abstract

Multibeam satellites make efficient use of the spectrum allocated for space communications by reusing these frequencies on adjacent beams. Cochannel interference is introduced by the antenna beam pattern. The carrier power levels are computed to minimize the interference-to-carrier-power ratios on all beams, which is shown to be equivalent to equalizing these ratios with one carrier power (or the sum of the carrier powers) preselected. An iterative procedure is proposed to determine these carrier power levels, which correspond to the eigenvector associated with the largest positive eigenvalue of the antenna beam pattern matrix. The procedure converges more rapidly than other methods now available.

Nomenclature

C_k	carrier power on k antenna beam
$E_k = (C_k/C_1)$	normalized carrier power level on k antenna beam with respect to C_1
$E_k(L)$	iterative variable corresponding to normalized carrier power level
E_{ko}	eigenvector corresponding to Perron eigenvalue R_o

Presented as Paper 74-470 at the AIAA 5th Communications Satellite Systems Conference, Los Angeles, Calif., April 22–24, 1974. This paper is based upon work performed in COMSAT Laboratories under the sponsorship of the International Telecommunications Satellite Organization (INTELSAT). Views expressed in this paper are not necessarily those of INTELSAT. The author wishes to thank Dr. K. Chitre and Dr. O. Shimbo for useful discussions and encouragement.

*Senior Staff Scientist, Systems Studies Division.

I_j	interference power on j antenna beam
N	number of spot beams
N_o	total thermal noise power on each beam normalized to carrier power C_1
R(L)	iterative variable corresponding to interference-to-carrier-power ratio
R_{jj}, R_j	interference-to-carrier-power ratio on j antenna beam
R_m	real eigenvalues of antenna beam matrix, other than R_o
R_o	Perron eigenvalue of antenna beam matrix
w_{jk}	matrix element of spot-beam antenna including effect of thermal noise
x_{jk}	matrix element of spot-beam antenna patterns

Introduction

Future satellites will be equipped with spot-beam antennas that concentrate the satellite transmitted power over separate small areas of Earth's surface. The angular separations at the satellite between these spot beams provide sufficient discrimination in both reception and transmission to enable the reuse of frequencies on adjacent beams.[1] In this way, heavy traffic links for points-to-point coverage can be implemented within the limited frequency bands allocated for space communications. Global and regional coverage antennas associated with most present-day satellites are more suited to broadcast and low-density multidestination traffic schemes in which frequency reuse occurs only to a limited extent, if at all.

Frequency reuse implies cochannel interference received at and transmitted from the satellite. It is possible to calculate the interference-to-carrier-power ratio on each spot beam from a knowledge of the carrier power levels received and transmitted and a matrix description of the spot-beam antenna gains in the direction of their adjacent spot beams. To minimize the interference-to-carrier-power ratios on all spot beams, with the spot-beam antenna patterns fixed, a unique choice of relative power levels for the carriers may be determined, applicable to Earth stations in the up-link and at the satellite for the down-link. The effect of thermal noise also is considered.

This paper is concerned with computing the relative carrier power levels that minimize the interference-to-carrier-power ratios on all spot beams.[2,3] The method proposed provides a solution to a set of simultaneous nonlinear equations by an iterative procedure that is unconditionally stable, rapidly convergent, and less complicated to implement than a variety of alternative techniques.[4,5] This program has been employed for 10 x 10 matrices with computer time reduced more than twentyfold over alternative conventional computational methods. The computer time saving increases as the matrices become larger.

Carrier Power Normalization

The interference power received in one satellite spot beam is the sum of the carrier powers aimed at the remaining spot beams, weighted in accordance with the particular spot-beam antenna pattern. These carrier powers may be summed since the signals on each beam are assumed to be statistically independent.

The interference-to-carrier-power ratio for the j spot beam is given by

$$R_{jk} = R_j = (I_j/C_j), \quad k = j$$
$$= 0 \quad , \quad k \neq j \tag{1}$$

where

$$I_j = \sum_{k=1, k\neq j}^{N} x_{jk} C_k \tag{2}$$

In matrix form Eq. (1) becomes

$$[0] = \{[-R] + [x]\} \quad [C] \tag{3}$$

where

$$x_{kk} = 0, \quad k = 1, N \tag{4}$$

A trivial solution to Eq. (3) is [C] = [0]. If one of the carriers, e.g., C_1, is selected arbitrarily on the basis of certain link parameters, we may write the interference-to-carrier-power ratio R_j in terms of normalized or relative carrier powers; then Eq. (1) becomes

$$R_j = \sum_{k=1}^{N} (x_{jk} E_k)/e_j \tag{5}$$

where

$$E_k = (C_k/C_1), \quad k = 1, N \tag{6}$$

The normalized version of Eq. (3) becomes

$$[0] = \{[-R] + [x]\} \quad [E] \tag{7}$$

where $E_1 = 1$ excludes the trivial solution associated with Eq. (3).

Selection Criterion for Normalized Carrier Power Levels

The criterion for selecting the relative carrier power levels E_k is to minimize the maximum interference-to-carrier-power ratio R_j, where $R_j \geq R_k$, $k = 1, N$. One method of solution would require a multidimensional search over all positive real values for the E_k, $k = 2, N$, which is a cumbersome computational procedure.

It can be shown (Appendix A) that minimizing the maximum R_j of R_k, $k = 1, N$, is equivalent to equalizing these interference-to-carrier-power ratios, i.e.,

$$R_k = R_o, \quad k = 1, N \tag{8}$$

R_o is then the unique positive eigenvalue of matrix [x] [the Perron root] the elements of whose eigenvector $[E_o]$ are all real and positive.[6] Other positive eigenvalues will not have all elements of their eigenvectors real and positive, which is required for a physically realizable solution. The values for E_k, $k = 2, N$, are derived from the reduced order [relative to Eq. (7)] matrix equation since, with R_o determined, there are only $N - 1$ unknowns E_k. Thus,

$$\begin{bmatrix} -x_{21} \\ -x_{31} \\ \cdot \\ \cdot \\ \cdot \\ -x_{N1} \end{bmatrix} = \begin{bmatrix} -R_o & x_{2,3} & \cdots & x_{2N} \\ x_{32} & -R_o & \cdots & x_{3N} \\ \cdot & \cdot & & \cdot \\ \cdot & \cdot & & \cdot \\ \cdot & \cdot & & \cdot \\ x_{N2} & x_{N3} & \cdots & -R_o \end{bmatrix} \begin{bmatrix} E_{2o} \\ E_{3o} \\ \cdot \\ \cdot \\ \cdot \\ E_{No} \end{bmatrix} \tag{9}$$

Hence, the normalized carrier power levels E_{ko}, $k = 2, N$, may be determined by first evaluating the largest real positive eigenvalue R_o of matrix [x], substituting this value R_o into the reduced order matrix Eq. (9), and finally inverting Eq. (9).[2] The uniqueness of the solution for the E_k, k = 2, N, is guaranteed by Perron's theorem.[6] The computational procedure is again cumbersome.

Iterative Solution of Nonlinear Simultaneous Equations

Equation (5) for the normalized carrier power levels E_k, k = 2, N, for equal interference-to-carrier-power ratios R on all beams may be rewritten in the form

$$E_j = \sum_{k=1}^{N} \frac{x_{jk}E_k}{R}, \quad j = 2, N \tag{10}$$

where

$$R = \sum_{k=2}^{N} x_{1k}E_k \tag{11}$$

This set of simultaneous nonlinear equations may be solved by an iterative procedure according to

$$E_j(L) = \frac{\sum_{k=1}^{N} x_{jk}E_k(L-1)}{\sum_{k=2}^{N} x_{1k}E_k(L-1)} \tag{12}$$

where L is the iterative index.

The number of iterations for convergence to within certain accuracy limits depends on the initial arbitrary values $[E_k(0), k = 2, N]$ selected for the normalized carrier power levels and the positive antenna gain matrix, elements $x_{ij}(E_1 = 1)$. The procedure approximates the unique set of positive values for the eigenvector $[E_o]$ associated with the Perron eigenvalue R_o of matrix [x] referred to previously.[6] Appendix B gives the Fortran Program for the iterative method.[4,5]

Convergence of the Iterative Procedure

Proof of convergence for the iterative procedure is given in Appendix C.

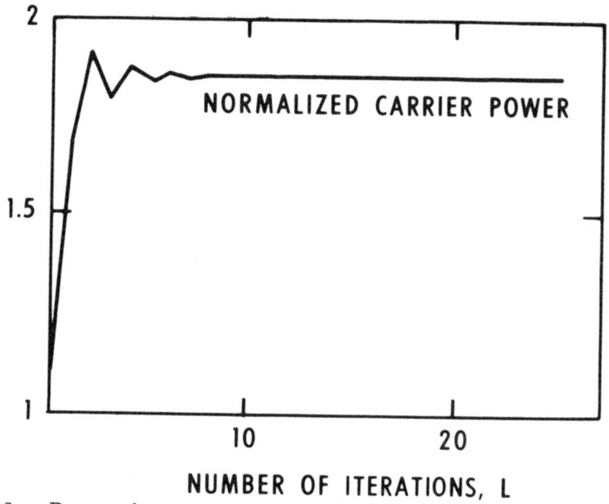

Fig. 1 Dynamic response of iterative procedure (normal initial conditions)

Figure 1 illustrates the dynamic response for one normalized carrier power for a system of four (N = 4) carriers. The initial values substituted into the program (Appendix B) were unity for all normalized carrier powers. It is interesting to note that, should the initial values for the normalized carrier powers correspond to another eigenvector associated with a real eigenvalue R_m of the antenna beam matrix [x], not the Perron eigenvalue, the dynamic response has an unstable equilibrium point at this critical value but eventually converges to the eigenvector associated with the Perron eigenvalue (Fig. 2).

Effect of Thermal Noise

Thermal noise affects the interference-to-carrier-power ratio in adding another term to the interference component. It has the same power on all spot beams and is statistically independent of all signal components. The interference-to-carrier-power ratio Eq. (5) is now given by a noise-to-carrier-power ratio

$$R_j = \sum_{k=1}^{N} (w_{jk}E_k)/E_j \tag{13}$$

where

$$w_{jk} = x_{jk}, \quad j, k = 1, N, \quad k \neq 1$$

$$w_{j1} = N_o + x_{j1}, \quad j = 1, N \tag{14}$$

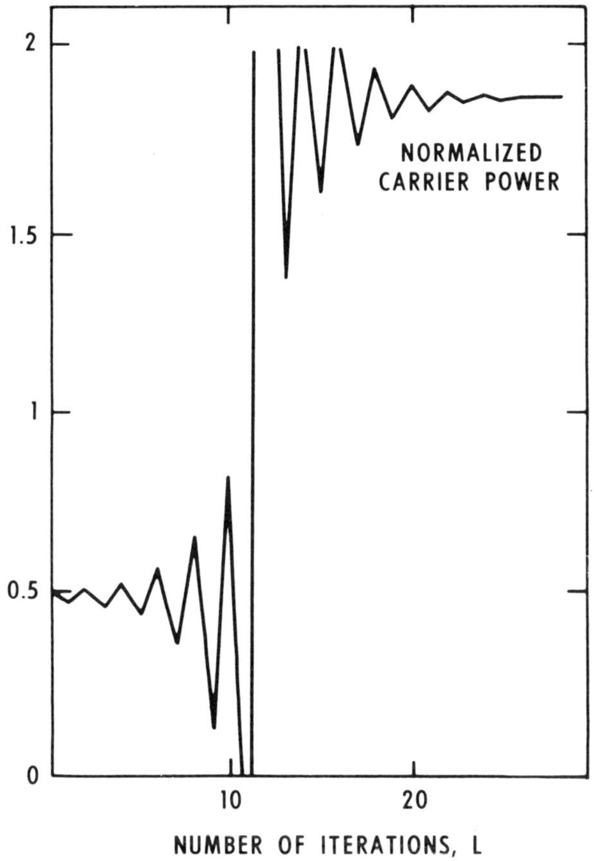

Fig. 2 Dynamic response of iterative procedure (special initial conditions)

and N_o is the same thermal noise power in each beam normalized to carrier power C_1.

The equation for the normalized carrier powers [E] in the presence of thermal noise N_o, corresponding to Eq. (7), is

$$[0] = \{[-R] + [w]\} \quad [E] \tag{15}$$

Appendix A shows that minimizing the maximum intereference-to-carrier-power ratio without thermal noise is equivalent to equalizing these ratios. The same result holds for the case with thermal noise.

With thermal noise included, the equations for the normalized carrier powers corresponding to Eq. (12) are

$$E_j(L) = \frac{\sum_{k=1}^{N} w_{jk} E_k(L - 1)}{\sum_{k=1}^{N} w_{1k} E_k(L - 1)} \tag{16}$$

R(I) = 0 in the computer program (Appendix B) is replaced by R(I) = N_o to take into account the thermal noise component.

Conclusions

The performance of multibeam satellites with frequency reuse can be optimized by determining the carrier power levels that yield minimum interference-to-carrier-power ratios on all beams. The evaluation of these levels involves the solution of a set of simultaneous nonlinear equations by an interative procedure which can be shown to be rapidly convergent. Hence, this method is computationally more efficient than alternative schemes that call for multidimensional searches or evaluation of the Perron eigenvalue of a matrix and the subsequent inversion of a related matrix.

The ability to determine these carrier powers will become important in the design of new satellite systems that incorporate multiple antenna beams on which the frequency is reused to provide the required communications capacity. The method proposed in this paper has been used in studies for the INTELSAT V series of satellites with great success.

Appendix A. Selection Criterion Equivalence

To show that minimizing the maximum interference-to-carrier-power ratio is equivalent to equating these ratios to the Perron eigenvalue, R_o, consider the simplest 3-carrier system, where

$$R_1 = x_{1\,2}\,E_2 + x_{1\,3}\,E_3 \tag{A1a}$$

$$R_2 = (x_{2\,1} + x_{2\,3}\,E_3)/E_2 \tag{A1b}$$

$$R_3 = (x_{3\,1} + x_{3\,2}\,E_2)/E_3 \tag{A1c}$$

$$R_1 = R_2 = R_3 = R_o > 0 \tag{A2}$$

then, from Eq. (9),

$$E_{2\,0} = (R_o x_{2\,1} + x_{2\,3}\,x_{3\,1})/(R_o^2 - x_{2\,3}\,x_{3\,2}) > 0 \tag{A3a}$$

$$E_{3\ 0} = (R_o x_{3\ 1} + x_{3\ 2}\, x_{2\ 1})/(R_o^2 - x_{2\ 3}\, x_{3\ 2}) > 0 \qquad \text{(A3b)}$$

Allowing small perturbations on E_2, E_3 about $E_{2\ 0}$, $E_{3\ 0}$, respectfully, all possible outcomes indicate the resulting values for R_1, R_2, R_3 will have at least one larger than R_o. Thus, for perturbations Δ_2, $\Delta_3 > 0$, and substituting into Eq. (A1),

a) $E_2 = E_{2\ 0} + \Delta_2$, $E_3 = E_{3\ 0} + \Delta_3$; $R_1 > R_o$ (A4)

b) $E_2 = E_{2\ 0} + \Delta_2$, $E_3 = E_{3\ 0} - \Delta_3$; $R_3 > R_o$ (A5)

c) $E_2 = E_{2\ 0} - \Delta_2$, $E_3 = E_{3\ 0} + \Delta_3$; $R_2 > R_o$ (A6)

d) $E_2 = E_{2\ 0} - \Delta_2$, $E_3 = E_{3\ 0} - \Delta_3$ (A7)

Then

$$R_2 = \frac{x_{2\ 1} + x_{2\ 3}(E_{3\ 0} - \Delta_3)}{E_{2\ 0} - \Delta_2} \qquad \text{(A8)}$$

$$\simeq R_o + \frac{\Delta_2 R_o}{E_{2\ 0}} - \frac{x_{2\ 3}\Delta_3}{E_{2\ 0}} \qquad \text{(A9)}$$

For

$$R_2 \leq R_o\ , \quad \frac{\Delta_3}{\Delta_2} \geq \frac{R_o}{x_{2\ 3}} \qquad \text{(A10)}$$

Similarly,

$$R_3 = \frac{x_{3\ 1} + x_{3\ 2}(E_{2\ 0} - \Delta_2)}{E_{3\ 0} - \Delta_3} \qquad \text{(A11)}$$

$$\simeq R_o + \frac{\Delta_3 R_o}{E_{3\ 0}} - \frac{x_{3\ 2}\ \Delta_2}{E_{3\ 0}} \qquad \text{(A12)}$$

For

$$R_3 \leq R_o\ , \quad \frac{\Delta_3}{\Delta_2} \leq \frac{x_{32}}{R_o} \qquad \text{(A13)}$$

However, from Eq. (A3),

$$R_o^2 > x_{23}\, x_{3\ 2} \qquad \text{(A14)}$$

Hence, R_o is the minimum value for the maximum of R_1, R_2, R_3. A similar reasoning may be applied to larger number ($N > 3$) carrier systems.

Appendix B. Iterative Procedure Fortran Program

```
Dimension E(N), E1(N), R(N), X(N,N)
Do 10 I = 1,N
E(I) = 1 (initial value of relative carrier power levels)
10 E1(I) = 1.1
Do 20 L = 1, 1000 (iterative index)
Do 30 I = 1,N
R(I) = 0
Do 30 J = 1,N
30 R (I) = R (I) + X(I,J)*E(J)
Do 40 I = 2,N
40 E(I) = R(I)/R(1)
Do 50 I = 2,N
If (ABS (E(I) - E1(I)). GE. .0001) Go to 35 (accuracy of
   procedure)
50 Continue
Go to 45
35 Do 60 I = 2,N
60 E1(I) = E(I)
20 Continue
45 write (6,5) [E(I), I = 2,N], R(1) (=Ro)
Stop
End
```

N = number of carrier powers
E(I) = I normalized carrier power level
R(I) = interference-to-carrier-power ratio for i beam

Appendix C. Convergence to the Perron Eigenvalue

A general proof[4,5] for convergence of the iterative procedure to the largest magnitude eigenvalue, eigenvector is as follows, based on the computer method in Appendix B:

$$[R(L)] = [x]\ [E(L - 1)] \tag{C1}$$

$$[E(L)] = [R(L)]/R_1(L) \tag{C2}$$

Hence,

$$[E(L)] = \frac{[x]\ [E(L-1)]}{R_1(L)}$$

$$= \frac{[x]}{R_1(L)} \frac{[x]^{L-1}}{\prod_{j=1}^{L-1} R_1(j)} [E(0)] \qquad (C3)$$

The initial vector [E(0)] may be written in terms of the eigenvectors $[E_m]$ of matrix [x], suitably normalized; that is,

$$[E(0)] = [E_o] + u_1[E_1] + u_2[E_2] + \ldots \qquad (C4)$$

The corresponding eigenvalues R_m of matrix [x] are such that

$$|R_o| > |R_1| > |R_2| \ldots \qquad (C5)$$

Hence,

$$[E(1-1)] = \frac{R_o^{L-1}}{\prod_{j=1}^{L-1} R_1(j)}$$

$$\cdot \left\{ [E_o] + u_1[E_1]\left(\frac{R_1}{R_o}\right)^{L-1} + u_2[E_2]\left(\frac{R_2}{R_o}\right)^{L-1} + \ldots \right\}$$

$$\simeq \frac{R_o^{L-1}}{\prod_{j=1}^{L-1} R_1(j)} [E_o] \text{ for } L >> 1 \qquad (C6)$$

and

$$[E(L)] \simeq \frac{R_o}{R_1(L)} \frac{R_o^{L-1}}{\prod_{j=1}^{L-1} R_1(j)} [E_o] \qquad (C7)$$

Near steady-state conditions

$$[E(L)] \simeq [E(L-1)] \simeq [E(L-2)] = \ldots \simeq [E_o] \qquad (C8)$$

and

$$R_1(L) \simeq R_1(L - 1) \simeq R_1(L - 2) = . . . \simeq R_o \qquad (C9)$$

References

[1]Kreutel, R. W., "Antenna Design for Frequency Reuse Satellite Communications System," National Telemetry Conference, Atlanta, Ga., 1973.

[2]Aein, J., "Power Balancing in Systems Employing Frequency Reuse," COMSAT Technical Review, Vol. 3, No. 2, Fall 1973, pp. 277–301.

[3]Meyerhoff, H. J., "Method for Computing the Optimum Power Balance in Multibeam Satellites," COMSAT Technical Review, Vol. 4, No. 1, Spring 1974, pp. 139–147.

[4]Pipes, L. A. and Hovanessian, S. A., Matrix Computer Methods in Engineering, Wiley, New York, 1969.

[5]Faddeev, D. K. and Faddeeva, V. N., Computational Methods of Linear Algebra, Freeman, 1963.

[6]Bellman, R., "On an Iterative Procedure for Obtaining the Perron Root of a Positive Matrix," Transactions of the American Mathematical Society, Vol. 125, 1955, pp. 719–725.

GENERALIZED DESIGN AND OPTIMIZATION TECHNIQUES FOR SATELLITE COMMUNICATIONS NETWORKS

Alan D. Levit*

International Business Machines Corporation,
Poughkeepsie, N. Y.

Abstract

The designer of a communication network can consider, among other factors, cost, reliability, availability and performance. Satellite technology introduces new opportunities to satisfy these considerations by taking advantage of characteristics inherent in both satellite and terrestrial environments. Processes of network design and optimization involve consideration of system cost and quality of service as a function of distance, traffic distribution, and geographic peculiarities. Techniques employed may be analytic or heuristic, depending upon degrees of complexity encountered. The procedure, being iterative, requires tradeoff decisions along the way as to degree of exactness vs the cost and time of analysis. The result is that the network designer combines techniques of programing analysis with human interaction and can orient his methods to the characteristics of the problem.

Presented as Paper 74-468 at the AIAA 5th Communications Satellite System Conference, Los Angeles, Calif., April 22-24, 1974. The author appreciates the work and advice of Barbara J. Gayhart. Also appreciated is the assistance of Peter A. Legault, Ernest V. Valera, Thomas M. Sullivan, Steve Meyers, Lin Woo, Robert S. Wilkov, Norman A. Vogel, Daniel P. White, Donald C. Flemming, John L. Dickinson, Robert L. West, Sally R. Noble, and Elizabeth M. Ragnone.

*Manager of System Support, Telecommunications Systems Technology, System Communications Division.

Introduction

To date, much work has been done to advance the state of the art of communication network design.[1,2,3] The process, however, is not routine. Peculiarities of each design case warrant the personalization of the design process. Terrestrial network design can be straightforward; however, when satellite links are interposed, special considerations must be included.

This paper will attempt to address some special design considerations, constraints, and techniques used in the design of integrated terrestrial/satellite networks.

Economics of Satellite/Terrestrial Networks

It is very difficult to generalize the justification of satellite communications. Misconceptions relating to geographic dispersions and lack of existing terrestrial plant often are made. It is easy to think that great distance between traffic sources justifies installing a satellite network, and yet it is not clear what the relationship between traffic intensity and mileage has to be before such a recommendation can be made. Australia, for example, is a country (continent) 2500 miles at its widest, and yet, because most of its traffic is concentrated in the eastern provinces, it might not be a very good candidate for a satellite network. It is therefore desirable to develop an educated "rule of thumb." This section will describe some techniques for preliminary evaluation.

Figure 1 shows the variations of terrestrial tariffs as a function of distance for approximately a full transponder's worth of traffic.[4] With this simple chart and an approximate cost of a hypothetical two-station satellite network, one can make some reasonable statements. Knowing the mix of terrestrial tariffs in the network being analyzed, one can go even farther toward stating the applicability of a satellite network.

Figure 2 illustrates this idea. Knowing the costs of the components of the satellite network and the makeup of terrestrial traffic, the distance between Earth stations 1 and 2 can be varied to determine when space segment costs offset terrestrial costs. Depending upon the nature of the traffic being analyzed, the minimum distance

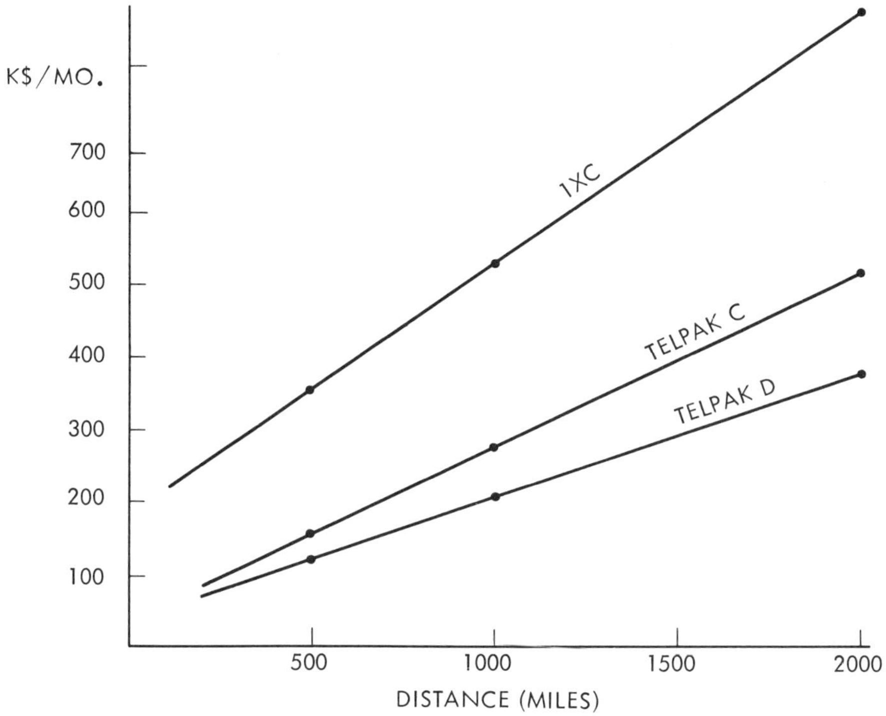

Fig. 1 Terrestrial tariffs as a function of distance.

to warrant the consideration of a satellite will be different. For example, reading from the graph of current tariffs shown in Fig. 1, and assuming a $250,000 space segment, only a 300-mile separation is required for a satellite network to equal the cost of an all terrestrial network. That is the distance from Los Angeles to San Francisco. Obviously, an all-leased-line network is unrealistic. But we could bound the problem by reading from the same graph that, in order to displace an all-TELPAK network, a separation of 900 miles would be required. New York to Chicago would certainly be a candidate.

Working this line of reasoning backwards, if one wanted to have a satellite network between Dallas and Chicago (936 miles), the satellite subsystem would have to cost between $200,000 and $500,000, depending on the nature of the traffic. This exercise comments only on traffic distances and terrestrial/satellite costs. It assumes a traffic volume that uses the full capacity of the equipment. In order to be more realistic, we have to consider traffic distribution more broadly.

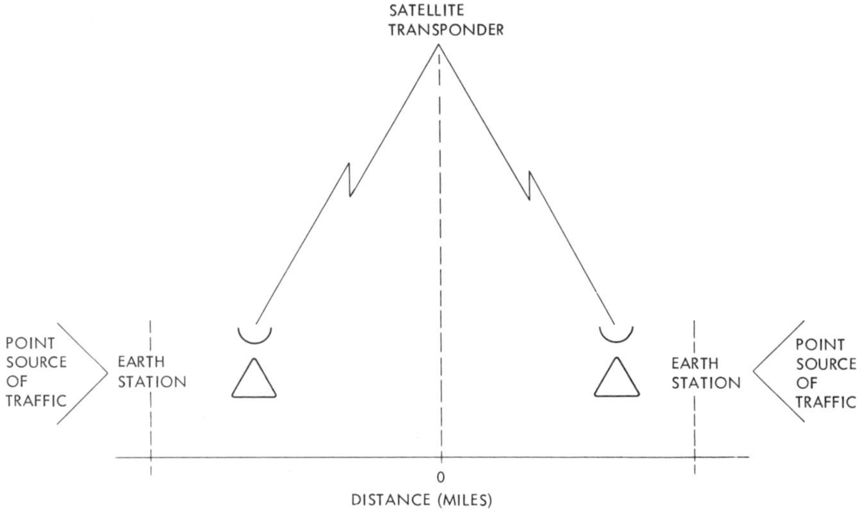

Fig. 2 Economics of satellite vs terrestrial communications: two-station example.

Consider the following hypothetical model: to approximate a terrestrial network, assume that all sites communicate through one large switch. Functionally equivalent networks can be configured for randomly distributed sites and site loads within a fixed geographical area, as illustrated in Fig. 3. Given the capacities and cost of a transponder and an Earth station and the distributions of traffic (the number of large sites compared to the number of small), minimum system costs for equivalent terrestrial and satellite networks can be calculated. By varying the number and loading of sites, it is possible to determine the amount of traffic necessary to make a satellite network cost less than an equivalent terrestrial network. Repeating the calculations for various grid sizes (geographic areas) leads to a graph as shown in Fig. 4. Figure 4 shows the relationship between geographical region size, traffic intensity, and the advisability of a satellite network. With a minimum of information about an existing terrestrial network, one can comment on its candidacy.

If the same or similar tariffs applied outside the bounds of the United States, then some of the following observations can be made from the same graph. For example, for a country the size of Nigeria (500 x 500 miles), a total of 50 sites including two large sites generates sufficient traffic to justify a satellite network. In the area of Brazil (1800 x 1800 miles), only 600 lines are needed for a satellite network to be cheaper than a terrestrial network. An interesting

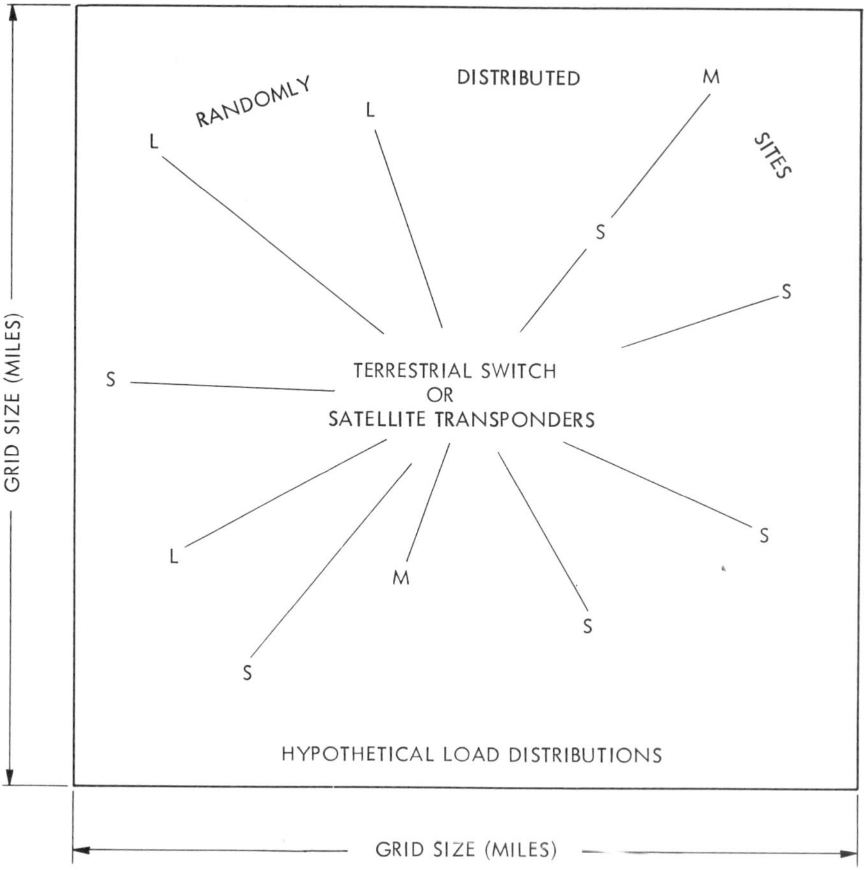

KEY: L = LARGE SITE
M = MEDIUM SITE
S = SMALL SITE

Fig. 3 Economics of satellite vs terrestrial communications: generalized example.

characteristic is that, once a network is justified for long-distance routes, shorter routes can be added without destroying the cost advantage. With coarse application of the foregoing techniques, the decision to proceed further along toward network design can be made.

Design Considerations

The general process of network design is a cost tradeoff between terrestrial and satellite facilities. Therefore, one of the most important factors to be considered is parameterization of system costs.

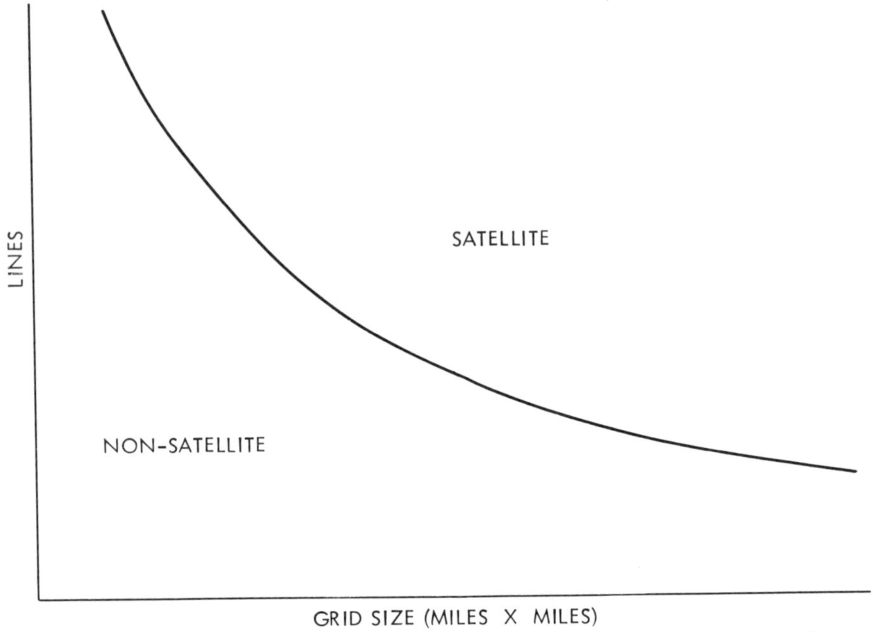

Fig. 4 Relationship of geographic region size, traffic intensity, and the advisability of a satellite network.

Without knowing the relationship of system element capacities to their cost, no meaningful tradeoffs can be made. Costs can be divided into two categories: terrestrial and space segment. Terrestrial costs are made up primarily of leased-line and interconnect elements. Space segment costs are determined primarily by transponder and earth station costs. These building blocks are shown in Fig. 5; the cost elements are shown in Fig. 6.

Interconnect costs are less predictable. These costs will vary as a function of transmission format and distance. Distance depends on freedom of choice of line concentrator and earth station location. Therefore, the cost of interconnect as a function of network design, as shown in Figure 6, is related to site eligibility. Frequency coordination difficulties, which are more severe at 4/6 GHz than at 12/14 GHz, as well as land availability, compound the problem. In light of these constraints, it is desirable to locate the satellite resource controller close to the RF station because of IF limitations. Alternatively, the line concentration function should be located near traffic sources to reduce line costs. If land availability correlates with traffic intensity, then this cost curve will reflect the

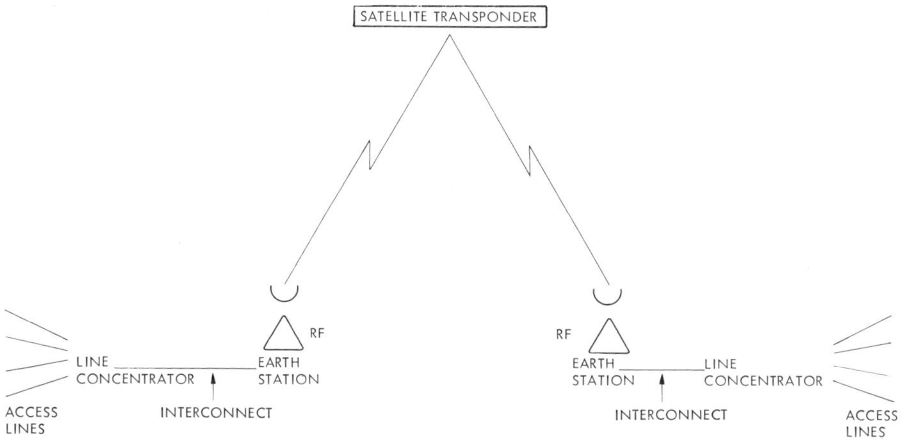

Fig. 5 Block diagram of a typical system configuration.

higher interconnect costs to bring smaller stations onto the network. At the higher frequencies, the process of site selection will more naturally follow traffic patterns.

Summing the component cost curves shown in Fig. 6, for a specific design case, yields a system cost curve from which an operating design range can be determined. Each of the three component curves may vary according to independent variables. For example, the line cost curve will change with tariffs; Earth station costs are a function of the design and technology; and interconnect costs are a function of Earth station location which may be geographically or politically limited. The system cost curve, therefore, will experience a minimum as a function of any of these independent variables, and it is important to parameterize the relationship. For example, the relationship of relative system cost to Earth station cost is shown in Fig. 7.

Design Techniques

So far, the discussion has been on cost parameterization: the analysis of system cost as a function of system constraint. There are other advantages to be gained by the further application of network design tools to optimize certain portions of the network. As the importance of cost parameterization has been emphasized, it is equally necessary to emphasize the importance of a matrix of traffic. Without a reliable data base, one could not progress beyond generalizations. The frequency coordination constraint that allows opera-

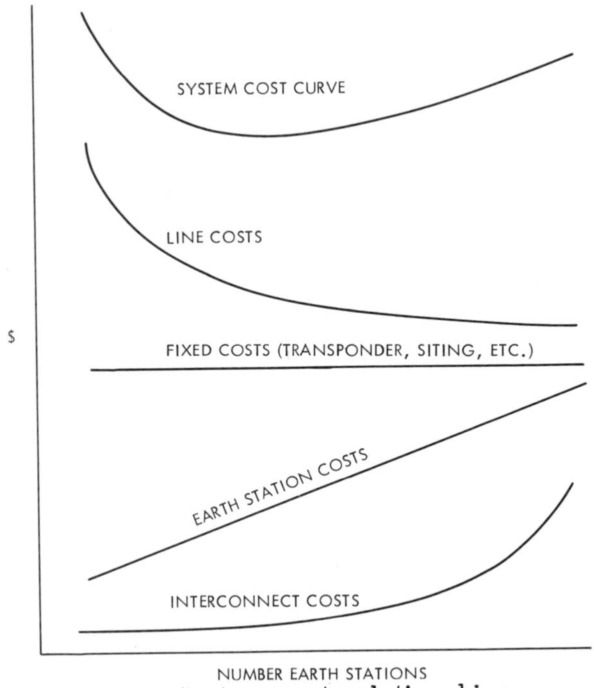

Fig. 6 System cost relationships.

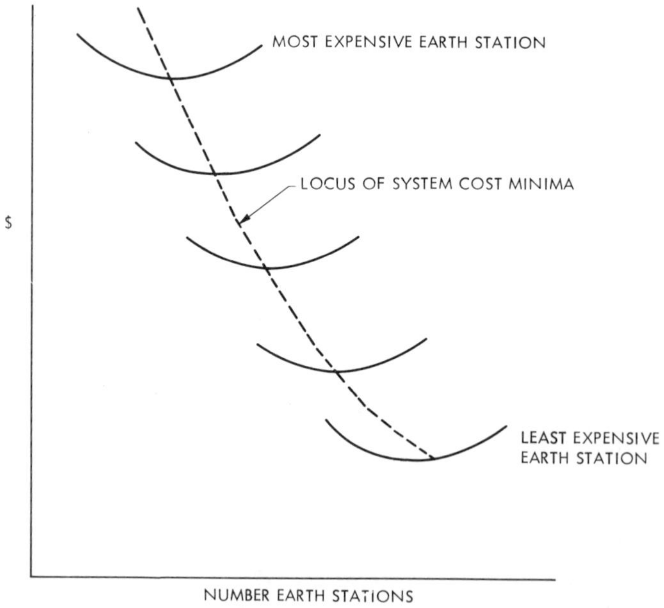

Fig. 7 Relative system cost.

tion at a more favorable position on the line cost curve would be meaningless if prediction of line cost variations, by analyzing a data base of total system traffic (n x n, point to point) and calculating its values for hypothetical numbers of Earth stations, was not possible. The absence of a data base also would prevent the investigation of the relationship of performance parameters such as blocking probabilities and reliability to system cost.

The data base of traffic, when factored by the number of Earth stations, becomes the definition of satellite eligible traffic. It then can be embodied in an m x m matrix representing the community of interest of the m Earth stations in the network. The total traffic, manipulated and processed by technology algorithms of demand assignment and the like, determines the number of transponders required in the system. The assumption is that transponder cost is a significant system cost element. If the number of transponders is greater than one, the efficient apportionment of traffic to transponder becomes another element of network design. It is this requirement for transponder netting that will determine which Earth stations will access only one transponder and which will access more than one, since these will bear the burden of extra cost. The recursive nature of this network design procedure becomes obvious immediately. In order to choose a candidate site as an Earth station, we must know the tradeoff of terrestrial/satellite costs, but transponder access must be an element of this cost, and that will not be known until transponder capacity is apportioned.

Elements of system cost can be estimated coarsely (Fig. 6). Based on these estimates, the best set of Earth station primary (the concentration nodes) and secondary (the traffic sources) sites can be determined. (Primary sites also can be sources.) Each candidate for system design then can be analyzed for transponder requirements. The ideal objective of optimizing transponder utilization is to minimize the number of transponder accesses required without jeopardizing expandability or flexibility. A goal is to equalize the grade of service for all users.

It is easiest to describe the idea of transponder netting with only two transponders. The complications incurred with increasing numbers of transponders can be extrapolated. The total number of Earth stations in the network will be divided into two groups: those assigned to transponder one, and those to transponder two. Traffic is of two types: intergroup or intragroup. Traffic can be subdivided further,

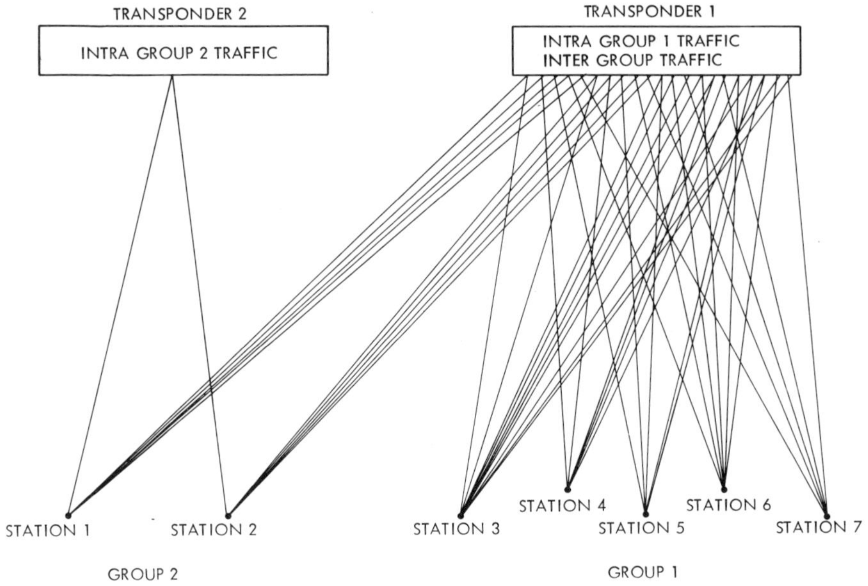

Fig. 8 Transponder netting.

but further subdivision will not be considered in this discussion. It can be seen from this example that one group can be made to access only one transponder. Cost considerations suggest that this group contain as many members as possible because of the added hardware required to communicate with two transponders from one Earth station.

The simplest approach would be to group the smallest stations until the remaining channel capacity is not enough to contain the traffic from a whole inter or intra station group. The remaining larger stations then would fall into the second transponder. The idea is shown graphically in Fig. 8.

The calculation of transponder capacities is a function of the blocking probability, Pb, allocated to this portion of the system, as well as the technology and algorithms (such as demand assignment) used in the satellite resource controller at the Earth station. Each can weigh heavily in determining the minimum number of transponders required for a given amount of traffic. The decision to quantize group traffic affects the bounds of transponder capacity. In other words, for a fixed value of Pb, a transponder is never exactly filled. For the example shown in Fig. 8, the process started by ranking the seven stations and assigning their traffic to transponder

1 starting with station 7. The interstation traffic among station 2 and stations 3-7 would not fit as a group in transponder 1 and had to be assigned to transponder 2. The excess capacity in transponder 1 will result in less blocking for the members of that group.

Quantum grouping results in discontinuities in the relationship of transponder cost as a function of the number of Earth stations and therefore introduces discontinuities in the total system cost curve which should really be depicted as shown in Fig. 9, rather than continuously, as shown in Fig. 6. This characteristic of transponder netting encourages the trading off of Earth stations if the minimum system cost occurs close to a discontinuity. Also, because this approach to transponder netting generally results in lower utilization of the second transponder, its cost justification can be weighed against the cost of the investment.

The process of "optimizing" combinations of Earth station assignments to share transponder bandwidth can be performed and cost

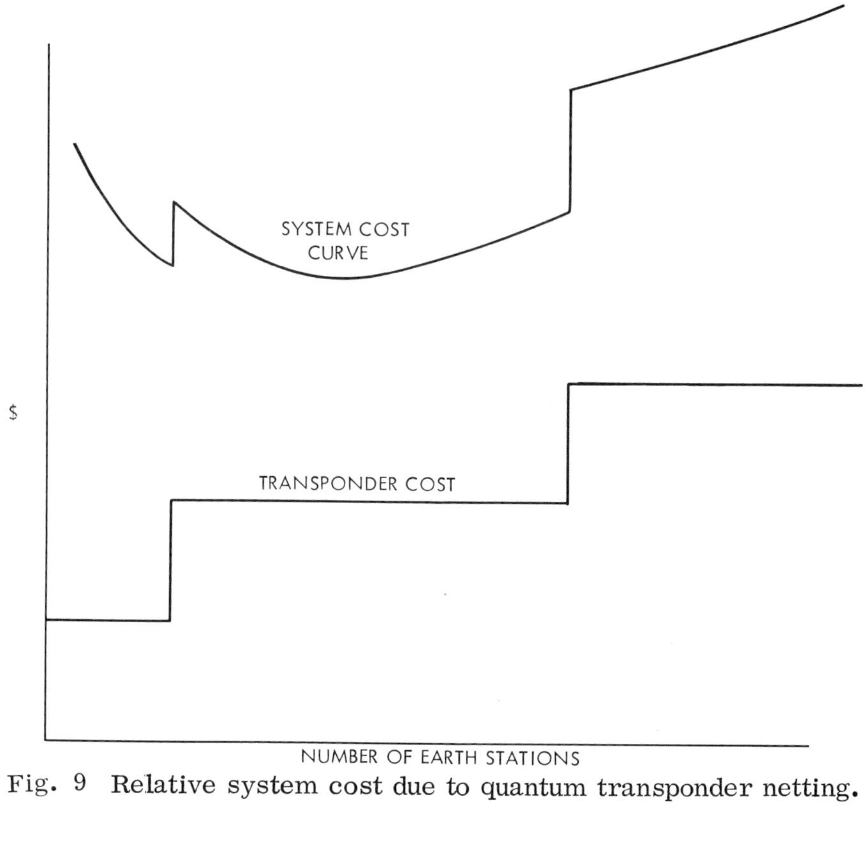

Fig. 9 Relative system cost due to quantum transponder netting.

personalization for each site developed as a function of the number of transponder accesses at each site, site preparation costs, and localized costs as a function of capacities. Figure 6 therefore contains generalized component costs but personalized system cost.

Programing Considerations

The development of automated network design tools is a key ingredient in the processes being discussed. The discipline of programing enables the processing and manipulation of large, cumbersome, and disjoint data into simpler forms that lend themselves to sensitivity and parameterization studies. These data can be, and often are, in different forms and sample sizes. The formation and processing of data bases therefore involves measurement tapes and punched cards for input and produces reams of output reports of point-to-point matrices, lists of cities, and indices. FORTRAN and PL/1 seem to handle this high-volume input/output task fairly well. The mathematics of the calculations is generally straightforward, so that utilization of a batch processor (or time-shared processor) is suitable.

Where complicated algorithms are involved, real-time debugging is necessary to shorten development cycles. For reliability calculations, for example, where vector lengths of lines and availability quickly get out of hand, an interactive language such as APL is advantageous. Debugging time is minimized, and the flexibility of algorithm development is enhanced.

However, reliability calculations, data base matrix manipulations, system cost personalization, and analysis all are interrelated. It would be foolish to design a system without considering all of these features. There are two alternative procedures: do the development of the network design tools in the most suitable language for the job, and then convert to a common-base language (acceptable if you have a dedicated facility); or, secondly, develop an approach based on a time-sharing system that supports APL, FORTRAN, assembly, and PL/1 (or similar programing languages), and rely on the systems file management capability for the integration of techniques and the sharing of data bases and intermediate results. The result will be that analyses can be performed in reasonable time.

Conclusion and Summary

An attempt has been made in this paper to address network design as applied to satellite communication networks. The importance of some basic factors in network design: parameterization, data base, automation (programing), and design interaction has been emphasized. The initial decision to consider satellite networks is usually made broad-brush with as little information and expenditure of resources as possible. Early in this process, recognition must be given to the distinguishing characteristics of satellite networks: frequency coordination, technology, and the need for parameterization of cost; for the process that is really being performed is that of cost tradeoff of satellite circuits for terrestrial circuits. At this point, one is forced to be more specific about the design candidates in terms of traffic intensities and geographic dispersions. Similarly, one has to know the characteristics of the transmission resources in order to complete the parameterization and configure a system to meet the loads. The recursive nature of the exercise quickly makes it apparent that network design tools have to be programed so that in reasonable time each subsystem can be estimated, and then optimized, and then reoptimized until solution is approached.

With greater degrees of sophistication, these network design tools can be refined to allow sensitivity and trade studies to be performed. The cost of various system components could be varied to measure the effect on network configuration. It is important to know the sensitivity of the design to each element. Similarly, the application of these tools can be used to determine the sensitivity of network availability to each component as well as specific architectures. To know the critical paths can enhance system design. These analyses can lead to families of curves that describe the system in terms of its component parts. The optimum design is not necessarily a minimum cost point. Rather it is a range of solutions for which tradeoff decisions weighing alternative designs and considerations will have to be made.

References

[1] Woo, L. S. and Tang, D. T., "Optimization of Teleprocessing Networks With Concentrators," National Telecommunications Conference, Nov. 1973, Atlanta, Ga.

[2]Greenberg, D.A., "A New Approach for the Optimal Placement of Concentrators in a Remote Technical Communications Network," National Telecommunications Conference, Nov. 1973, Atlanta, Ga.

[3]Pan, G.S., "Design of a CCSA Nationwide Telephone Network, Telecommunications, Vol.39, No. 9 , Sept. 1972, pp. 27-32

[4]Mirowsky, B.J., "Design of a Cost Effective Satellite-Terrestrial Communications Network," EASCON, 1973, Washington, D.C.

Bibliography

Abramson, N. and Kuo, F., eds., Computer Communications Networks, Prentice Hall, Englewood Cliffs, N.J., 1973.

Kiesling, J.D., Elbert, B.R., Garner, W.B., "A Technique for Modelling Communications Satellites," Comsat Technical Review, Vol. 2, No. 1, Spring, 1972, pp. 73-103.

Northrop, G.M., "Aids for the Gross Design of Satellite Communications Systems," IEEE Transactions on Communication Technology, Vol. COM-14, No. 1, Feb. 1966, pp. 46-56.

CONUS AERONAUTICAL RADIONAVIGATION BY SATELLITE

Irvin G. Stiglitz*

Massachusetts Institute of Technology, Lexington, Mass.

Abstract

Over the past half decade, a variety of satellite system concepts have been promulgated as solutions to the CONUS ATC problems. By categorizing these, it is possible to draw some generally valid observations about the characteristics of each of them. By selecting system architectures, representative of each category, key technical aspects of systems within each category can be explored. Critical aspects illuminated include avionics complexity, required number of satellites, system vulnerability, capacity, required ground processing, and accuracy.

I. Introduction

A variety of satellite system concepts have been advanced as candidates to provide Air Traffic Control (ATC) surveillance, communication, and/or navigation service over the continental United States (CONUS).[1-9] Each has its advantages and disadvantages. All employ position location service by multilateration using a constellation of satellites. This paper summarizes the highlights of a technical assessment of satellite techniques performed by Lincoln Laboratory as part of the Advanced Air Traffic Management System (AATMS) study.[10-14]

No attempt has been made in the work summarized here to assess the broad spectrum of operational or economic implications of employing these techniques in the national airspace

---

Presented as Paper 74-439 at the AIAA 5th Communications Satellite Systems Conference, Los Angeles, Calif., April 22-24, 1974. This work has been supported by the Transportation Systems Center, Department of Transportation, Cambridge, Mass.

*Leader, Advanced Techniques, Surveillance and Control Division, Lincoln Laboratory.

system. No consideration has been given to such factors as 1) the cost effectiveness of satellites as an element in the CONUS ATC system, 2) the manner by which any of these satellite techniques might evolve from present-day aircraft surveillance, navigation, and/or communication systems, 3) the detailed operational requirements on such a system, or 4) the implications of sharing system resources with other non-ATC functions. Instead, the work has focused entirely on technical issues.

The results of this assessment of the key technical issues have verified that satellite-based techniques for CONUS ATC could be developed without reliance on high-risk technology. No one particular technique has emerged as optimum; instead, several viable alternatives have been identified.

II. Focus of the Program

The study objective has been to make technical assessments of several possible techniques for utilizing satellites in air traffic control. These techniques can be categorized as follows:

1) Coordinated aircraft-to-satellite techniques (CAST).

Aircraft respond to discretely addressed interrogation by satellite. Aircraft responses are repeated by the visible satellites to a central facility for processing.

2) Random access aircraft-to-satellite techniques (RAST).

Aircraft transmit periodically in an independent, uncoordinated fashion. These signals are repeated by the visible satellites to a central facility for processing.

3) Satellite-to-aircraft technique (SAT).

Transmitted signals from the visible satellites are used by aircraft to compute their position. These position data are displayed to the pilot and can be transmitted over a line-of-sight link to the ground.

The method used to assess the foregoing techniques was to select and analyze a "baseline configuration" for each. It was not the intent of the baseline to be a "best possible" implementation of the techniques. Instead, the intent was to insure that all major technical issues were addressed.

One of the primary attractive attributes of satellites is their inherent ability to provide broad coverage of low-altitude airspace. General aviation aircraft are predominant users of low-altitude airspace. Hence, a central issue throughout this study has been the complexity of general aviation avionics required for satellite operation.

A. Coordinated Aircraft-to-Satellite Techniques (CAST)

Aircraft interrogations originate at a centralized ground facility. These are relayed through a satellite repeator to the aircraft. Interrogations are received using top-mounted aircraft antennas. In general, the interrogation waveform contains aircraft identification and a digital message, e.g., for intermittent positive control, traffic advisories, weather data, vectoring, etc. Aircraft respond (using the same top-mounted antenna) by the transmission of a timing pulse and a digital message, e.g., confirming receipt of the uplink message, reporting status, requesting new vectors, etc. These signals are received by at least four visible, widely separated satellites for relay to a centralized ground facility for processing.

At this facility, signal time of arrival is determined for each of these satellites, and aircraft position is determined by multilateration. Included messages are decoded, and the surveillance data base is updated. Thus, CAST can provide both surveillance and digital communications service. Aircraft are interrogated periodically according to a schedule (continually updated by the ground facility) designed to eliminate mutual interference between aircraft replies. The requirement that aircraft transmissions not interfere is the factor that ultimately limits capacity, i.e., the number of aircraft that can be handled per second. The Astro-DABS [5, 8] concept advanced by the MITRE Corporation is an example of CAST.

B. Random Access Aircraft-to-Satellite Techniques (RAST)

Aircraft transmissions are initiated by the aircraft itself. Each aircraft employs a top-mounted antenna for periodic transmission of a unique identifying signature waveform consisting of a string of pulses. These signals are received by a constellation of satellites, at least four of which are in view. The received waveforms are then relayed to a central facility for processing. The signals from each satellite are processed to detect aircraft presence and estimate arrival time. Aircraft position, determined by multilateration, is used to update the surveillance data base.

Since the times at which aircraft transmit are uncoordinated (random), the transmissions from different aircraft can mutually interfere at the receiver. This effect, called multiple-access noise, complicates the ground data processing problem and eventually limits system performance for large numbers of aircraft.

A variety of concepts employing RAST have been advanced. These include the North American Rockwell AATMS system A concept,[7] the RCA SATAN concept,[2] the Boeing satellite concept,[3] and the TRW LIT concept.[1]

C. Satellite-to-Aircraft Techniques (SAT)

Timing signals are transmitted by at least four visible, widely separated satellites. Periodically, each satellite transmits a timing pulse and an accompanying digital message. In general, the message includes satellite identification and location and special broadcast messages, e.g., severe weather advisories. These signals are received at the aircraft using top-mounted antennas. Aircraft position is determined aboard each aircraft by multilateration, using the signal time-of-arrival data and the satellite location. The calculations are performed using a microprocessing computer (available today for less than $1000). The computed aircraft position then is displayed to the pilot as navigation information.

These position data also could be relayed over the DABS (Discrete Address Beacon System) data link to the ground for inclusion in the ATC data base. Thus, SAT can provide navigation to the pilot and dependent three-dimensional surveillance to the ground. A number of concepts employing SAT previously have been advanced in the military sector. These include the Defense Navigation Satellite System (DNSS), TIMATION, system 621B, and the Global Positioning System.[4, 6]

III. Key Technical Features

A. Avionics Features

The range between an aircraft and a satellite is typically 100 times larger than that between an aircraft and a line-of-sight ground station (e.g., 22,000 miles vs 220 miles). Consequently, signals for satellite service are attenuated much more severely than those for ground-based

Table 1 Representative avionics characteristics

	CAST	RAST	SAT	ATCRBS
Transmitter	Coherent 500 w of power	Coherent 500 w of power, power control	Not required	Incoherent 200 w of rf power
Frequency stability	$1:10^6$	$1:10^6$	$1:10^6$	$1:10^3$
Receiver noise figure	11 db	Not required	11 db	15-19 db
Receiver processor	Coherent matched filter	Not required	Coherent matched filter	Leading-edge detector
Antenna	Top-mounted crossed dipole	Top-mounted crossed dipole	Top-mounted crossed dipole	Blade
Data processing	Special-ized logic	Not required	Micro-processing computer	Minimal

systems. As exemplified in Table 1, avionics for satellite service must satisfy more stringent performance requirements and hence costs more than today's Air Traffic Control Radar Beacon System (ATCRBS) avionics.

The avionics transmitter with CAST and RAST must put out several times more power than today's typical ATCRBS transponder. Moreover, coherent transmitted signal waveforms are required to permit significant receiver processing gain. By contrast, today's ATCRBS transponder employs a simple incoherent cavity oscillator that is pulsed on and off. The transmitters intended for use with RAST must, in addition, be carefully power-controlled to minimize multiple-access degradation of disadvantaged aircraft due to signal overlap at a satellite repeator.

A frequency drift of several mega-Hertz is permissible with ATCRBS transponders. Frequency drift for satellite avionics must, however, be maintainted to within 1 kHz. To realize the full potential of receiver processing gain, coherent matched filter receivers also are required. By way of contrast, ATCRBS transponders employ simple leading-edge detectors. Avionics receivers for CAST and SAT must employ low-noise front ends; these also are not required for today's ATCRBS service.

The top-mounted antennas must have nearly uniform upper hemispheric coverage. One candidate is a crossed dipole antenna fabricated using inexpensive stripline techniques. Today's ATCRBS antenna is an even simpler bottom-mounted blade.

The small avionics computer required to perform the position-determination computations for SAT could be realized today with a (single circuit board) microprocessing computer at an OEM (original equipment manufacturer) cost under $1000. Even lower costs may be achieved with custom integrated circuits. The overall cost of the minimum package for RAST, CAST, or SAT is estimated to be two to four times that of an ATCRBS transponder.

B. Satellite Antenna

The high attenuation that signals experience in traveling between satellites and aircraft can be offset partially by use of large-aperture, high-gain, multibeam antennas at the satellites. Use of such antennas significantly reduces the requirements on transmitter power and receiver sensitivity and therefore reduces avionics costs.

In the assumed 1535- to 1660-MHz aeronautical radio navigation band, the use of a 30-ft antenna (e.g., similar to that deployed on ATS-6) could provide coverage zones of hundreds of miles (as illustrated in Fig. 1). For the indicated 10 beams, a 10-db power saving results (relative to a single CONUS coverage antenna). With such an antenna, a 1-kw satellite power amplifier operating with an avionics receiver having an 11-db noise figure front end would perform as well as a 5-kw satellite power amplifier and a CONUS coverage antenna working with an 8-db noise figure avionics receiver. Similarly, a 500-w aircraft transmitter would suffice with a high-gain satellite antenna, whereas 5 kw would be required for a CONUS coverage antenna.

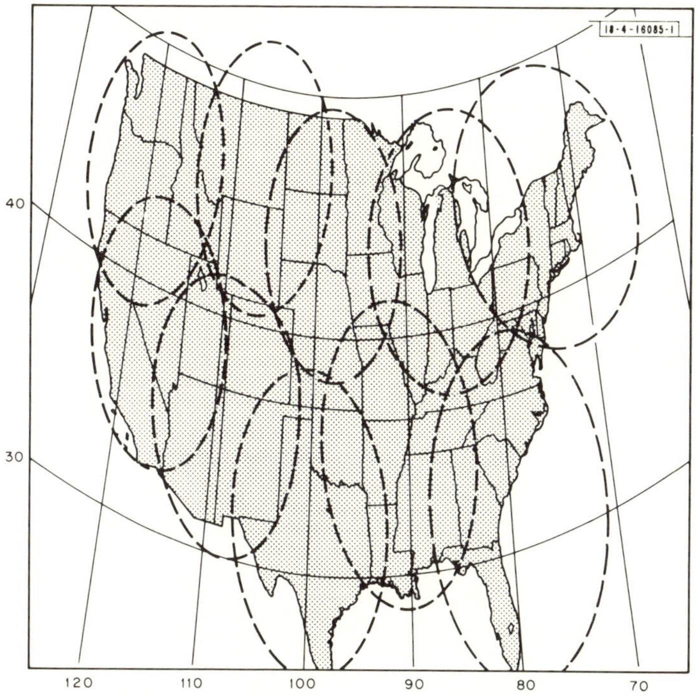

Fig. 1 Typical coverage zones for a large-aperture satellite antenna.

With CAST, an added benefit to CONUS interrogation efficiency accompanies the use of a multibeam antenna. It is evident that the interrogation algorithms for nonoverlapping coverage zones can be made largely independent, hence permitting high surveillance update rates. With RAST, a similar benefit derives from the ability of aircraft within different beams to operate with little mutual interference. With SAT, the use of a large-aperture satellite antenna permits use of impulse-like signals and a relatively simple avionics receiver for time-of-arrival estimation.

C. Computation Requirements

In CAST and RAST, the aircraft positions are computed in a large, centralized data-processing facility located on the ground. By contrast, in SAT the aircraft positions are calculated in a distributed manner; each aircraft calculates its own position. Although all tasks have not been sized, it has been estimated that the computations required for position determination alone substantially exceed that for NAS Stage A. In the case of CAST, the computation problem is complicated by the need to schedule aircraft interrogations. In the case of RAST, the problem is complicated by the need to sort the incoming pulses (which become interleaved due to the random transmission times). The data-processing requirement for CAST and RAST is evidently huge. Because of the "iceberg" nature of large automated systems, it is difficult to assess the true scope of this development; hence, the ground data-processing facility presently is rated as an area of significant technical uncertainty for CAST and RAST.

D. Vulnerability to Intentional Disruption

A system that employs CAST or RAST depends upon the aircraft-to-satellite uplink for proper operation. The high mobility of aircraft precludes use of an antenna that must be pointed toward each of the satellites. Hence, the aircraft antenna must be nondirectional, i.e., it must cover nearly the entire upper hemisphere. By contrast, a jammer fixed on the ground can focus a highly directional antenna toward each satellite. Thus, a "toaster-powered" jammer can "swamp out" signals transmitted from aircraft to a particular satellite.

For purposes of illustration, assume that the aircraft transmits a 500-w, 500-TW (time-bandwidth) product pulse through a top-mounted, crossed dipole (2.5-db gain) aircraft antenna and that adequate performance requires a 10-db signal-to-noise ratio. If the jammer uses a 6-ft (8° beamwidth, 26.5-db gain) antenna, then the jammer need transmit only 100 w of rf power into the antenna in order to be a threat (27 dbw + 27 db + 2.5 db - 10 db - 26.5 db = 20 dbw). To disrupt normal operation, the jammer would have to point beams toward several different satellites. Because of the multiple satellite beams covering CONUS, several different locations (most probably in regions with multiple-beam overlap) would have to be used to completely disrupt CAST and RAST service.

CAST and RAST also require the use of a large central facility (for processing aircraft replies) whose disruption could have serious consequences. An independent on-line backup facility might be employed for guaranteed performance.

By contrast, SAT is comparatively jam-proof because of the absence of an aircraft-to-satellite uplink. Vulnerability of the dependent surveillance is comparable to that of the present ATC system, i.e., local jamming for short time periods.

E. Improving Accuracy

The so-called Geometric Dilution of Precision (GDOP) conventionally has been used to assess the accuracy afforded by a multilateration system. In effect, GDOP is an error magnification factor that specifies how much the basic ranging error is magnified by a particular transmitter-receiver geometry. Clearly, the smaller the GDOP, the better.

A new procedure for determining GDOP was developed at Lincoln Laboratory as part of the AATMS study.[15, 16] The new procedure differs from previous methods in that it provides considerable insight for the selection of good geometries. Basically, the method reduces the problem of calculating GDOP to that of determining the moments of inertia of a mass configuration easily obtained from the satellite-aircraft geometry.

The insight provided by the method indicated that satellite constellations with significantly improved GDOP(s) could be designed. Accordingly, a family of new constellations was designed to determine the practical extent to which GDOP could be improved or, equivalently, the required number of satellites could be reduced. The constellations include large (15-satellite), medium (10-satellite), and small (7-satellite) constellations. In Table 2, the performance of one of the resulting constellations, LL-10,[17, 18] is compared to that of the previous RCA-8[2] and NAR-15[3] constellations. The entries are based upon GDOP calculations over a 119-point CONUS grid. Evidently, the performance of the 10-satellite constellation (LL-10) is comparable to that of the 15-satellite constellation (NAR-15) and significantly better than that of the eight-satellite constellation (RCA-8). The work also showed that, although small (seven-satellite) constellations

Table 2 Accuracy performance comparison for three different satellite constellations

Constellation	Nominal level flight		Single satellite failure	30° aircraft bank		
	Average GDOP	Rms deviation	Percent GDOP>10	Average GDOP<10	Percent GDOP>5	Percent GDOP>10
RCA-8	3.8	0.38	42.0	7.6	38	17.0
LL-10	2.4	0.42	0.8	2.7	13	9.5
NAR-15	2.4	0.35	0	2.8	11	0.3

can produce reasonably low values of GDOP during level flight, the GDOP(s) are highly sensitive to satellite failure and aircraft maneuvers. Thus, if high accuracy is to be maintained following a satellite failure and during maneuvers, larger constellations are necessary. Analyses of a variety of constellations indicated that 10 satellites should suffice.

IV. Principal Conclusions

No significant technological advances are essential for developing a system employing any one of the techniques. Both CAST and RAST provide surveillance as a primary service. Two-way digital communication is especially easy to accommodate with CAST. Both techniques can be configured to support tens of thousands of aircraft at an average surveillance update rate of once every 10 sec. Navigation is the natural service with SAT. Surveillance (dependent on the navigation data) could employ the DABS transponder for line-of-sight digital communications with ground stations.

Because signals between satellites and aircraft travel tens of thousands of miles, they are attenuated severely when they arrive at their destination. A combination of techniques is needed to combat this power loss. On the satellite, large antennas are required both to collect a useful fraction of the power transmitted by the aircraft and also to transmit focused signals to them. In addition, special avionics designs with coherent high-power transmitters and sensitive receivers with high signal processing gain should be employed. The required avionics for satellite service is expected to be several times more costly than today's ATCRBS transponders.

Systems employing either CAST or RAST would be vulnerable to disruption by hostile groups. In particular, both techniques rely upon an aircraft-to-satellite uplink that could be disrupted by several inexpensive, easily transportable, low-power jammers. In addition, both techniques require a large centralized computing facility, the disruption of which would disable all service.

Regardless of which technique is employed, at least 10 satellites are required for realistic "down to the ground" CONUS satellite service. In contrast, today's ground-based system employs hundreds of ground sites and does not provide this "down to the ground" coverage everywhere. However, satellite-based systems also require the support of extensive ground facilities. Both CAST and RAST require a large central processing facility considerably more complex than today's NAS stage A system. SAT has no such requirement; rather, it employs a network of relatively simple ground sites for line-of-sight aircraft-ground digital communications.

Although there are no major technical obstacles to the development of systems employing any one of these techniques, the associated cost to the government will be considerable. For example, estimates of engineering, development, facilities, and equipment cost range from a few hundred million dollars for SAT to nearly a billion dollars for CAST and RAST.

References

[1] Otten, D.D., "Study of a Navigation and Traffic Control Technique Employing Satellites," Group Rept. 08710-6012-12000, Vols. I-IV, Dec. 1967, TRW Systems, Redondo Beach, Calif.

[2] "Fourth Generation Air Traffic Control Study," Rept. DOT-TSC-304-1, Vols. I-IV, June 1972, North American Rockwell Corp., Anaheim, Calif.

[3] "Study and Concept Formulation of a Fourth Generation Air Traffic Control System," Rept. DOT-TSC-506-1, Vols. I-V, April 1972, Boeing Co., Renton, Wash.

[4] Diamond, P.M., "The Potential of a System of Satellites as a Part of an Air Traffic Control System," <u>AGARD Conference Proceedings 105</u>, June 1972, p. 20-1.

[5] Schuchman, L., "The ASTRO-DABS Concept," MTR-6287, Jan. 1973, MITRE Corp., McLean, Va.

[6] McDonald, K.D., "A Survey of Satellite-Based Systems for Navigation, Position Surveillance, Traffic Control, and Collision Avoidance," <u>Journal of the Institute of Navigation</u>, Vol. 20, No. 4, Winter 1973-1974, pp. 301-320.

[7] "Concept for a Saturn Based Advanced Air Traffic Management System," Rept. DOT/TSC-OST-73-29, Vols. I-X, Feb. 1974, Rockwell International Corp., Autonetics Division, Anaheim, Calif.

[8] "An Advanced Air Traffic Management Concept Based on Extension of the Upgraded Third Generation ATC System," Rept. FAA-EM-73-10A, March 1974, MITRE Corp., McLean, Va.

[9] "Advanced Air Traffic Management System Study Executive Summary," Rept. DOT/TSC-OST-75-1, Jan. 1975, U.S. Department of Transportation, Cambridge, Mass.

[10] Stiglitz, I.G., Beusch, J.U., Eckberg, A., and Schneider, K.S., "Concept Formulation Studies of the Surveillance Aspects of the Fourth Generation Air Traffic Control System," Project Rept. ATC-7, DOT/TSC-241-1, Sept. 21, 1971, Lincoln Laboratory, Massachusetts Institute of Technology, Lexington, Mass.

[11] Schneider, K.S., Stiglitz, I.G., and Eckberg, A.E., "Surveillance Aspects of the Advanced Air Traffic Management System," Project Rept. ATC-10, DOT/TSC-241-2, June 22, 1972, Lincoln Laboratory, Massachusetts Institute of Technology, Lexington, Mass.

[12] Orr, R.S. and Schneider, K.S., "Technical Assessment of Satellites for CONUS Air Traffic Control, Vol. I: Coordinated Aircraft-to-Satellite Techniques," Project Rept. ATC-26, DOT/TSC-241-4, Jan. 31, 1974, Lincoln Laboratory, Massachusetts Institute of Technology, Lexington, Mass.

[13] Schneider, K.S. and Orr, R.S., "Technical Assessment of Satellites for CONUS Air Traffic Control, Vol. II: Random Access Aircraft-to-Satellite Techniques," Project Rept. ATC-26, DOT/TSC-RA-3-8-2, Feb. 1974, Lincoln Laboratory, Massachusetts Institute of Technology, Lexington, Mass.

[14] Lee, H.B. and Goode, B.B., "Technical Assessment of Satellites for CONUS Air Traffic Control, Vol. III: Satellite-to-Aircraft Techniques," Project Rept. ATC-26, DOT/TSC-RA-3-8-3, Feb. 1974, Lincoln Laboratory, Massachusetts Institute of Technology, Lexington, Mass.

[15] Lee, H.B., "Accuracy Limitations of Hyperbolic Multilateration Systems," TN 1973-11, DOT/TSC-241-3, March 22, 1973, Lincoln Laboratory, Massachusetts Institute of Technology, Lexington, Mass.

[16] Lee, H.B., "Accuracy Limitations of Range-Range (Spherical) Multilateration Systems," TN 1973-43, DOT/TSC-RA-3-8-1, Oct. 11, 1973, Lincoln Laboratory, Massachusetts Institute of Technology, Lexington, Mass.

[17] Lee, H.B. and Wade, A.E., "Improved Satellite Constellations for CONUS ATC Coverage," Project Rept. ATC-23, DOT/TSC-RA-3-8-4, May 1, 1974, Lincoln Laboratory, Massachusetts Institute of Technology, Lexington, Mass.

[18] Lee, H.B. and Wade, A.E., "New Satellite Constellations for CONUS ATC Coverage," AIAA Paper 74-430, April 1974, Los Angeles, Calif.; also published elsewhere in this volume.

NEW SATELLITE CONSTELLATIONS FOR CONUS ATC COVERAGE

H. B. Lee* and A. E. Wade*

Massachusetts Institute of Technology, Lexington, Mass.

Abstract

This paper describes several new constellations capable of providing surveillance and/or navigation coverage over the continental United States. Position determination by hyperbolic multilateration is presupposed. The constellations are suggested by a novel procedure for calculating the geometric dilution of precision (GDOP). Constellation performance is characterized not only by reduced GDOP(s) but also by reduced sensitivity to single satellite failure and aircraft maneuvers.

1. Introduction

A number of methods have been described for providing an aircraft surveillance/navigation service over the continental United States (CONUS) using a constellation of orbiting satellites.[1-9] Basically, the methods treat the satellite positions as known and determine the aircraft positions relative to the satellites by multilateration using signal time-of-arrival measurements.

This paper summarizes a recent study of the problem of designing the associated satellite constellations.[10] It is assumed throughout that the aircraft positions are determined

---

Presented as Paper 74-430 at the AIAA 5th Communications Satellite Systems Conference, Los Angeles, Calif., April 22-24, 1974. This work was supported by the Transportation Systems Center, Department of Transportation, Cambridge, Mass. It is a pleasure to acknowledge several beneficial interchanges with Ron Loeliger of the Autonetics Division of Rockwell International Corporation early in the work. The authors also wish to acknowledge the encouragement and support of Irvin Stiglitz of Massachusetts Institute of Technology, Lincoln Laboratory.

*Staff Members, Lincoln Laboratory.

by means of hyperbolic multilateration using all satellites visible at elevation angles exceeding a prescribed minimum angle.

The study has approached the problem of constellation design by asking the following questions:

1) What is the best performance that can be obtained given N satellites?

2) How closely can the theoretical optimum performance be achieved by real constellations?

Although this approach is widely used in engineering design, it apparently has not been used previously in the design of constellations for navigation/surveillance applications over CONUS.

It should be noted that the constellation design problem addressed here aims at providing multilateration service to a small portion of the Earth's surface (i.e., CONUS). Thus, the problem differs from that of providing a global multilateration service and also from that of providing either a local or a global communication service.

The main results are as follows:

1) Comprehensive analyses of three "baseline" constellations are summarized. The constellations are representative of previous small, medium, and large constellations. The analyses include:

a) Calculation of the geometric dilution of precision (GDOP) during level flight (basically, GDOP is an error magnification factor that indicates how much the basic ranging error is magnified by the constellation-aircraft geometry; see Sec. 2).

b) Calculation of GDOP after a key satellite has been deleted (to assess the impact of satellite failure).

c) Calculation of GDOP for aircraft making a 30° bank on test headings of 0°, 10°,...350° (to assess the impact of aircraft maneuvers).

The calculations are performed at each point of a 119-point CONUS grid. In addition, the calculations are performed under the separate assumptions of "visibility down to aircraft wingtips," "visibility down to 15° above the wingtips," and "visibility down to 30° above the wingtips." Comparison of the resulting average GDOP(s) with the theoretical minimum GDOP(s) shows that the actual GDOP(s) vary between 2.2 and 4.5 times the minimum attainable GDOP. Consequently, the constellations allow considerable room for accuracy improvement.

2) A new procedure for calculating GDOP is described. The procedure "explains" the (comparatively high) GDOP(s) provided by the baseline constellations. The procedure also indicates that substantially improved GDOP(s) can be obtained by employing synchronous retrograde orbits in place of the previously used posigrade orbits. (A retrograde orbit is one for which the rotational sense of the satellite is opposite to that of Earth. A posigrade orbit is one for which the rotational sense is the same as that of Earth.)

3) A variety of methods for utilizing retrograde orbits in constellations are examined. One method emerges as the best of those considered. The method involves using retrograde orbits to approximate an idealized constellation wherein the satellites are distributed uniformly within a "viewing cone."

4) New large, medium, and small constellations are designed by the foregoing method. The resulting constellations are subjected to the same analyses as the baseline constellations. In almost all cases, the new constellations exhibit smaller GDOP(s) that vary between 1.6 and 2.3 times the optimum GDOP values.

5) The analyses of both the baseline constellations and the new small (seven-satellite) constellations indicate that small constellations can produce useful values of GDOP (e.g., three) during level flight. However, the GDOP(s) are highly sensitive to satellite failure or aircraft maneuvers. For example, during banking GDOP exceeds 10 on more than 50% of the test headings if visibility down to 30° above the wingtips is assumed. Accordingly, use of more than seven satellites is indicated if high accuracy is to be maintained following a satellite failure and during aircraft maneuvers. The analyses of medium-sized constellations suggest that 10 satellites is a reasonable number.

2. Accuracy Measure Used

For present purposes, GDOP is used as a measure of constellation accuracy. Basically, GDOP is an error magnification factor that indicates how much the basic ranging error is magnified by the constellation-aircraft geometry. GDOP is defined as the ratio

$$\mathrm{GDOP} = \sigma_a/\sigma_r \tag{1}$$

where σ_a denotes the rms error in the calculated position of the aircraft, and σ_r denotes the rms ranging error.

In general, the positional error σ_a depends upon the algorithm used to calculate the aircraft position. The positional error σ_a here is taken to be that made by the (optimal) linearized least-squares procedure when data from all visible satellites are utilized. In addition, it is assumed that the ranging errors for the different satellite-aircraft links can be modeled as zero-mean, equal variance, uncorrelated random variables. This assumption parallels that made in earlier studies.[2,3,8] The zero-mean assumption presupposes that the timing signals have been corrected for the mean delay introduced by the ionosphere. The equal-variance assumption treats as omnidirectional the errors due to the residual ionospheric delay, satellite tracking errors, and errors due to receiver noise. Section 18 discusses relaxation of the equal variance assumption. See Eq. (I-13) in Ref. 2 for the detailed expression used to evaluate GDOP.

3. Constellation Design Problem

The constellation design problem is taken to be that of assigning satellites to physical orbits such that the following conditions are satisfied:

1) Low values of GDOP are obtained at all times throughout CONUS.

2) The GDOP(s) are relatively insensitive to satellite failure (dropout).

3) The GDOP(s) are relatively insensitive to typical aircraft maneuvers.

The accuracy of any candidate constellation depends, of course, upon the number of measurements utilized or, equivalently, upon the number of visible satellites. The question of visibility is resolved by treating as visible any satellite that satisfies the following two conditions:

1) The satellite is above the local Earth horizon.

2) The satellite is within a "viewing cone" of half-angle ϕ centered about the aircraft vertical, as shown in Fig. 1.

This approach makes it possible to draw conclusions that are independent of any particular technology. For example, the results do not depend upon any assumptions concerning "reasonable" transmitter power levels. At the same time, the approach makes it possible to account for the cutoff characteristic of realistic antennas by varying the half-angle ϕ of the viewing cone shown in Fig. 1.

4. Satellite Orbits Used

All satellite constellations considered employ synchronous orbits with perigee at the southernmost orbital point. Again, this decision follows Refs. 2, 3, and 8. The motivation for restricting attention to synchronous orbits is that the design objective calls for coverage only to a small portion of the Earth's surface (i.e., CONUS). The reason for placing the perigees at the southernmost orbital point is to maximize the time of satellite visiblity from the northern hemisphere (specifically from CONUS).+

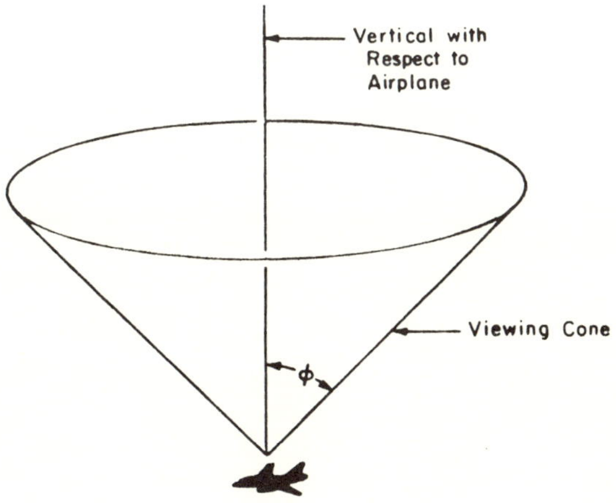

Fig. 1 Viewing cone of half-angle ϕ.

+Note that these restrictions would not be called for if global coverage were the design objective.

Four parameters suffice to describe satellite orbits of the foregoing kind. These are as follows:

- e = orbit eccentricity
- i = inclination of the orbital plane with respect to the equatorial plane
- γ_p = longitude of the subperigee point
- T_p = time of passage through perigee

To reduce the temporal variations of GDOP, the orbits in most cases are configured such that 1) the satellites traverse a common ground track, and 2) the subsatellite points are spaced equally "in time" along the ground track.

5. Baseline Constellations

To provide a baseline for assessing the new constellations, the following previous constellations were examined: 1) the RCA-8 constellation as modified by Autonetics (see p. B-32, Appendix B, Vol. III, Ref. 3); 2) the LL-1 constellation (see pp. 131 and 134, Ref. 2); and 3) the Hybrid constellation (see Refs. 3 and 8, and note that the constellation is called "the System-A Constellation" in Ref. 8, pp. 2-57 to 2-75, Vol. 2, Book 1). The constellations are representative of previous small, medium, and large constellations. All three constellations employ combinations of equatorial orbits and posigrade orbits (or orbits with an inclination i smaller than 90°). Detailed descriptions of these constellations can be found in Ref. 10 or in the original sources.

The constellations were examined over a grid extending from 60° to 140° W long, and from 25° to 55° N lat. The grid encompasses the continental United States (CONUS) and its oceanic approaches. Adjacent grid points are separated by 5° in longitude and/or latitude. Thus, the grid consists of 119 points. The following analyses were performed for each constellation:

1) To assess accuracy during level flight, GDOP was calculated at each grid point.

2) To assess the impact of satellite failure, GDOP was recalculated at each grid point after a key satellite was deleted.

3) To assess the impact of aircraft maneuvers on accuracy, GDOP was calculated for aircraft in a 30° bank at test headings of 0°, 10°, 20°, . . ., 350° at each grid point.

The analyses were performed for cone half-angles of $\phi = 90^{\circ}$, 75°, and 60°. Note that a half-angle of $\phi = 90^{\circ}$ corresponds to visibility "down to the horizon." All results were generated using variants of a computer program made available by Autonetics (see Ref. 10, Appendix A).

6. Performance of the Baseline Constellations

The small baseline constellation (RCA-8) utilizes eight satellites. Two of the satellites are in circular equatorial orbits; the remaining six satellites are in elliptical orbits of eccentricity 0.25 inclined at 80°. The elliptical orbits are selected so that the satellites traverse a common "figure-eight" ground track.

GDOP statistics for the three viewing cones are shown in Table 1 for time t = 0 hr. The entries in the second column are the average numbers of satellites visible at time t = 0 over all of CONUS. The entries in the third column are the averages of the GDOP(s) smaller than 10. As expected, the average number of visible satellites decreases and GDOP increases as the cone half-angle is made smaller. The entries in the fourth column are the rms deviations of the GDOP(s) smaller than 10. The entries in the fifth column are explained in Sec. 7. Clearly, the nominal performance of the constellation is quite reasonable.

Dropout results are summarized in Table 2 for failure of the northernmost satellite. It is evident from the first row of the table that the failure significantly impacts accuracy even for a cone half-angle of 90°. GDOP more than doubles as the average number of visible satellites decreases by one. The second row shows that the impact is more substantial for a cone half-angle of 75°. Specifically, GDOP at 42% of the grid points exceeds 10, with GDOP at the remainder averaging 7.6. The third row shows that the impact is severe for a

Table 1 Level flight (baseline small)

Cone half-angle, deg	Average number visible	Average GDOP < 10	Rms dev.	Excess
90	7.0	3.5	0.04	3.0
75	6.5	3.7	0.37	2.8
60	4.9	4.7	0.98	2.6

Table 2 Dropout results (baseline small)

Cone half-angle, deg	Average number visible	Percent GDOP > 10	Average GDOP < 10	S (sens.)
90	6.0	0	7.7	8.5
75	5.5	42	7.6	6.7[a]
60	4.1	86	7.4	3.4[a]

[a]Larger if GDOP(s) exceeding 10 are included.

Table 3 30° bank results (baseline small)

Cone half-angle, deg	Average number visible	Percent GDOP > 10	Average GDOP < 10
90	6.5	2	3.8
75	5.6	17	4.5
60	4.3	54	5.5

cone half-angle of 60°. GDOP at 86% of the grid points exceeds 10, with GDOP at the remainder averaging 7.4. The entries in the "Sensitivity" column are the ratios of the resulting fractional changes in GDOP to the fractional changes in the numbers of visible satellites. Thus, for example, with $\phi = 90^{\circ}$, the satellite failure increases the average GDOP by 121%, while decreasing the average number of visible satellites by 14%, so that Sensitivity = 8.5 (= 121/14).

The bank analysis results are summarized in Table 3. The results cover aircraft making 30° banks at test headings of 0°, 10°, ... , 350° at each grid point within CONUS. The first row of the table shows that, in the case of a viewing cone half-angle of $\phi = 90^{\circ}$, the impact upon accuracy is modest. The banking aircraft "see" 6.5 satellites on the average. Moreover, only 2% of the banking aircraft have GDOP(s) in excess of 10. The average GDOP is 3.8 for aircraft with GDOP < 10. The second and third rows of the table show that banking impacts accuracy substantially in the case of smaller viewing cones. For $\phi = 75^{\circ}$, the banking aircraft sees 5.6 satellites on the average. Approximately 17% of the aircraft have GDOP(s) in excess of 10. For $\phi = 60^{\circ}$, the average number of visible satellites falls to 4.3 or just 0.3 above the minimum number required. Approximately 54% of the banking aircraft have GDOP(s) in excess of 10.

To summarize, the baseline small constellation produces a highly uniform GDOP over all of CONUS. Accuracy is highly sensitive to failure of a single satellite, however. Moreover, accuracy degrades substantially during aircraft maneuvers (e.g., banking), particularly for the smaller viewing cones.

The baseline medium constellation (LL-I) contains 12 satellites rather than 8. All satellites are placed in elliptical orbits of eccentricity 0.4 and inclination 63.4°. The orbits are configured so that 6 of the satellites traverse one ground track, whereas the remaining 6 satellites traverse a second ground track. The two ground tracks are identical (degenerate) figure eights offset 35° in longitude so as to provide improved east-west coverage.

GDOP statistics for the constellation are given in Table 4. Again the average number of visible satellites decreases as the cone half-angle is reduced from 90° to 60°, and GDOP increases correspondingly.

Dropout results are summarized in Table 5 for failure of a satellite in an extreme southwesterly position in the viewing cone. The average number of visible satellites decreases by almost unity for $\phi = 90^{\circ}$, 75°, since the satellite deleted is visible over most of CONUS for these cone half-angles. The decrease is smaller (0.5) for $\phi = 60^{\circ}$, in which case the satellite deleted is visible over approximately half of CONUS.

The bank analysis results are given in Table 6. The first row shows that, for a viewing cone half-angle $\phi = 90^{\circ}$, banking does not degrade accuracy significantly. Specifically, the average GDOP increases by 25% compared to level flight. The second and third rows show that banking degrades accuracy more substantially for smaller viewing cones. For example, for $\phi = 75^{\circ}$, 1% of the banking aircraft have GDOP(s) in excess of 10. For $\phi = 60^{\circ}$, 22% of the aircraft have GDOP(s) in excess of 10.

In summary, the baseline medium constellation provides reasonably uniform GDOP over CONUS for a viewing cone half-

Table 4 Level flight (baseline medium)

Cone half-angle, deg	Average number visible	Average GDOP < 10	Rms dev.	Excess
90	9.9	2.8	0.56	3.1
75	9.1	4.5	2.4	4.2
60	8.2	6.5	2.5	4.5

Table 5 Dropout results (baseline medium)

Cone half-angle, deg	Average number visible	Percent GDOP > 10	Average GDOP < 10	S (sens.)
90	9.0	0	3.7	3.0
75	8.4	0	6.5	5.6
60	7.7	14	8.5[a]	5.0[a]

[a]Larger if GDOP(s) exceeding 10 are included.

Table 6 30° bank results (baseline medium)

Cone half-angle, deg	Average number visible	Percent GDOP > 10	Average GDOP < 10
90	9.4	0	3.6
75	8.7	1	4.7
60	7.3	22	6.2

angle of $\phi = 90^{\circ}$. As the size of the viewing cone is reduced, GDOP increases significantly and loses its uniformity. Again, accuracy is sensitive to failure of a single satellite and also to aircraft maneuvers.

The baseline large constellation (Hybrid) contains 15 satellites. The constellation consists of 6 satellites in circular equatorial orbits and 9 satellites in elliptical orbits of eccentricity 0.25 and inclination 80°. Once again, the inclined orbits are configured so that the satellites traverse a common "figure-eight" ground track.

GDOP statistics for the baseline large constellation at time t = 0 hr are given in Table 7. Again, the average number of visible satellites decreases as the cone half-angle is reduced. The unit reduction from 10.8 to 9.8 is due to masking of one or another of the equatorial satellites by the 75° viewing cone. The reduction from 9.8 to 8.0 is due to further masking of one or more equatorial satellites and one or more nonequatorial satellites. It is evident that the constellation provides uniform and comparatively low GDOP(s) over all of CONUS. For example, for $\phi = 90^{\circ}$, the average GDOP is 1.8, whereas the rms GDOP variation is 0.25.

Table 7 Level flight (baseline large)

Cone half-angle, deg	Average number visible	Average GDOP < 10	Rms dev.	Excess
90	10.8	1.8	0.25	2.2
75	9.8	2.4	0.35	2.5
60	8.0	3.8	1.1	2.9

Table 8 Dropout results (baseline large)

Cone half-angle, deg	Average number visible	Percent GDOP > 10	Average GDOP < 10	S (sens.)
90	10.4	0	2.2	4.5
75	9.1	0	2.8	1.9
60	7.6	0	4.1	1.4

Table 9 30° bank results (baseline large)

Cone half-angle, deg	Average number visible	Percent GDOP > 10	Average GDOP < 10
90	9.9	0	2.3
75	8.7	0.3	3.2
60	6.8	11.0	4.6

Dropout results for the constellation are summarized in Table 8 for failure of an equatorial satellite near the southwestern cone horizon. Clearly, the constellation is relatively insensitive to single-satellite failure.

The results of the bank analysis are given in Table 9. The table shows that degradation is not significant for the viewing cone half-angles $\phi = 90^{\circ}$ and $\phi = 75^{\circ}$. For $\phi = 60^{\circ}$, degradation becomes more noticeable. Here 11% of banking aircraft have GDOP(s) in excess of 10.

To summarize, the baseline large constellation provides low and relatively uniform GDOP(s) over CONUS. Accuracy is not seriously degraded either by failure of a single satellite or by aircraft maneuvers.

7. Comparison with Optimum GDOP

It has been shown elsewhere[11] that the minimum value of GDOP, which can be attained by (optimally) positioning N satellites within a viewing cone of half-angle ϕ, is given by the expression

$$\mathrm{GDOP} = \frac{1}{\sqrt{N}} \frac{4}{\sqrt{1 + \cos\phi}\,(\sqrt{5 - 3\cos\phi} - \sqrt{1 + \cos\phi})} \qquad (2)$$

The entries in the excess column of Tables 1, 4, and 7 compare the actual GDOP(s) produced by the baseline constellations with the minimum value given by Eq. (2). Specifically, the excess is defined as the ratio of the actual (average) GDOP to the minimum value given by Eq. (2). Thus, for example, for $\phi = 90^{\circ}$, the average GDOP for the baseline large constellation is 1.8, whereas that given by Eq. (2) is 0.85; therefore, the associated excess = 2.1 (= 1.8/0.85). Reference to the excess columns of Tables 1, 4, and 7 shows that the actual GDOP(s) are 2.2 to 4.5 times the minimum attainable GDOP(s). Consequently, the baseline constellations allow considerable room for GDOP improvements. The following section describes a recently developed procedure[11] for calculating GDOP. The procedure suggests how such GDOP improvements can be achieved.

8. An Alternate Procedure for Calculating GDOP

The new calculation procedure consists of three basic steps. First, a simple configuration of masses is constructed from the satellite-aircraft geometry. Next the moments and products of inertia of the mass configuration about its center of mass are calculated. Finally, GDOP is calculated from the inverse of the moment-of-inertia matrix.

The procedure for constructing the mass configuration is quite simple. In the case of N visible satellites, the configuration is constructed as follows:

1) Draw a sphere of unit radius centered at the aircraft.

2) Place unit masses $m_1, \ldots, m_N$ at the N subsatellite points on the unit sphere.

The next step consists of calculating the moment-of-inertia matrix of the mass configuration about its center of mass (CM). To perform this step, it is necessary to locate the center of mass and set up a Cartesian coordinate system (X, Y, Z) there. Then the moment of inertia matrix $\underline{L}$ is calculated from the expression

$$\underline{L} = \begin{bmatrix} \sum_{j=1}^{N} X_j^2 & \sum_{j=1}^{N} Y_j X_j & \sum_{j=1}^{N} Z_j X_j \\ \sum_{j=1}^{N} X_j Y_j & \sum_{j=1}^{N} Y_j^2 & \sum_{j=1}^{N} Z_j Y_j \\ \sum_{j=1}^{N} X_j Z_j & \sum_{j=1}^{N} Y_j Z_j & \sum_{j=1}^{N} Z_j^2 \end{bmatrix} \tag{3}$$

where X_j, Y_j, and Z_j denote the coordinates of mass m_j. [Note that the diagonal elements of Eq. (3) correspond to the moments of inertia of the unit masses $m_1, m_2, \ldots, m_N$

projected into the X-Y, X-Z, and Y-Z planes; the off-diagonal terms correspond to the associated products of inertia.]

The final step consists of calculating GDOP from the expression

$$GDOP = \sqrt{\text{trace } (\underline{L}^{-1})} \tag{4}$$

9. Attributes of a "Good" Constellation

By working backward through the steps in Sec. 8, it is possible to deduce attributes of a constellation exhibiting low values of GDOP. Specifically, GDOP, as given by Eq. (4), can be small only if the trace of $\underline{L}^{-1}$ is small. But the trace of $\underline{L}^{-1}$ is small only if that of $\underline{L}$ is large. Accordingly, to obtain a (comparatively) small value of GDOP it is necessary that the moments of inertia L_{xx}, L_{yy}, and L_{zz} be large. This conclusion suggests that the following is a good strategy for obtaining low GDOP(s).

Strategy: Select the satellite orbits so that the configuration of masses $m_1, m_2, \ldots, m_N$ exhibits large moments of inertia in all three directions.

The strategy obviously produces large values of L_{xx}, L_{yy}, and L_{zz}. It also tends to make $\underline{L}^{-1}$ have small diagonal elements, so that GDOP as calculated from Eq. (4) is small.

10. Spreading Out the Ground Tracks

It is instructive to examine the baseline constellations from the foregoing viewpoint. Figure 2 depicts the placement of mass on the unit sphere for the baseline large constellation as viewed from a grid point in central CONUS at time $t = 0$. Each integer denotes the mass location for the correspondingly numbered satellite. Only the visible satellites are represented. An asterisk indicates multiple masses just to the left of the asterisk. The reason for the (comparatively) high GDOP(s) of the baseline constellations is apparent immediately from Fig. 2. Specifically, the masses very nearly lie in a plane. Consequently, the vertical moment arms are small, so that the moment of inertia L_{zz} is small and the Z-Z element of $\underline{L}^{-1}$ is correspondingly large, with the result that GDOP as calculated from Eq. (4) is large.

It is evident from these observations how reduced GDOP(s) can be achieved. To reduce GDOP, it is necessary only to spread out the masses, particularly in the vertical direction.

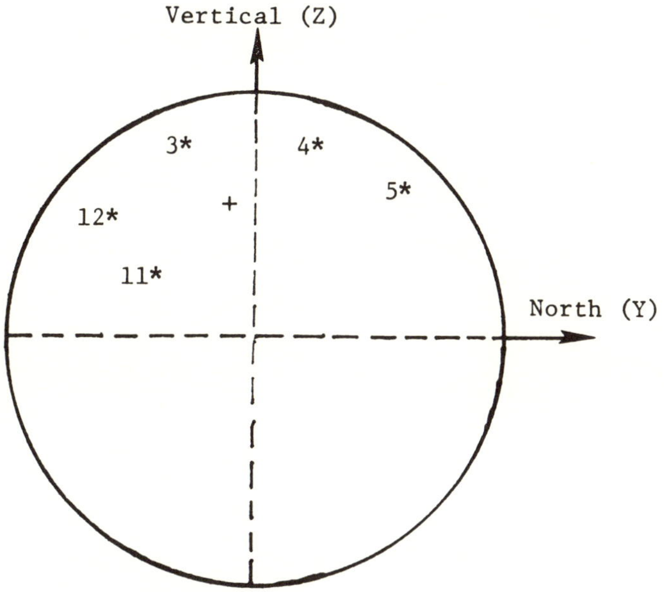

Fig. 2 Placement of mass on the unit sphere for the baseline large constellation.

That is, it is necessary only to spread out the subsatellite points on the unit sphere. To understand how this can be done, observe that, if the unit sphere is scaled up to the size of the Earth, then the scaled-up subsatellite points closely correspond to the subsatellite points on the Earth. Consequently, reduced GDOP(s) can be obtained by spreading out the subsatellite points on the Earth or, equivalently, by spreading out the ground track.

Several methods for spreading out the ground track can be devised. The method that is explored here is to employ retrograde orbits in place of the previous posigrade orbits. Satellites in retrograde orbits are visible from central CONUS throughout their northern passage, provided that the orbital inclination is not excessive (see Sec. 7.2, Ref. 10).

11. Use of Retrograde Orbits

One effective strategy for employing retrograde orbits is to use them to fill the "viewing cone" with satellites in an approximately uniform manner. Although a "uniform" constellation is suboptimum, it is close to optimum.[11] In addition, uniform constellations are relatively insensitive to satellite failure and aircraft maneuvers. Three different methods for approximating uniform constellations are considered here.

The first method involves placing satellites in all three types of orbits: retrograde, posigrade, and equatorial. The satellites in retrograde orbits provide coverage near the rim of the viewing cone, particularly toward the north. The satellites in posigrade orbits provide coverage overhead, whereas those in equatorial orbits provide coverage to the south.

The second method utilizes satellites in retrograde and equatorial orbits only. The retrograde orbits are configured so that the satellites traverse a single ground track. The retrograde satellites are used to provide coverage overhead, along the northern, northeastern, and northwestern horizons, and also in the southeast and southwest. The equatorial satellites provide additional coverage near the southern horizon.

The third method also utilizes satellites in retrograde and equatorial orbits only. Here, however, the retrograde orbits are configured so that the satellites traverse two identical ground tracks offset in the east-west direction. Again the retrograde satellites provide coverage overhead and along the northern horizon. The reason for offsetting the ground tracks is to obtain improved coverage near the eastern and western horizons. Once again, the equatorial satellites provide coverage near the southern horizon.

To determine the relative merits of the three methods, one or more constellations were constructed using each. Fifteen satellites were used for this purpose, and visibility "down to the horizon" (i.e., $\phi = 90^{\circ}$) was assumed. Each constellation was designed in two steps. First, a "ballpark" set of parameters was selected. Then the constellation was "tuned" by repeatedly analyzing it and varying the constellation parameters. The parameter values that produced the smallest average GDOP for level flight were selected as the final values.

The retrograde orbits for each initial constellation were selected to have their apogees on the northern rim of a viewing cone located in Kansas. The parameter γ_p initially was taken to be 100° W long (except for Method 3). The tuning step involved varying γ_p and the orbital inclinations in increments of 5° (or smaller) and varying the orbital eccentricities in increments of 0.05. The GDOP results for level flight are summarized in Table 10. It is evident from the table that the second method is the best of those consid-

Table 10 "Uniform" constellations

Method	Constellation	Retrograde	Posigrade	Equatorial	Avg. GDOP
1	U1	6	6	3	1.46
2	U2	12	0	3	1.35
	U3	13	0	2	1.35
	U4	5	0	10	1.35
	U5	12	0	3	1.35
	U6	6	0	9	1.37
3	U7	6+6	0	3	1.38

ered. Additional analyses of satellite failure and aircraft banking indicated that the design method implicit in Constellation U2 is the best of those considered. The procedure can be summarized as follows.

The Best Design Procedure

1) Place approximately 4/5 of the available satellites in retrograde orbits of modest eccentricity, such that the subsatellite points are spaced evenly in time along a single ground track, and such that a satellite at apogee is just above the northern rim of the viewing cone.

2) Place the remaining satellites in circular equatorial orbits uniformly distributed between 20° and 180°W long.

Figure 3 depicts a vertical-north projection of the mass configuration for Constellation U2. Clearly, the constellation eliminates the near planarity of the masses previously noted in Fig. 2. The result in terms of GDOP is evident from Tables 7 and 10. For $\phi = 90^{\circ}$, the U2 constellation produces an average GDOP of 1.35, whereas the baseline-large constellation produces an average GDOP of 1.8.

12. Design of the New Constellations

The foregoing procedure was used to design large, medium, and small constellations for viewing cone half-angles of $\phi = 90^{\circ}$, 75°, and 60°. Fifteen satellites were used in each

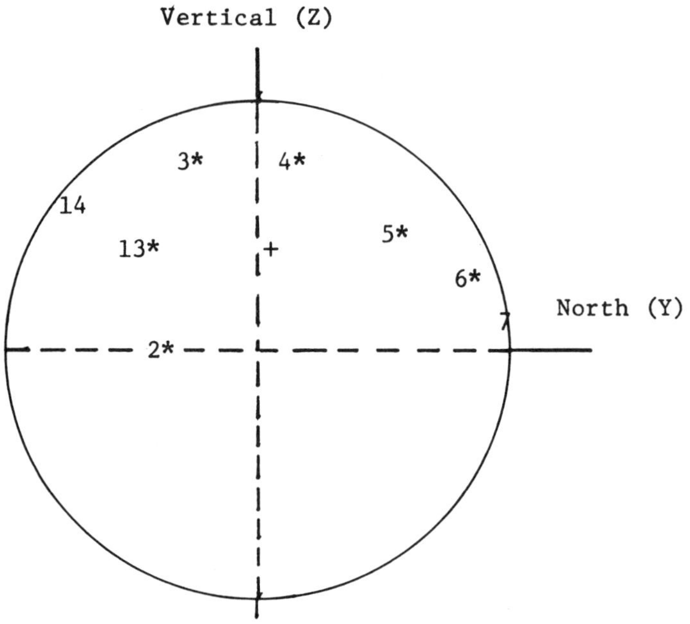

Fig. 3 Mass configuration for Constellation U2.

large constellation. Ten satellites were used in each medium constellation and 7 satellites in each small constellation. The resulting constellations were named according to the convention

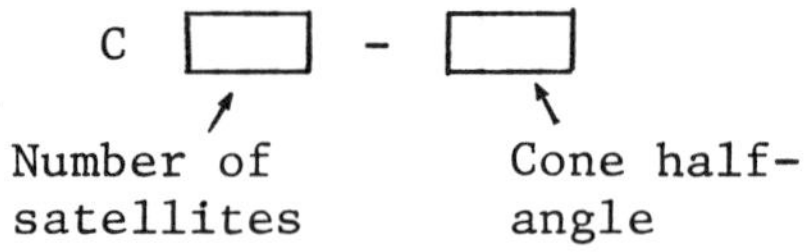

Thus, for example, C10-75 is the medium-sized constellation designed for a viewing cone half-angle of $\phi = 75^{\circ}$. Each constellation was designed in two steps. First a "ballpark" constellation was designed using the final procedure of Sec. 11. Then the constellations were "tuned" by varying 1) the relative numbers of satellites in retrograde and equatorial orbits, 2) the inclination of the retrograde orbits, 3) the eccentricity of the retrograde orbits, and 4) the longitudinal separation of the equatorial satellites. The parameter values that produced the lowest average GDOP during level flight then were taken to be the final constellation parameters.

The tuning step was simplified greatly by the discovery that the orbital parameters, e and i, have largely independent effects upon the average GDOP. This frequently made it possible to tune a constellation by finding the best value of i for a given value of e, and then finding the best value of e for the resulting i.

13. New Large Constellations

The C15-90 constellation (see Ref. 10 for detailed orbital parameters) consists of 3 satellites in circular equatorial orbits and 12 satellites in retrograde orbits of eccentricity 0.5 and inclination of 116.6°. (In cases where the optimum inclination was within a few degrees of 116.6°, the latter inclination was selected with a view toward orbital stability.) The retrograde orbits are configured so that the 12 satellites traverse a common ground track at 2-hr intervals. The first rows of Tables 11-13 summarize the performance of C15-90. Reference to Tables 7-9 shows that, for $\phi = 90^\circ$, C15-90 compares favorably to the baseline-large constellation in all categories. For example, C15-90 produces an average GDOP of 1.4 (vs 1.8), with an rms deviation of 0.11 (vs 0.25). Moreover, the sensitivity to satellite failure is only 0.3 (vs 4.4). Finally, the average GDOP during banking is 1.7 (vs 2.3).

The orbital parameters for C15-75 are identical to those of C15-90, except that the inclinations of the retrograde orbits are reduced from 116.6° to 105°. This reduction is necessary for the retrograde satellites to remain visible over the northern rim of the (smaller) viewing cone. The second rows of Tables 11-13 summarize the performance of C15-75. Reference to the corresponding entries in Tables 7-9 shows that, for $\phi = 75^\circ$, the C15-75 constellation compares favorably with the baseline constellation in all categories.

Table 11 Level flight (new large)

Constellation	Average number visible	Average GDOP < 10	Rms dev.	Excess
C15-90	12.5	1.4	0.11	1.6
C15-75	11.2	1.8	0.20	1.8
C15-60	9.1	2.9	0.67	2.2

Table 12 Dropout results (new large)

Constellation	Average number visible	Percent GDOP > 10	Average GDOP < 10	S (sens.)
C15-90	11.5	0	1.4	0.30
C15-75	10.2	0	1.9	0.67
C15-60	8.3	0	3.2	1.00

Table 13 30° bank results (new large)

Constellation	Average number visible	Percent GDOP > 10	Average GDOP < 10
C15-90	9.9	0	1.7
C15-75	9.8	0	2.3
C15-60	8.0	2	4.0

The orbital parameters of the constellation C15-60 also are similar to those of C15-90. Minor changes involve reducing the inclination of the retrograde orbits to 99° and reducing the longitudinal separation of the equatorial satellite to 30°. Again, the changes are necessary to keep the majority of the satellites above the rim of the (still smaller) viewing cone. The third rows of Tables 11-13 summarize the GDOP performance of C15-60. Note that, on the average, C15-60 provides two fewer visible satellites than C15-75. This is because either near-apogee satellites or equatorial satellites, but not both, generally are visible in C15-60. By contrast, both generally are visible in C15-75. Although GDOP(s) for C15-60 are well above those for C15-90 and C15-75, the GDOP(s) still are quite reasonable. Reference to Tables 7-9 shows that, for a viewing cone with $\phi = 60^{\circ}$, constellation C15-60 compares favorably with the baseline large constellation in all categories.

14. New Medium-Sized Constellations

Except for minor deviations, the new medium-sized constellations are scaled-down versions of the large constellations. Accordingly the resulting GDOP(s) for level flight are roughly $\sqrt{15/10}$(=1.23) times larger than those for the large constellations. (Scaling the size of a constellation by a factor k tends to reduce GDOP by the factor $1/\sqrt{k}$. See Sec. 6.7 of Ref. 10.) Also, the GDOP(s) are more sensitive to satellite failure and aircraft banking than those for the large constellations.

The C10-90 constellation consists of 8 satellites in retrograde orbits of eccentricity 0.5 and inclination $^{\circ}$110, and 2 satellites in circular equatorial orbits located at 25° and 175°W long. Again the retrograde orbits are configured so that the satellites traverse a common ground track. The first rows of Tables 14-16 summarize the GDOP performance of C10-90. Clearly, C10-90 provides low GDOP(s) not only during level flight (Table 14) but also following a satellite failure (Table 15) and during aircraft banking (Table 16).

The C10-75 constellation consists of 7 satellites in retrograde orbits of eccentricity 0.35 and inclination 100.0°, and 3 satellites in circular equatorial orbits. The reduced inclination of the retrograde orbits is necessary to keep the satellites above the rim of the viewing cone at most grid points. Note that C10-75 has one more equatorial satellite and one less retrograde satellite than C10-90. In addition, the retrograde orbits have reduced eccentricities (0.35 vs 0.5). The second rows of Tables 14-16 summarize the performance of C10-75. Note that performance still is quite reasonable. Nonetheless, comparison of the first two rows of Table

Table 14 Level flight (new medium)

Constellation	Average number visible	Average GDOP < 10	Rms dev.	Excess
C10-90	8.0	1.7	0.10	1.7
C10-75	6.9	2.4	0.42	2.0
C10-60	6.6	3.7[a]	0.93	2.3

[a]Larger if GDOP(s) greater than 10 are included.

Table 15 Dropout results (new medium)

Constellation	Average number visible	Percent GDOP > 10	Average GDOP < 10	S (sens.)
C10-90	7.5	0	2.0	2.5
C10-75	5.9	1	2.7[a]	0.8[a]
C10-60	6.0	6	4.0[a]	0.8[a]

[a]Larger if GDOP(s) exceeding 10 are included.

Table 16 30° bank results (new medium)

Constellation	Average number visible	Percent GDOP < 10	Average GDOP > 10
C10-90	6.9	0.05	2.2[a]
C10-75	5.9	10	3.0[a]
C10-60	5.6	37	4.2[a]

[a]Larger if GDOP(s) greater than 10 are included.

14 shows that the rms deviation of GDOP for C10-75 is more than four times that of C10-90. Similarly, the C10-75 entries of Tables 15 and 16 show that a small fraction of the GDOP(s) now exceed 10 following a satellite failure or during aircraft banking. The reduced GDOP performance of C10-75 compared to C10-90 and the "C15" constellations is due to the following factors:

1) Fewer satellites are available in favorable positions at any one time (because of the reduced total number of satellites).

2) The satellites in favorable positions influence GDOP over smaller geographical areas (because of the reduced size of the viewing cone).

The constellation C10-60 consists of 8 satellites in retrograde orbits of eccentricity 0.5 and inclination 95°, and 2 equatorial satellites separated by 20° in longitude. Again, the reduced inclination of the retrograde orbits and the reduced separation of the equatorial satellites are necessary to insure visibility at a majority of CONUS grid points. The third rows of Tables 14-16 summarize the performance of C10-60. The effect of further reducing ϕ has been to increase GDOP and the fluctuations in GDOP. The average GDOP is more than 50% greater than that for C15-75. In the case of satellite failure, an increased percentage (6%) of GDOP(s) exceed 10. In the case of banking, more than one-third of the aircraft have GDOP(s) greater than 10. These developments again are consequences of factors 1 and 2.

Comparison of the performance of all three medium-sized constellations with that of the baseline medium constellation shows that, for each value of cone angle ϕ, the new constellations perform better in all categories. Indeed, comparison of Tables 14-16 with Tables 7-9 shows that the performance of the new medium-sized constellations is comparable to that of the baseline-large constellation.

15. New Small Constellations

Again, except for minor perturbations, the constellations are scaled-down versions of the large constellations. The average GDOP(s) for level flight are approximately $\sqrt{15/7}$ (=1.46) times those for the large constellations. The GDOP(s) following satellite failure or during banking are significantly higher than those predicted by the scaling argument of Ref. 10, however. This is because the average number of visible satellites closely approaches the minimum number of 4.

The constellation C7-90 consists of 5 satellites in retrograde orbits of eccentricity 0.70 and inclination 116.6°, and 2 satellites in circular equatorial orbits. Like C10-90, constellation C7-60 is very nearly a scaled-down version of C15-90. The GDOP statistics for C7-90 are given in Tables 17-19, along with those for C7-75 and C7-60. On the average, 5.8 of the 7 satellites are visible over CONUS. The only satellite not generally visible is that at perigee. Considering its small size, the constellation performs well following satellite failure. Table 18 shows that only 2.5% of the GDOP(s) exceed 10 when a key satellite fails, and the remaining GDOP(s) average 3.1. Performance is more sensitive to aircraft banking. Table 19 shows that, for 16% of the selected headings at CONUS grid points, GDOP exceeds 10, with the GDOP(s) averaging 2.7 on the remaining headings.

Table 17 Level flight (new small)

Constellation	Average number visible	Average GDOP < 10	Rms dev.	Excess
C7-90	5.8	2.0	0.19	1.6
C7-75	5.8	3.0	0.40	2.1
C7-60	5.1	4.1	0.38	2.2

Table 18 Dropout results (new small)

Constellation	Average number visible	Percent GDOP > 10	Average GDOP < 10	S (sens.)
C7-90	4.8	2.5	3.1[a]	3.4[a]
C7-75	4.9	11	3.9[a]	1.4[a]
C7-60	4.4	36	4.5[a]	0.7[a]

[a]Larger if GDOP(s) exceeding 10 are included.

Table 19 30° bank results (new small)

Constellation	Average number visible	Percent GDOP > 10	Average GDOP < 10
C7-90	5.0	16	2.7[a]
C7-75	4.9	28	3.4[a]
C7-60	3.7	76	4.5[a]

[a]Larger if GDOP(s) greater than 10 are included.

The constellation C7-75 consists of 5 satellites in retrograde orbits of eccentricity 0.5 and inclination 95°, and 2 satellites in circular equatorial orbits separated 48° in longitude. The GDOP statistics for C7-75 are given in Tables 17-19. Note that, although C7-75 and C7-90 provide approximately the same numbers of visible satellites during level flight, GDOP is significantly higher for C7-75. This is a direct consequence of the higher elevation angles of the C7-75 satellites. Table 18 indicates an increased sensitivity to satellite dropout. Specifically, failure of a key satellite causes GDOP to exceed 10 at more than 10% of the grid points. Table 19 indicates an increased sensitivity to aircraft banking, with GDOP exceeding 10 at 28% of the test headings.

The constellation C7-60 consists of 5 satellites in posigrade orbits of eccentricity 0.5 and inclination 87°, and 2 satellites in circular equatorial orbits. For the case of 7 satellites and $\phi = 60^\circ$, the "tuning" step showed that retrograde orbits do not provide lower GDOP(s) than highly inclined posigrade orbits. Accordingly, posigrade orbits were used.

The GDOP statistics are given in Tables 17-19. Tables 18 and 19 show that performance is quite sensitive to satellite dropout and aircraft banking. For example, failure of a key satellite causes GDOP to exceed 10 at over one-third of the grid points. Table 19 shows that during aircraft banking fewer than the minimum number of satellites are visible on the average. Consequently, GDOP exceeds 10 over more than 75% of the test headings.

In spite of reduced performance compared to the medium and large constellations, the new small constellations compare quite favorably to the baseline-small constellation (see Tables 1-3).

Tables 1-3 and 17-19 collectively suggest that, regardless of the design method used, 7 satellite constellations are highly sensitive to satellite failure and aircraft maneuvers. Accordingly, more than 7 satellites are indicated if high accuracy is to be maintained following a satellite failure or during aircraft maneuvers.

16. Further GDOP Improvements

The constellations of Secs. 13-15 are the best that the authors have devised. Although the GDOP(s) improve upon those for the baseline constellations, they do not closely approximate the theoretical GDOP limit, Eq. (2). Instead, the constellation excesses are in the range

$$1.6 \leq \text{excess} \leq 2.3 \tag{5}$$

so that the optimum GDOP(s) are 40% to 60% smaller than the actual GDOP(s). Therefore, it is relevant to ask whether further reductions in GDOP can be achieved. It is the authors' opinion that, although small (10%) reductions in GDOP may be possible, large reductions (30% or more) will be difficult in the context of the problem statement of Sec. 3. The reasons for this are as follows:

1) The satellites do not occupy fixed positions but must traverse real physical orbits.

2) Low GDOP(s) are required not just at a single point but throughout CONUS.

3) Low GDOP(s) are required not just at one instant but at all times.

Certainly satellites can be placed in orbits so that at a given time the optimum GDOP, Eq. (2), is obtained at one location in CONUS. The optimum GDOP will not be obtained anywhere else in CONUS, however. Nor will the optimum value be obtained a short time later. Consequently, some departure from the optimum value is unavoidable. The authors' experience suggests that the departure may be as great as 30%.

17. Other Possible Analyses

Almost all of the foregoing analyses are for time t = 0 hr. Thus, the results represent a sample of the performance of each constellation. Had additional resources been available, it would have been desirable to examine the time dependence of GDOP for each constellation over a complete period. Such analyses would have provided useful information on the time variation of GDOP for each constellation. Time averaging of the data also would have smoothed the sensitivity data. Although additional analyses would have refined the results, it is doubtful that they would have altered any of the major conclusions. (See Sec. 8.8 of the full study report.[10])

18. Unequal Ranging Errors

The study has made the simplifying assumption that the ranging errors for all signal paths are uncorrelated and have equal variances (see Sec. 2). The same basic approach can be used in cases where the ranging errors have unequal variances, provided that the errors are uncorrelated. Equations (2-4) can be generalized easily to the unequal variance case. Once this has been done, the moment-of-inertia viewpoint can be used as in Secs. 9 and 10 to assist in design. The same basic approach also can be used when range-range multilateration is used in place of hyperbolic multilateration.[12]

References

1 Otten, D.D., "Study of a Navigation and Traffic Control Technique Employing Satellites," Group Rept. 08710-6012-12000, Vols. I-IV, Dec. 1967, TRW Systems, Redondo Beach, Calif.

2 Stiglitz, I.G., Beusch, J.U., Eckberg, A., and Schneider, K.S., "Concept Formulation Studies of Surveillance Aspects of the Fourth Generation Air Traffic Control System," Project Rept. ATC-7, DOT/TSC-241-1, Sept. 21, 1971, Lincoln Laboratory, Massachusetts Institute of Technology.

3 "Fourth Generation Air Traffic Control Study," Final Rept. C71-61/301, Nov. 1971, Rockwell International Corporation, Anaheim, Calif.

4 "Study and Concept Formulation of a Fourth Generation Air Traffic Control System," Rept. DOT-TSC-3601-1, Vols. I-V, April 1972, Boeing Company, Renton, Wash.

5 Diamond, P.M., "The Potential of a System of Satellites as Part of an Air Traffic Control System," AGARD Conference Proceedings No. 105, June 1972, p. 20-1.

6 Schuchman, L., "The ASTRO-DABS Concept," TR MTR-6287, Nov. 1972, MITRE Corporation, McLean, Va.

7 McDonald, K.D., "A Survey of Satellite-Based Systems for Navigation, Position Surveillance, Traffic Control and Collision Avoidance," Addendum to Proceedings of ION National Aerospace Meeting, March 1973, Institute of Navigation, Washington, D.C.

8 "Advanced Air Traffic Management System," Final Rept. C72-1206/201, April 1973, Rockwell International Corporation, Anaheim, Calif.

9 "Advanced Air Traffic Management-System B Summary Report," MTR-6419, Series 1, FAA-EM-73-10, June 1973, MITRE Corporation, McLean, Va.

10 Lee, H.B. and Wade, A.E., "Improved Satellite Constellations for CONUS ATC Coverage," Rept. ATC-23, May 1, 1974, Lincoln Laboratory, Massachusetts Institute of Technology.

[11] Lee, H.B., "Accuracy Limitations of Hyperbolic Multilateration Systems," TN 1973-11, DOT/TSC-241-3, March 22, 1973, Lincoln Laboratory, Massachusetts Institute of Technology.

[12] Lee, H.B., "Accuracy Limitations of Range-Range Multilateration Systems," TN 1973-43, DOT/TSC-RA-3-8-(1), Oct. 11, 1973, Lincoln Laboratory, Massachusetts Institute of Technology.

EFFECT OF MULTIPLE-BAND/SERVICE SATELLITES ON ORBIT AND SPECTRUM UTILIZATION

W. G. Long Jr.*

Computer Sciences Corporation, Falls Church, Va.

Abstract

Various possible combinations of bands and services allocated for satellite communications are outlined, and initial analyses of the effect of some possible multiple-band configurations on a single spacecraft are presented. Relationships among the principal parameters of a satellite communications system such as antenna characteristics, noise temperature and margins, and the information transfer per unit of orbital arc and rf bandwidth (orbit and spectrum utilization) are developed. Both homogeneous and heterogeneous aspects are introduced. Initial results indicate that use of multiple-band satellites and intersystem heterogeneity may reduce orbit and spectrum utilization. However, additional study is necessary before the definitive conclusions may be made.

I. Introduction

As a result of the 1971 World Administrative Radio Conference on Space Telecommunication and previous space conferences, many allocations are available, in a variety of services, for space telecommunications. The way in which these band allocations and services are combined on individual spacecraft may have an effect on the utilization of the geostationary orbit. The nature of this effect is dependent on the disparity of system characteristics among the allocated bands and services that utilize a single satellite. With the

---

Presented as Paper 74-433 at the AIAA 5th Communications Satellite Systems Conference, Los Angeles, Calif., April 22-24, 1974.

*Senior Scientist; presently a member of the staff of the Military Satellite Communications Systems Office of the Defense Communications Agency.

objective of efficient use of the geostationary orbit, the various possible and probable combinations of bands and services should be analyzed with respect to any deleterious effects on orbit and spectrum utilization. Orbit and spectrum utilization is defined as the information transfer between Earth stations per unit of orbital arc and rf bandwidth.

II. Multiple-Band/Service Satellites

For purposes of definition, a multiband satellite refers to a satellite that utilizes more than one up- or downlink frequency allocation in the same service allocation, in this case the fixed satellite service. Within this definition, there are a number of permutations, which can include 1) more than one "normal" uplink and downlink band pair (6/4 and 14/11 GHz), 2) more than one uplink allocation (6/4 and 14/4 GHz), or 3) more than one downlink allocation (6/4 and 6/11 GHz). A multiservice satellite is defined as a satellite that utilizes overlapping frequency bands for more than one service such as fixed satellite and broadcasting satellite in the 11.7 to 12.2-GHz band. A multiband and service satellite is defined as a satellite that utilizes nonoverlapping frequency allocations for different services. Included in this definition are the cases where a fixed satellite service allocation may be utilized in support of another service, such as the use of 6/4 GHz between a fixed Earth station and a satellite with an L-band aeronautical mobile allocation between an aircraft and the satellite.

In order to demonstrate the possibilities involved with multiband and service satellites, Table 1 lists the common worldwide fixed satellite allocations and the different types of service allocations. "Normal" band pairs are indicated in the table. It is obvious that a large number of permutations are possible. From the list of services, it would appear that the most probable services that would be utilized with the fixed satellite service are broadcasting satellite and mobile satellite. Mobile satellite and radionavigation satellite also would tend to be associated. Large disparities in the characteristics associated with these services are quite probable. For instance, a large difference in Earth antenna gains would exist between fixed stations and mobile stations and between fixed stations and direct broadcast receiving Earth stations. Large differences in system characteristics tend to result in large differences in the separation required for an acceptable interference level. Considerable differences in the service areas associated with different services

Table 1 Allocations and services

Fixed satellite allocations, GHz		Satellite services
3.700 - 4.200	6/4	Fixed
5.925 - 6.425		Mobile
7.250 - 7.750	8/7	Aeronautical mobile
7.900 - 8.400		Maritime mobile
10.95 - 11.20	14/11	Broadcasting
11.45 - 11.70		Radionavigation
14.00 - 14.50		Aeronautical radionavigation
17.70 - 21.20	30/20	Maritime radionavigation
27.50 - 31.00		Earth exploration
40.00 - 41.00		Meteorological
50.00 - 51.00		Amateur
92.00 - 95.00		Standard frequency
102.0 - 105.0		Time signal
140.0 - 142.0		Space research
150.0 - 152.0		Space operation
220.0 - 230.0		Intersatellite
265.0 - 275.0		

also could exist which could limit adversely the usable arc of the orbit and thus reduce the satellite repositioning flexibility. Thus it may be inferred that the use of multiband/service satellites might result in a reduction of the potential utilization of the geostationary orbit. Conversely, there are a number of incentives for utilizing multiband/service satellites. A prime incentive appears to be related to economics; i.e., it is generally more cost-advantageous to place a number of repeaters on the same satellite rather than on separate satellites, from both the space segment and Earth segment standpoints. However, economic incentives do not necessarily correspond to efficient orbit and spectrum utilization. Another incentive is the ability to cross-connect the various uplink and downlink frequency bands utilized on a satellite. This capability might allow better mating of the orbit and spectrum to disparities in the traffic node requirements, and thus it is possible that a multiband satellite might result in better orbit and spectrum utilization than a group of single-band satellites fulfilling the same requirements.

The preceding paragraphs are intended to provide a gross framework for further studies and analyses with respect to the effect of multiband/service satellites on orbit and spectrum utilization. In the following section, a preliminary analysis is initiated with respect to the use of multiband satellites in the fixed satellite service and some numerical results presented with respect to the various combinations involved in the nominal 6-, 14-, and 30-GHz uplink frequencies and 4-, 11-, and 20-GHz downlink frequencies.

III. System Parameters

In order to estimate the effects of using multiband satellites in the fixed satellite service, it is necessary to assume certain parametric relationships with respect to the significant system characteristics. These include Earth station antenna gains, noise temperatures, operating margins, satellite spacing, and bandwidth expansion.

Earth Station Antenna Gains

The Earth antenna on-axis gain has a very significant effect on the orbit and spectrum utilization of the geostationary orbit when considered with respect to the Consultative Committee International Radio (CCIR) reference radiation[1] diagram for off-axis gain for large antennas. In the design of Earth station antennas, both technical and economic factors influence the achievable gain. For satellites with multiple frequency band repeaters, differences in Earth antenna gains at the different frequency bands may exist. In order to estimate the gain-frequency function of Earth antennas, the results obtained by Pope[2] based on surface tolerance and costs for exposed parabolic reflectors are utilized. With an aperture efficiency of 57%, the gain-frequency function is

$$G = 18 + 20 \log fD - 4.58 \times 10^{-7} f^2 D^3 \text{ (db)} \quad (1)$$

where D is in meters and f in gigaHertz.

The diameter D_o at which maximum gain is achieved is then

$$D_o = 185/f^{2/3} \quad (2)$$

The diameter corresponding to the optimum cost gain ratio D_c is

$$D_c = 32\,[1 - (D/D_o)^3] \tag{3}$$

where D/D_o is related to frequency by

$$f \approx 13.9 \left[\frac{(D/D_o)}{1 + (D/D_o)^3}\right]^{3/2} \tag{4}$$

In the following analyses, a reference gain of 59.1 db at 4 GHz, along with three antenna gain-frequency functions, is utilized. These functions and the resultant antenna gains are shown in Table 2 for the nominal 6/4-, 14/11-, and 30/20-GHz fixed-satellite bands.

Noise Temperatures

The noise temperatures of the satellite and Earth stations affect the radiated power requirements and thus affect orbit/spectrum utilization. System noise temperatures result from a number of discrete noise sources, but for clear air transmission the first amplifier noise tends to dominate. For Earth stations, which normally employ low-noise amplifiers, the system noise temperature generally increases with frequency. This trend may be approximated by a function of the form ($A + Bf^x$) over the range of frequencies of interest. Using a reference value of G/T of 40.7 db with a G of 59.1 db at 4 GHz results in a system noise temperature T of 70^oK. Assume that 50^oK represents the amplifier noise and 20^oK represents antenna noise, and an exponent of 0.8 results in a noise temperature T of

$$T = 20 + 16.7f^{0.8} \, (^oK) \tag{5}$$

where f is in gigaHertz. With the preceding assumptions for antenna gains and system noise temperatures, the Earth terminal G/T's as a function of frequency can be computed. The G/T values for the 4-, 11-, and 20-GHz bands are shown in Table 2. It is assumed that the satellite noise temperature is independent of frequency.

System Margins

As operating frequencies are increased above about 10 GHz, atmospheric absorption through fog and rain increases very rapidly and it becomes necessary to increase the operating margins and perhaps limit the satellite elevation angles and

Table 2 Assumed system parameters

			D = D_c			D = 0.4 D_o			G_o = 59.1 db		
Frequency, GHz	Margin, db	T, °K	D, m	G, db	G/T, db	D, m	G, m	G/T, db	D, m	G, db	G/T, db
3.950	...	70	30.0	59.1	40.7	30.0	59.1	40.7	30.0	59.1	40.7
6.175	0	...	30.0	62.7	...	30.0	62.7	...	30.0	62.7	...
11.325	4	138	25.0	65.9	44.5	15.0	62.2	40.7	10.4	59.1	37.7
14.250	6	...	25.0	67.4	...	15.0	64.1	...	10.4	61.1	...
19.450	10	200	20.0	68.2	45.2	10.4	63.8	40.7	6.1	59.1	36.1
29.250	12	...	20.0	70.0	...	10.4	67.0	...	6.1	62.1	...

utilize diversity reception. Since margin affects power levels and thus orbit and spectrum utilization, some assumptions on practical margins are necessary for the following analyses. Margin differences rather than absolute values are significant, and assumed values also are shown in Table 2.

Satellite Spacing Requirements

The basic methodology for computing angular separation requirements given in Appendix 29 of Ref. 3 is utilized. It is assumed that the modulation characteristics and the satellite antenna coverage are identical for all satellite systems. Antenna gains, margins, and noise temperatures are variables. Of interest is the ratio of angular separations required for the same interference level as a function of the different parameters utilized with different frequency bands as indicated by θ and θ_1 in Fig. 1. Utilizing the CCIR off-axis gain slope for large Earth antenna of $G(\theta) \propto \theta^{-2.5}$ and the methodology of Appendix 29 of Ref. 3 results in an angular separation ratio (θ_1/θ) of

$$\frac{\theta_1}{\theta} = \left[\frac{M_1\,(G/T)}{M\,(G/T)_1} \cdot \frac{(\gamma G_{d1}/\gamma_1 G_{u1} + 1)}{(G_d/G_u + 1)}\right]^{0.4} \tag{6}$$

and

$$\gamma = T_e/\mu T_s \tag{7}$$

where

μ = downlink to uplink noise
no subscript = reference band pair
subscript 1 = other band pair
subscript e = Earth
subscript s = satellite

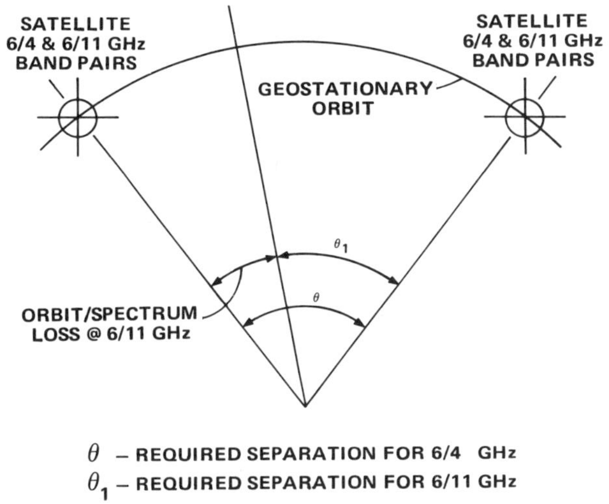

Fig. 1 Homogeneous model.

subscript u = uplink
subscript d = downlink

When only downlink interference exists,

$$\frac{\theta_1}{\theta} = \left[\frac{M_1 (G/T)}{M\,(G/T)_1}\right]^{0.4} \tag{8}$$

And when only uplink interference exists,

$$\frac{\theta_1}{\theta} = \left[\frac{M_1\,(G/T)\,(\gamma G_{d1}/\gamma_1 G_{u1})}{M\,(G/T)_1\,(G_d/G_u)}\right]^{0.4} \tag{9}$$

When $T_1 = T$, $\gamma_1 = 8$, and $M_1 = M$,

$$\frac{\theta_1}{\theta} = \left[\frac{(1/G_{d1} + 1/G_{u1})}{(1/G_d + 1/G_u)}\right]^{0.4} \tag{10}$$

With the assumption of identical modulation, these expressions relate the ratio of the angular separations required between two single repeater satellite systems having different characteristics, or, conversely, the ratio is a measure of the orbit and spectrum utilization of the system associated with the repeater requiring the least separation when both repeaters are on the same satellite; i.e., there is a loss in utilization for values of (θ_1/θ) greater or less than one.

Bandwidth Expansion

Another factor implicit in orbit and spectrum utilization is the information transfer per unit of occupied rf bandwidth, i.e., the bandwidth expansion utilized in the modulation. Under the proper set of conditions, the bandwidth expansion in one repeater can be adjusted to compensate partially for the loss in orbit and spectrum utilization incurred when $\theta_1/\theta \neq 1$ on a multiple repeater satellite. These conditions are met when the system associated with smaller angular separation requirement concurrently requires higher margins. For FM modulation, the relative power levels P between two different carriers for the same performance can be approximated as a function of the rms modulation index m as

$$P_1/P = (m/m_1)^{2.8} \text{ for } 0.4 \leq m \leq 4 \tag{11}$$

A change in modulation index results in a change in the information bandwidth B per unit of rf bandwidth in the approximate ratio of

$$B/B_1 = (P_1/P)^{0.285} \tag{12}$$

The relative angular separation change from θ_1 to θ_{11} for the CCIR antenna sidelobe envelope slope and for a fixed interference level in the baseband is

$$\theta_{11}/\theta_1 \propto (P_1/P)^{0.4} \tag{13}$$

Thus, the relative orbit and spectrum utilization U is improved by the ratio B/B_1 and is

$$U = (\theta_1/\theta)(P_1/P)^{0.285} \text{for } (\theta_1/\theta)(\theta_{11}/\theta_1) \leq 1 \tag{14}$$

When $(\theta_1/\theta)(\theta_{11}/\theta_1) = 1$, the angles are equalized, resulting in optimum compensation and a resulting utilization U_o of

$$U_o = (\theta_1/\theta)^{0.287} \tag{15}$$

The preceding values of U are for the band in which orbit and spectrum are not fully utilized. The total relative utilization U_T for a multiband case can be expressed as

$$U_T = \frac{1 + (b_1/b)\,U_1 + \cdots + (b_n/b)U_n}{1 + (b_1/b) + \cdots + (b_n/b)} \tag{16}$$

where n is the number of repeaters, $b_1, \ldots, b_n$ are the respective rf occupied bandwidths, and b is the reference bandwidth. For the case of digital modulation, the bandwidth expansion is not easily changed, and therefore optimization may be difficult. However, if the bandwidth expansion can be changed to equalize the angular separation, orbit and spectrum utilization generally will be improved.

IV. Orbit and Spectrum Utilization

Homogenous Model

Utilizing the preceding analyses, estimates of the relative orbit and spectrum utilization can be made for a homogenous model. It is assumed that two adjacent satellites have the same repeater combinations and that the systems associated with each repeater in each satellite are identical. It also is assumed that interference occurs between identical band pairs on each satellite but not between different band pairs. If a common frequency exists between different band pairs as shown in Fig.1, it is assumed that other forms of isolation exist; i.e., a 6/4 repeater does not interfere with a 6/11 repeater.

Table 3 shows the results of computations for the antenna models considered for the equal bandwidth expansion and the FM index optimized cases. This table is arranged in a matrix form, with band pairs ordered in increasing margin requirements. The relative orbit and spectrum utilization for all possible combinations and number of bands per satellite can be derived from this table. To determine the utilization, the lowest-order pair in the frequency column is entered, and the value given in that row indicates the relative utilization of the row band pairs. For example, if each satellite has a 6/4, 14/11, and 30/20 repeater, the relative orbit and spectrum utilization is 1 for the 6/4, 0.57 for the 14/11, and 0.46 for the 30/20 band for equal bandwidth expansion. The equal bandwidth expansion values given in this table may be indicative of utilization with digital modulation where the bandwidth expansion is not adjusted easily. However, it also should be noted that the required angular separations are not necessarily the same for FM and digital modulations. In the more severe cases indicated in this table, about one-half of the usable arc in a band could be lost by multibanding. However, this loss in utilization can be reduced under certain conditions by changing the bandwidth expansion. In this case, the changes in power levels required to equalize the separa-

Table 3 Multiband utilization factors: homogenous case ($D = D_c$)

Frequency Pairs, GHz	Relative orbit and spectrum utilization U							
	6/11	14/4	14/11	6/20	14/20	30/4	30/11	30/20
6/4	1)a 0.73	0.92	0.57	0.69	0.51	0.89	0.53	0.46
	2)a 0.92	0.98	0.85	0.90	0.82	0.97	0.83	0.80
	3)a 0.83	0.94	0.67	0.83	0.64	0.95	0.67	0.62
6/11	1)	0.79	0.79	0.95	0.71	0.81	0.73	0.64
	2)	0.79	0.92b	0.99	0.91	0.81b	0.91	0.88
	3)	0.79b	0.82	0.98	0.77	0.81b	0.81	0.76
14/4		1)	0.63	0.75	0.56	0.98	0.58	0.50
		2)	0.63b	0.92	0.75b	0.99	0.86	0.75b
		3)	0.63b	0.87	0.68	0.99	0.70	0.66
14/11			1)	0.83	0.90	0.64	0.92	0.80
			2)	0.83b	0.97	0.64b	0.98	0.94
			3)	0.83	0.95	0.64b	0.97	0.93
6/20				1)	0.75	0.77	0.77	0.67
				2)	0.75b	0.77b	0.87b	0.76b
				3)	0.75b	0.77b	0.80	0.76
14/20						1) 0.57	0.97	0.90
						2) 0.58b	0.97b	0.97
						3) 0.57b	0.97b	0.91
30/4						1)	0.59	0.52
						2)	0.59b	0.52b
						3)	0.59b	0.52b
30/11							1)	0.87
							2)	0.87b
							3)	0.87b

a 1) $D = D_c = 0.4\ D_o$, same bandwidth expansion, G = 59.1 db; 2) $D = D_c$, FM index adjusted; 3) $D = 0.4\ D_o$, FM index adjusted.

b Cannot be optimized fully.

tions are computed and are compared with the assumed margin differences. If the margin difference is equal to or greater than the required power level difference, the angular separations can be equalized. The cases where FM index adjustment cannot optimize the utilization fully (where the angles cannot be equalized) are identified in the table footnotes. In some cases, no optimization is possible. The case of G = 59.1 db is intrinsically angle equalized. The case of $D = D_c$ is utilized as a reference in this table.

It is not obvious from the values given in this table that any general guidelines exist in terms of particular band pairs on a multiband satellite relative to orbit and spectrum utilization. Using the limited number of samples given in Table 3 and assuming that each sample is equally probable, Fig. 2 shows the general trend of the percentage of cases which do not exceed a given orbit and spectrum utilization loss for two-band satellites.

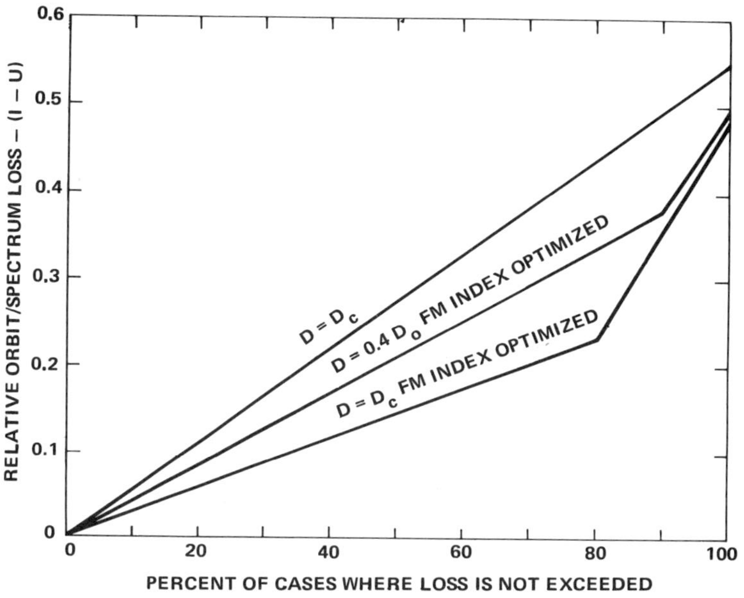

Fig. 2 Distribution of orbit and spectrum loss for two-band pairs: homogenous case.

It is noted that a flat probability density function appears to exist for the equal bandwidth expansion case and for most of the range for the FM index adjusted cases. The orbit and spectrum loss can be reduced by a factor of 2 by FM index optimization for the $D = D_c$ case. The effect of antenna gain also is indicated in the figure.

Orbit and Spectrum Utilization vs Number of Band Pairs. The total orbit and spectrum utilization U_T can be computed for all of the frequency pair combinations for two to eight pairs on each satellite. The average value of U_T for two pairs, three pairs, etc., was computed and is shown in Fig. 3. The worst and best case values also are shown. In general, the orbit and spectrum decreases as the number of frequency band pairs increases. However, the worst combinations appear to be relatively independent of the number of pairs for the $D = D_c$ case. The values of U_T represent fractional utilization of the orbit and spectrum available which is proportional to the number of band pairs used. Figure 4 shows the relative loss of orbit and spectrum in bandwidth-angle units as a function of the number of band pairs. The actual orbit and spectrum loss tends to increase in proportion to the number of band pairs but for the best cases is relatively small.

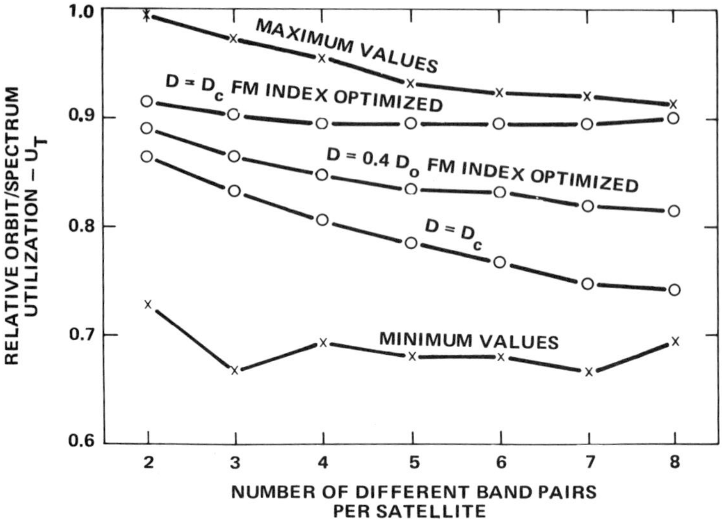

Fig. 3 Total utilization vs number of band pairs per satellite: homogenous case.

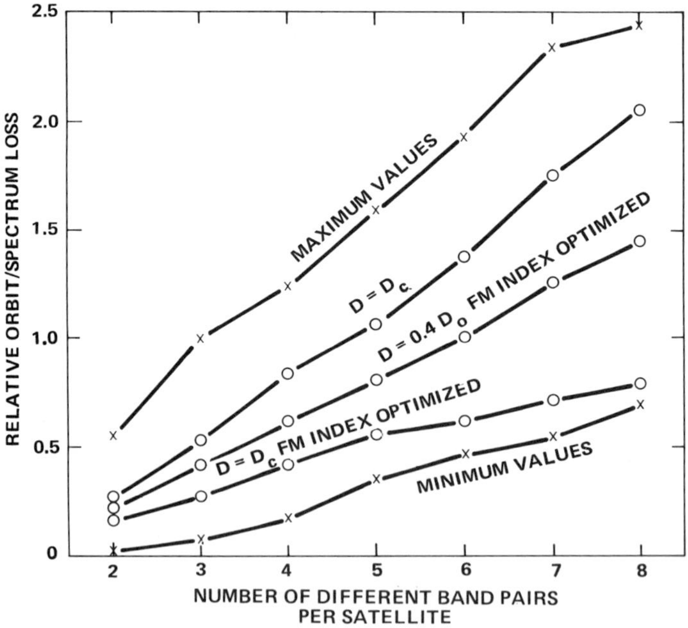

Fig. 4 Orbit and spectrum loss vs number of band pairs per satellite: homogenous case.

Orbit and Spectrum Utilization vs Number of Repeaters. The effect of the number of repeaters utilized on a satellite also may be estimated utilizing the values given in Table 3. For this case, a sample set was developed using only normal up-down frequency pairs and combinations for n = 2 to 8, where n is the number of repeaters on the satellite. The average values of U_T were computed for these combinations. The results of these computations are shown in Fig. 5. The average value of U_T decreases slowly with increasing numbers of repeaters. The worst case shown, $D = D_c$, also decreases with increasing numbers of repeaters, whereas the best case values show little loss. The relative orbit and spectrum loss in bandwidth angle units is shown in Fig. 6. Again, the loss tends to increase with number of repeaters, with a wide variation between worst and best cases.

Orbit and Spectrum Utilization with Mixed Uplinks and Downlinks

Another model can be considered in which the frequency bands can arbitrarily be used as uplinks or downlinks. Table 4 shows values of (θ_1/θ) for various normal and mixed uplinks and downlinks for two band pair satellites. The values are referenced to the normal uplink and downlink pairing. For the

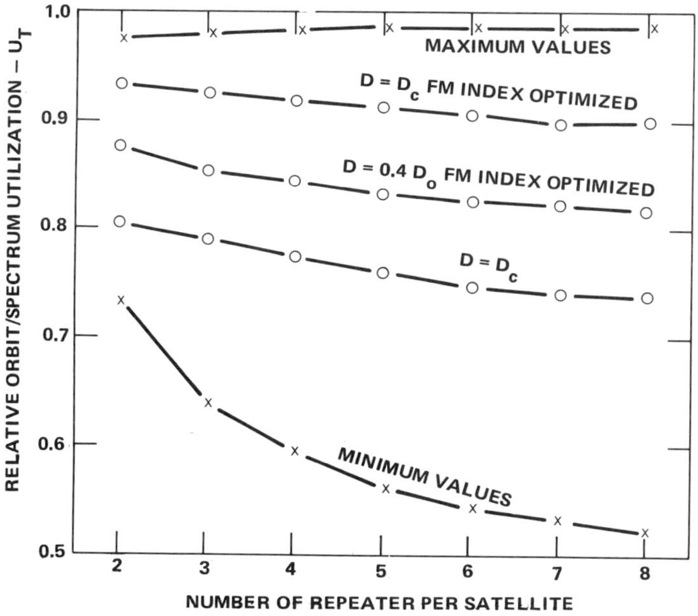

Fig. 5 Total utilization vs number of repeaters per satellite: homogenous case.

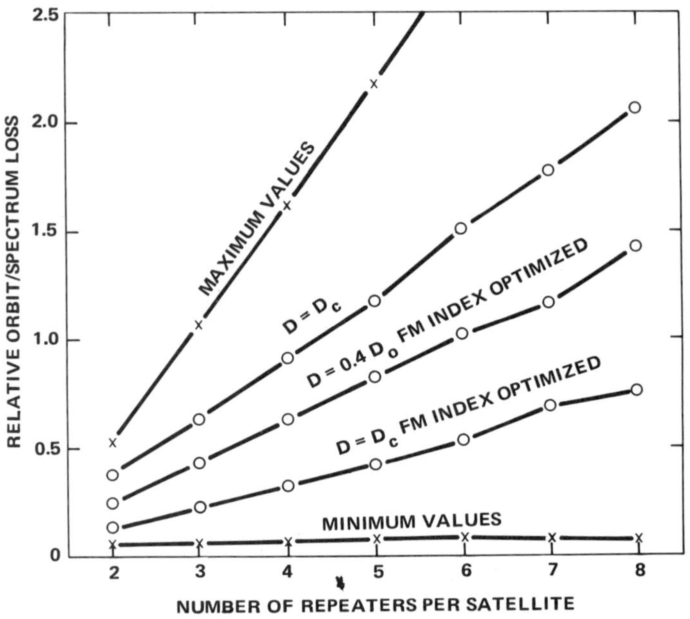

Fig. 6 Orbit and spectrum loss vs number of repeaters per satellite: homogenous case.

examples shown, the mixed band pairs show a higher degree of orbit and spectrum utilization than the normal band pairing arrangements.

V. Conclusions

Multiband Satellites in the Fixed Satellite Service

It must be emphasized that the analyses presented in Sec. IV are based on assumptions in which there is a relatively high degree of similarity among the system characteristics, and therefore any conclusions drawn from this analysis do not necessarily extend to the more general case of a multiband/ service satellite. From this analysis involving the various potential combinations of a multiband satellite in the fixed satellite service utilizing 6-, 14-, and 30-GHz uplinks and 4-, 11-, and 20-GHz downlinks, the following general conclusions may be drawn:

1) Losses in orbit and spectrum utilization range from negligible to quite significant values. There does not appear to be any obvious criterion for determining low or high loss combinations, but, in general, losses increase as the number of bands and/or repeaters is increased.

Table 4 Multiband factors with mixed uplinks and downlinks ($D = D_c$)

Satellite 1 frequency pairs, GHz	Satellite 2 frequency pairs, GHz	(θ_1/θ)	Total loss in angle units
6/4 14/11	6/4 14/11	1 0.57(0.85)a	0.43(0.15)a
14/4 11/6	14/4 11/6	0.91 0.73	0.00
14/6 11/4	14/6 11/4	0.70(0.84)a 0.93	0.09(+0.05)a
6/4 30/20	6/4 30/20	1 0.46(0.80)a	0.54(0.20)a
30/4 20/6	30/4 20/6	0.89 0.69	+0.02
30/6 20/4	30/6 20/4	0.66(0.79)a 0.91	0.07(+0.06)a
14/11 30/20	14/11 30/20	1 0.80(0.94)a	0.20(0.06)a
30/11 14/20	30/11 14/20	0.92 0.90	+0.14
30/14 20/11	30/14 20/11	0.84(0.95)a 0.97	0.07(+0.04)a

[a] Bracketed values are FM index optimized.

2) Optimization of the bandwidth expansion can reduce the loss significantly in orbit and spectrum utilization. Although this is quite feasible for FM, it may not be practical with digital modulations.

3) There appears to be no advantage, from an orbit and spectrum utilization standpoint, to normal band pairing (6/4) as compared to non-normal pairing (14/4). It may be that non-normal pairing may provide some advantage.

4) The angular spacing requirements using reversed up- and downlink bands are the same as with the normal up and down bands.

5) Multiband satellites which utilized mixed up and down bands (some normal and some reversed) may provide better orbit and spectrum utilization than either all normal or all reversed used.

Multiband/Multiservice Satellites

Considering the general case of multiband/multiservice satellites as described in Sec. II, the number of potential combinations of services and frequency band which might be accommodated on a single satellite is quite large. The disparity in system characteristics associated with a multiservice satellite can also be quite large, such as between the fixed satellite and mobile satellite services, and thus potentially large orbit and spectrum utilization losses may occur.

References

[1] "Generalized Earth Station Antenna Radiation Pattern for Use in Interference Calculations, Including Coordination Procedures, in the Frequency Range 2-10 GHz," CCIR Recommendation 465, XIIth Plenary Assembly, 1970, New Delhi, India; ITU.

[2] Pope, D. L., "Parametric Representation of Ground Antennas for Communications System Studies," Bell System Technical Journal, Vol. 47, No. 10, Dec. 1968, pp. 2145 - 2168.

[3] "Addition of a New Appendix (Appendix 29)," Final Acts of the World Administrative Radio Conference on Space Telecommunication, Annex 19, 1971, Geneva; ITU.

CHAPTER IV—ADVANCED CONCEPTS

Communication satellite technology has matured rapidly over the past decade, and it is now possible to provide new services to an ever-increasing community of users. The new services include wideband digital service, color television, secure voice, rapid and flexible extension of service. The new users served include mobile platforms (ships and aircraft), remote terminals, and small direct-access users. This is a greatly expanded role over the large fixed terminal point-to-point deployment that has been the mainstay of INTELSAT. As stated previously, the main deterrent to the expanded role has been the development of the political machinery to regulate these services. The U.S. domestic regulatory policy is now developing rapidly, and it is expected that this policy will permit a more innovative environment than in the INTELSAT arena. The ability of the international political framework to provide special services to mobile users has been slow, but it is believed that the threshold has been overcome and systems of this nature will be deployed in increasing numbers. The rapidly evolving policy in all these areas is strongly driven by the technical feasibility of providing these services; therefore it is extremely important to press forward with new technology and the advanced concepts that capitalize on these new technologies.

Improvements in component reliability have made it feasible to increase the complexity in the satellite and at the same time reduce the size of the crew that operates and maintains the terminals. To allow many more users to obtain direct access to these new services requires that the terminals be economical from both the initial investment and the O/M standpoints. Terminals can be made simpler and more economical by placing an increased burden on the spacecraft (i.e., increased satellite radiated power and onboard processing). This seems to be the logical trend for the future, and the new generations of satellites, now in the planning stages, are making provision for narrower beam antennas and onboard processing. The paper by Schmidt discusses several alternatives for achieving switching at the satellite.

The concept of a large distributed network of satellite terminals reduces the dependence on the record carriers and also reduces the costs of access lines. The concept of a large private line service is discussed in the paper by Metz.

Communication satellites provide the most flexible means for establishing effective command and control systems. Command and control normally requires the gathering of information, the distribution of this information to the correct decision centers, the decision process, and the distribution of the control messages to the controlled units. Satellites can greatly assist all of these steps except the actual decision process. However, if the decision processor is distributed for survivability or economic reasons, satellite links will be valuable for tying the components of the processor together. Command and control is an important part of any organization, whether it is the DoD, FAA, a large multinational corporation like General Motors, or any of a number of life insurance companies. The most difficult future problems of command and control will be in developing methods for obtaining consensus to take action on important issues when many nations are impacted. It is believed that communication satellites will play a large role in shaping the actual structure of international organizations.

One of the first federal agencies to be impacted by communication satellites will be the FAA. The command and control of aircraft is greatly enhanced by satellite navigation and communications. Future concepts will combine these functions and also will permit dynamic routing and rerouting, which will alter airline operations. One of the advanced system concepts that eventually will provide the capability to alter operations is ASTO-DABS, as described by Schuchman. A similar concept has been tested and is reported on by Courtney and Kadar. Communication satellites are expected to have a large impact on other institutions, such as banking and education. Some of the advanced technology required to support these advanced system concepts is discussed in Volume 42.

SATELLITE-SWITCHED TDMA: TRANSPONDER-SWITCHED OR BEAM-SWITCHED?

William G. Schmidt*

Computer Sciences Corporation, Falls Church, Va.

Abstract

For satellite communications in the 1980's, there is a strong trend toward wide-band communications packages utilizing a multiplicity of narrow high-powered beams. There are numerous advantages to this trend; a disadvantage, however, is the burden this technology places on the interconnection flexibility of the satellite. To restore this interconnection flexibility, an onboard real-time time-division multiple access (TDMA) communications system linking the stations has been proposed. This approach has been broadly termed satellite-switched TDMA (SS-TDMA) and is based upon the switching occurring between multiple transponders, each with fixed antenna pointing. Another class of SS-TDMA incorporates individual transponders with variable antenna pointing (beam-switched). This paper explores and develops the conceptual configuration and technology base for both types of SS-TDMA, discusses hybrid approaches wherein both types are interconnected, considers the use of onboard regeneration, and closes with a treatment of the advantages of SS-TDMA to mobile and military satellite communications.

Presented as Paper 74-460 at the AIAA 5th Communications Satellite Systems Conference, Los Angeles, Calif., April 22-24, 1974. This paper is based partially upon the author's participation in the Intelsat research and development program. The opinions expressed are not necessarily those of Intelsat.

*Operation Director, Satellite Communications and Transmission Systems Operation.

I. Introduction

The first generations of communications satellites were power-limited and used FM to trade bandwidth for C/N improvement by having available relatively large amounts of satellite bandwidth. As the power limitations lessened in subsequent satellites and the number of FM carriers increased, the intermodulation products of these carriers became significant factors in the capacity equations; this resulted in "channelized" satellites with 12- and 24-transponder communications packages. With still higher performance payloads now available, the major limitation is the rf spectrum allocation. To utilize the allocations more fully, methods of frequency reuse are being investigated with a view toward possible implementation in the next generation of communication satellites. Orthogonal polarization and multiple spot-beams (space-division multiple access) appear attractive, the latter being a lower risk and more cost-effective route at this time.[1] The new allocations in the 14/11-GHz band and 30/20-GHz bands also would help in handling the very heavy links or in penetrating those geographical areas where interference with terrestrial microwave links is a problem.

The advantages of narrow beams are not confined strictly to spectrum reuse. The higher gains of the satellite antennas allow the Earth station antenna gains to drop proportionately while maintaining the capacity of the link; this is especially important to mobile applications. Furthermore, the highly directive narrow beams tend to reduce significantly the prospect of interference to military communications systems. However, the price that one pays for this also must be considered – the loss of interconnectivity. Whereas the interconnection of two users is relatively simple when Earth-coverage beams are employed, it is a bit more difficult when spot-beams are used, since they are transmitting blind (i.e., being able to receive and monitor their own downlink signals) and transmitting to a station in another beam zone which may have a different frequency plan. It becomes significantly more complex when trying to accomplish this with some 50 or more Earth stations located within eight or more beam zones. A particular burden falls on the spacecraft itself, which, when operated in an FDMA (frequency-division multiple-access) mode, must make provisions for filtering and frequency translating the many multidestinational carriers of present and future size. The weight and power consumption of this equipment becomes a major element in the satellite design.

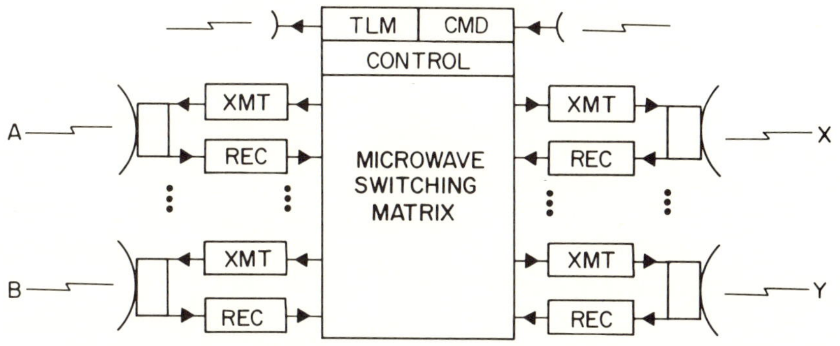

Fig. 1 Transponder-switched satellite diagram.

The use of time-division multiple-access (TDMA), when coupled to an onboard real-time time-division microwave switching matrix, appears to be an attractive solution to the problem of restoring interconnectivity to the multiple-beam satellite.[2,3] The impact of satellite-switched TDMA, sometimes called SS-TDMA or SDMA/TDMA, upon the capacity of a satellite has been recognized by some satellite communications operators who are funding research and development programs in this area with a view toward incorporating this technology within their later generation satellites.

There are two distinct types of satellite-switched TDMA. The first, such as is described in Refs. 2-4, may be called transponder-switched because the antenna beams are fixed, and it is only the "half-repeaters" that are interconnected as is shown in Fig. 1. Normally this approach is associated with a multiple transponder satellite configuration. The second type, which may be called beam-switched, makes use of one or more antennas whose beams may be redirected quickly. Within this type are two general configurations: the one-antenna (simultaneous transmit and receive) configuration, which employs onboard store-and-forward message switching; and the two-antenna version,which does not require onboard message storage. A satellite could carry multiple one-antenna or two-antenna repeaters but usually is not envisaged as carrying more than a few such repeater configurations. One also can give thought to a satellite that would carry both types of switching, and this is referred to in this paper as hybrid satellite-switched TDMA.

II. Transponder-Switched TDMA

The basic idea of transponder-switched TDMA is simply to take a fixed antenna farm configuration and, by means of the microwave switching matrix, make the transponder ensemble appear to have beam-hopping capability. Actually, each beam is never turned off and can carry traffic most of the time. Hence, the existing fixed-beam antenna technology and conventional transponders are retained, with the incorporation of advanced technology being in the form of the switching matrix (operated at rf with nanosecond switching times) and the distribution control unit, which is command-programable to operate the switching matrix in accordance with the requisite traffic pattern.[5]

As an indication of the interest in this technology, development began in 1973 of engineering models of a strip-line 16 x 16 pin-diode microwave switching matrix and a CMOS/LSI implementation of the distribution control unit. The diode switch operated with 50-nsec switching times and 50-db isolation specified between the on and off paths. The development of second-generation implementations is well underway, with significant advances planned in the areas of reliability and implementation. For example, the distribution control unit will be constructed with only five integrated circuits and contains logic to bypass failed microwave switching matrix junction drivers.

The decision to use TDMA as the multiple-access technique has two implications; the first is that the transponders will be as wide-band as possible, since channelization is not necessary to combat intermodulation interference; the second is that a synchronization technique is available to operate in the postulated environment. The former leads to modems operating up to 1 Gbit/sec, whereas the latter results in questions such as, "Shall the satellite act as a reference station? What is the most reliable and economical approach to achieving and maintaining network synchronization?"

Among the variety of synchronization approaches possible, one attractive candidate is to take a short segment of the SS-TDMA frame[6] (and switching matrix frame) and use it to "loop back" to the zone of origin; i.e., the uplink of zone A is connected to the downlink to zone A. Hence, for that short time period the transmissions are not "blind." This loopback switch segment, timed by the satellite clock, then is used by all stations in each zone to

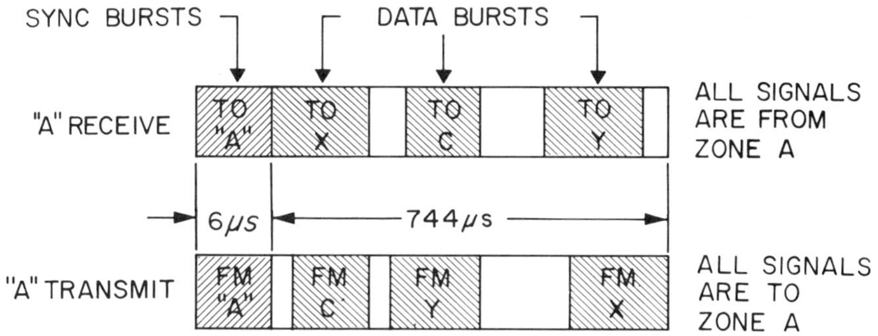

Fig. 2 Sample SS-TDMA signal format through transponder A for a 750-μsec TDMA frame.

achieve and maintain their burst (switch) synchronization in the subsequent "blind" transmissions in the remainder of the TDMA frame. The possible formats of such transmissions are indicated in Fig. 2.

Another synchronization method would have the satellite generate a synch signal which is "broadcast" into all zones simultaneously and rely on there being one segment of the onboard switch which "loops back" to the zone of origin, during which the more conventional TDMA burst synchronization could take place. The satellite-generated synch is then the reference synch burst. However, this method requires the inclusion of additional frequency generation equipment, high-speed logic, and a modulator. This technique appears more suitable for the period of regenerative repeaters that will be discussed later.

Another element in the evolution of transponder-switched TDMA is the "zone-shared transponder" approach. In the early thinking about satellite-switched TDMA, it was recognized that the full-time allocation of a complete transponder in a satellite implies a capacity requirement of, let us say, 1 Gbit/sec. While some beam zones do require this capability, most do not and, therefore, some advantage could be taken of that fact. The result is a pre-/post-matrix switching arrangement, shown in Fig. 3, which appears to significantly reduce the number of transponders (but not antennas) required. A twofold reduction seems likely as based on some typical examples drawn from operational scenarios.

Transponder-switched TDMA has the merit of providing the enhanced capacity associated with multibeam satellites with little added onboard weight and power. No demodulation to baseband and message-storage is required, and, with the exception of the microwave switching matrix, the current state-of-the-art satellite technology is applicable. Some further developments and testing are required of ultra-high-speed TDMA terminals and their synchronization; technical feasibility could be provided at lower speeds, however. This technique, because it is based on fixed and limited numbers of antenna beams, is not amenable to servicing mobile stations, such as in maritime and aeronautical applications. For these situations, the precious Earth coverage bandwidth allocations must be used, or beam-switched TDMA must be employed.

III. Beam-Switched TDMA

Within the category of beam-switched TDMA there are, as suggested earlier, two classes of operation; the distinguishing feature is whether one or two antennas are employed. The one-antenna approach, as exemplified by the diagram of Fig. 4, uses the same antenna for simultaneous reception and transmission. This implies the existence of a store-and-forward onboard capability since, in general, only one station or zone can be serviced at one time. Intrabeam traffic is usually insignificant. Furthermore,

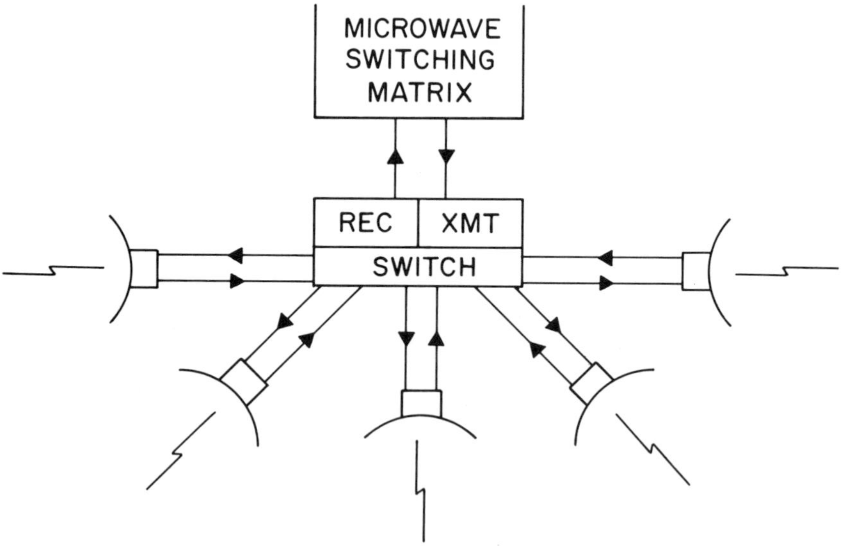

Fig. 3 A transponder-switched TDMA system with a zone-shared repeater.

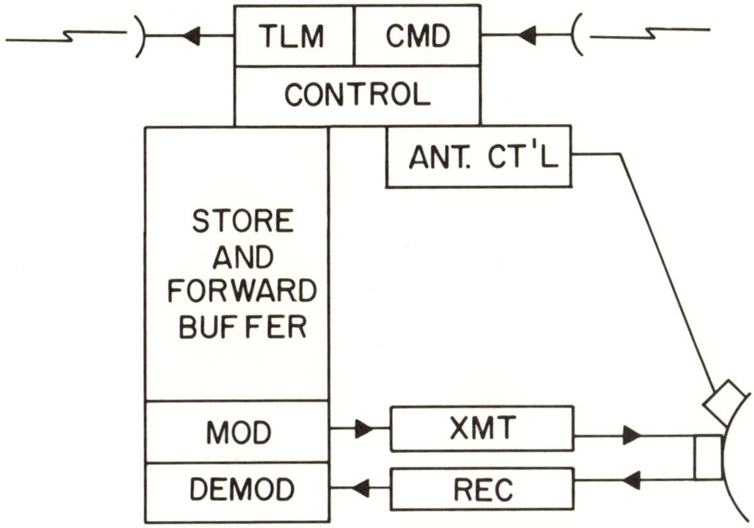

Fig. 4 Single-antenna beam-switching.

since the storage function is at baseband, onboard demodulation and remodulation also follow. The capacity of this system is determined largely by the store-and-forward processor and its associated memory, as well as the link bit rate that can be supported. Because of the onboard store-and-forward delay, the use of such a system for voice channels must be generally ruled out.

One of the more intriguing problem areas is that of synchronizing the uplink burst transmissions, since such synchronization will increase the efficiency by a large measure. If the satellite asynchronously (or quasi-synchronously) "polls" the associated Earth terminals by transmitting "start now" command pulses when the beam has switched to point at that station, about 0.5 sec is lost before data start to be received. Hypothesizing a 30-station network with an average of 300 bit/sec per station leads to a 150-sec polling frame to maintain a 90% uplink efficiency; the onboard storage then would require a memory of some 1.35 Mbit. If a voice channel (64 kbit/sec) per station were proposed, then a storage of 288 Mbits would be required. The 150-sec frame would present an obvious conversational problem.

If a terminal knows the exact relative location (time delay) to the satellite, as well as receives some sort of periodic beam-switching framing signal reference from the satellite, a much higher level of efficiency can be achieved, with a concomitant lower onboard storage requirement. Assuming that the switching of the beam, clock stabilities, ranging uncertainties, etc., lead to a guard time of 200 nsec, with an identical period for burst preambles, the previously proposed voice network then would probably use a 125-μsec polling frame and become more feasible. The result would still be an efficiency of about 90% but an onboard storage requirement of 240 bits, or six orders of magnitude less than the asynchronous polling method. Since each application of beam-switching TDMA will be somewhat unique, i.e., different traffic levels, Earth terminal complexities, etc., the optimum synchronization approach probably will be customer-tailored to the application. The point to be made is that a single onboard antenna with variable pointing can provide narrow-beam communications to any location within normal Earth coverage, with such capacity as the cost of onboard remodulation and storage processing permits.

The second class of operation is the two-antenna method in which one variable-pointing antenna is reserved for reception and the other is reserved for transmission (Fig. 5). The control logic that manipulates the pointing plan of these antennas is programed and verified by means of the command telemetry links of the satellite. No store-and-forward capability is required on the satellite, thereby permitting much higher network and link capacities. Another version of this class of operation would have the two antennas, but each has receive-transmit capability; in fact, the antennas may function in different frequency bands.

The vital element of these two beam-switching approaches is the antenna. Phased-array antennas require too much power at this state in their development, but the lens antenna, as exemplified by the unit designed for LES-7, appears to be an attractive candidate.[7] The lens structure is fed by 19 2-in.-diam conical feed horns. Each of 19 3° beam positions is achieved by illuminating a single horn to the center of the lens, and mutual coupling between feeds is greater than 42 db down. The logic controlling the feeds is solid state and in a tree configuration. Switching times are less than 100 nsec, with driver power at a 2.5-w level. Hence, the overall power consumption per antenna is higher than with transponder-switching but not significantly so. More recent work on multiple-beam antennas is expected to disclose other attractive alternatives.

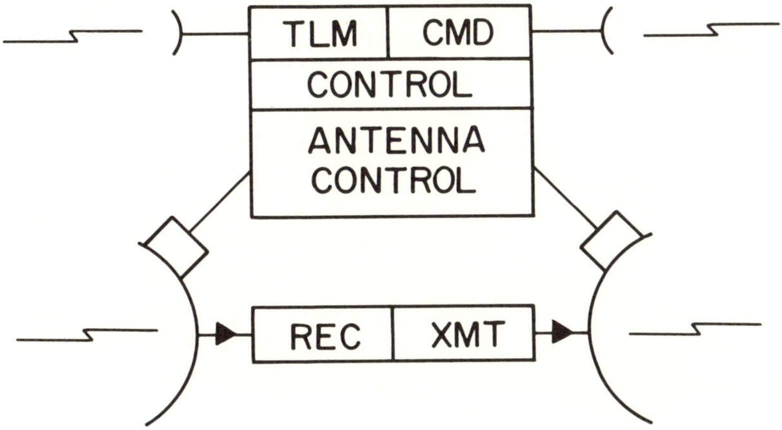

Fig. 5 Dual-antenna beam-switching.

IV. Hybrid Techniques

The application area of beam-switching is such that it is difficult to envisage complete antenna farms of such antennas, since "fixed" antennas can be repositioned mechanically while the satellite is in space and fixed antennas are more attractive for the predominant modes of traffic, i.e., between fixed stations. One easily can relate to a fixed-antenna farm that is made more flexible in responding to certain light-traffic requirements by the addition of one or more beam-switching antennas. This section will explore two hybrid configurations in which beam-switching and transponder-switching are mixed to achieve an attractive overall communications package for a satellite.

Figure 6 shows the use of a beam-switching antenna in conjunction with the switching matrix of the transponder-switching ensemble, much in the fashion of the zone-shared repeater. For example, a two-way link could be established between stations in zones A and X for a period of time, then switched by matrix switch and beam switch, to provide a link between a zone C station and a zone Y station, etc. This approach appears more attractive than zone-sharing transponders in that the multiple low-usage antennas are replaced by one beam-switching antenna. The problem comes when zone X needs a link to zone Y. Although such a link cannot be provided on a single-hop basis, the connection can be established on a double-hop basis by linking X to A, and A to Y.

Note that no onboard store-and-forward capability is required. Although this appears to be desirable, one of the more interesting aspects of having store-and-forward capability is the rate-changing possibilities afforded, which may be a very definite asset in some applications. For example, if a G/T-limited station in zone X wants a connection to a high-G/T station in zone A, the time used by the high-G/T station receiving/transmitting the long low-rate burst from/to the small station would penalize the satellite and the large station. This means that capacity has three dimensions: time, power, and bandwidth. The latter two are usually operated at maximum capability in TDMA systems; hence, time in TDMA is directly proportional to potential capacity. During a certain time period, a station could handle a 2-channel link to a small station, or a 200-channel link to a large station. If the 2-channel link is served, it represents a 198-channel loss of potential capacity because of uplink/downlink imbalance.

Onboard store-and-forward switching can limit that imbalance by processing the transmissions between zone X and the satellite at, for example, 64 kbit/sec, while the satellite and link between the zone A operates at 64 Mbit/sec. This approach is outlined in Fig. 7. The control unit is preprogramed to employ certain combinations of

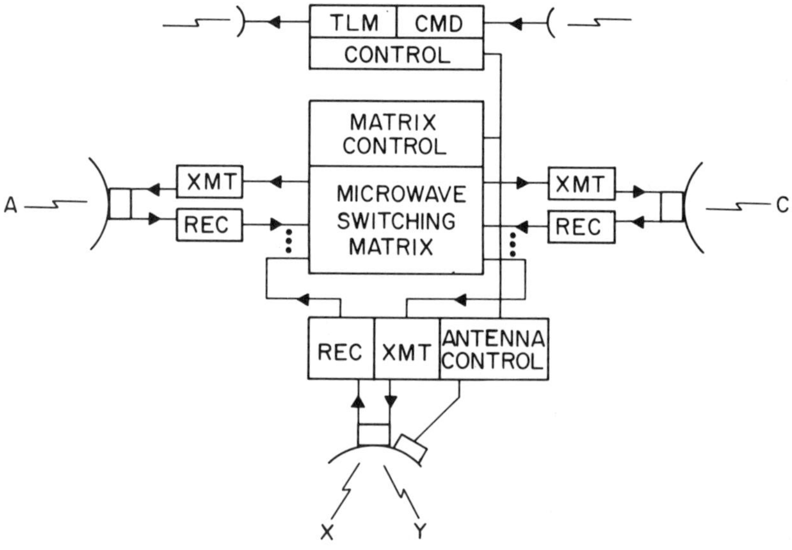

Fig. 6 A beam-switched transponder connected to a transponder-switched transponder ensemble.

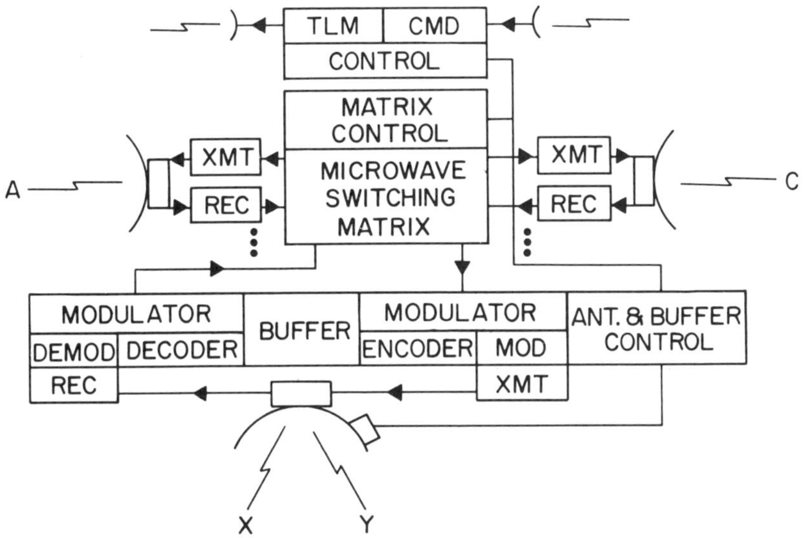

Fig. 7 A beam-switched store-and-forward transponder connected to a transponder-switched transponder ensemble.

modulation and coding for each link. The fixed antenna links are assumed to operate at a standardized rate, which is probably the highest rate in the system.

V. Onboard Regeneration

Nosaka and Muratani[4] have raised the very interesting question of whether regenerative repeaters should be incorporated in TDMA-oriented satellite transponders. There are several advantages to this approach. One is the lessening of the impact of transponder nonlinearities upon the modulation scheme, particularly the higher-order but "softer" digital modulation methods, such as 8- and 16-phase Phase Shift Keying (PSK). Currently, such nonlinearities severely penalize the use of higher-order digital modulation schemes in hard-limited situations. The degradations introduced by AM-PM, AM-AM, polarization interference, etc., can all be minimized by onboard regeneration.

Another strong advantage is the elimination of the multiplicity of traveling-wave tube (TWT) amplifiers, thereby saving significant weight and power in a satellite. A high-powered oscillator with outputs to the multiple digital modulators would replace it. The use of the millimeter wave region of the spectrum enhances the influence of rain on the propagation path; however, onboard regeneration will lessen its effects. Note also that the onboard transponder switching will not be done at rf but, rather, at baseband using conventional digital integrated circuits.

The question is whether the incorporation of a regenerative repeater will provide the improved reliability and increased capacity promised while still maintaining its attractiveness as a satellite-borne subsystem. Indications are that it will, but it also is clear that multirate modulation systems will not be able to use this capability until later in the technology evaluation. The development of onboard regeneration repeaters will be one of the areas requiring more effort in the future.

Nosaka and Muratani proposed the use of double differentially encoded Phase Shift Keyed (DPSK) modulation, citing the rationale of easier and more reliable implementation as well as the carrier recovery problems of Coherent Phase Shift Keyed (CPSK) modulation. Since the downlink is still the weaker of the two links, one may want to consider the use of DPSK on uplink but CPSK on the downlink, where its 2.5-db or so improvement would provide a more balanced communications system. The implementations of the onboard modulator for DPSK or CPSK are not different and so the onboard "penalty" does not change. Furthermore, since all of the downlink TDMA carriers now would have the same reference carrier, i.e., the high-power oscillator that replaces the TWT amplifier, then the carrier recovery problem of the CPSK demodulator is alleviated greatly. A simplified diagram of such a transponder ensemble is shown in Fig. 8.

VI. Mobile and Military Applications

It has been suggested earlier that the use of satellite-switched TDMA would be of significant advantage to mobile communications and would possess merits for military communications. Exploring the mobile communications aspect first, the coupling of spot-beam technology to the utilization of small Earth stations is a natural step, since the ground antenna gain is being traded directly off against satellite antenna gain. Where one is concerned with hundreds of small stations, the ability to specify a substantially lower Earth station G/T has a major economic impact on the communications

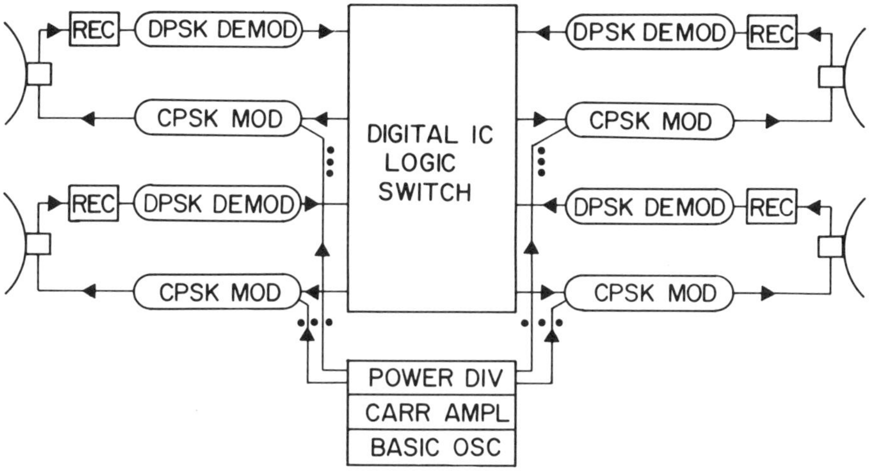

Fig. 8 Onboard regeneration.

system viability, and the attractiveness of that tradeoff is obvious. Nonetheless, the interconnectivity problem must be solved, and this paper already has devoted some time to that area. There are, however, two other facets of this technology which require addressing before such mobile systems become truly viable.

The first effort is the development of low-cost TDMA terminals for the mobile site. Most of the experimental work done in the field of TDMA has been directed toward stations with G/T of 30 db/°K or more and with requirements for 100 voice channels or more. Furthermore, a relatively autonomous station role has been assumed. In the mobile application, the most generally seen network is a hub configuration wherein the remote sites communicate only with the central station, this central station being much larger and more complex than any of the outlying stations. Advantage can be taken of this network configuration to develop lower-cost remote TDMA terminals by having the central terminal somewhat more complex. The plans to use TDMA for some maritime operations should provide incentive for development of low-cost TDMA terminals.

Another technology that must be developed more fully is that of retrodirective antennas. By this is meant a technique whereby an antenna in an Earth-coverage mode can detect a low-level signal and improve the effective antenna gain in the direction of that signal source, thereby enhancing the performance of that link. A number of variations of this approach are attractive to the mobile communications environment, particularly those that include the ability to track continuously the individual signals being received.

In the realm of military communications, what was said for the mobile environment applies to the military environment, which also requires highly mobile, and flexible, station and network operations. In addition to these requirements, military systems must have built-in protection against sophisticated intentional interference. Although there is a wide assortment of antijamming modulation forms, the ability of these techniques to rebuff a high-powered jammer is limited. Especially vulnerable are severely power-limited uplinks, and it is a major advantage of narrow-beam satellite communications that they have the potential of being virtually jam-proof. Before this potential is realized, however, the state of the art of satellite communications must be developed to be capable of beams and associated satellite stabilization brought to a level wherein a 50-100 mile "footprint" is achievable. In the meantime, the spot-beams can be time-hopped and frequency-hopped, which,along with conventional antijam modulation, will make things difficult but not impossible for the intentional interferer.

The hybrid scheme of Fig. 7 is appropriate to the tendency of military systems to operate in multiple frequency bands. One could envisage transponder-switching in one "trunking" band and one or more beam-switching transponders operating other bands selected by reason of existing user-equipment, intersatellite link considerations, covert or antijam signaling requirements, etc. The use of onboard regeneration is attractive because it would embody antijam demodulation and more efficient remodulation for the critical downlink.

Considering the above paragraphs, it appears that the increased onboard complexity inherent in military communications spacecraft will continue to represent a significant problem and compromises will have to be made. The era of the space shuttle and tug will lead to the repairable space communications stations of the future in which complexity and economy are more effectively joined.

VII. Conclusions

There seems to be little doubt that the next generations of communications satellites will start to explore and, subsequently, exploit the technological promise of satellite-switched TDMA. Since most satellites are faced with a wide range of types of traffic, the servicing of which is vital to their own economic viability, it appears that a hybrid form of satellite switching will prove to be most attractive.

As a prerequisite to this, an effort must be made to develop and space-qualify switching matrices and store-and-forward processors, as well as regenerative repeaters. Low-cost TDMA terminals and improved synchronization techniques must evolve from the experimental work of today. The quest for more sophisticated onboard processing will yield substantial rewards for mobile and military communications.

References

[1] Pritchard, W. L. and Bargellini, P. L., "Trends in Technology for Communications Satellites," Astronautics & Aeronautics, Vol. 10, No. 4, April 1972, pp. 36-42.

[2] Schmidt, W. G., "An Onboard Switched Multiple-Access System for Millimeter-Wave Satellites," Proceedings of the 1969 Intelsat/ IEE International Conference on Digital Satellite Communications, 1969, London, pp. 388-407.

[3] Schmidt, W. G. and Cooperman, R. S., "A Satellite-Switched SDMA/TDMA System for a Wideband Multibeam Satellite," 1973 International Conference on Communications, 1973, Seattle, Wash., pp. 12-7 to 12-12.

[4] Nosaka, K. and Muratani, T., "New Satellite Transponder with Regenerative Repeater and Higher Order DPSK for TDMA/ SDMA System," Proceedings of the Second International Conference on Digital Satellite Communications, 1972, Paris.

[5] Cooperman, R. and Dobyns, T., "A Distribution Control Unit for Satellite Switched Communications," International Telemetry Conference, Oct. 1972, Log Angeles, Calif.

[6] Withers, D. J. and Jefferis, A. K., "A Time Division Multiple-Access System for Intelsat," AIAA Progress in Astronautics and Aeronautics: Communications Satellite Systems,

Vol. 33, edited by P. L. Bargellini, MIT Press, Cambridge, Mass., pp. 393-410.

[7] Dion, A. and Ricardi, L., "A Variable-Coverage Satellite Antenna System," Proceedings of the Institute of Electrical and Electronics Engineers, Feb. 1971, pp. 252-262.

AN EXPERIMENT IN DOMESTIC SATELLITE UTILIZATION

R. J. Metz*

IBM Corporation, Poughkeepsie, N. Y.

Abstract

This experiment in utilization of domestic satellite communications was designed to evaluate performance of various techniques under development to support efficient, flexible use of a full transponder of a geostationary repeater satellite similar to those to be deployed for U.S. domestic service by 1975. The techniques under development are in the following areas generic to satellite communication technology: 1) multiple-access digital communication, 2) network and resource assignment control, and 3) encoding of traffic for digital transmission. The results of the experiment concern the characterization of the equivalent IF channel made available by the vendor services, the operability of various algorithms and techniques for utilization of the satellite path channel capacity, and performance when handling real and simulated traffic while interconnected with IBM's private line network.

Presented as Paper 74-444 at the AIAA 5th Communications Satellite Systems Conference, Los Angeles, Calif., April 22-24, 1974. Several people were instrumental in actually doing the work herein presented: James W. Kennedy, who scheduled and carried out the measurement program in his key role as test conductor; Charles A. Kunzinger, who dealt in the mysterious and wonderful black art of high rate burst Modems both of his own design and execution as well as those of others; Brian E. Parker, who conceived the architecture of the DAMA engineering model and drove the team that managed to keep it working; Ernest Valera, who executed the software to feed the network simulation processor; and William G. Crouse, of the SDD Laboratory at Research Triangle Park, N. C., who designed and implemented the voice-processing chain engineering model, particularly with respect to the digital echo control scheme.

* Senior Engineer.

I. Introduction

The IBM Corporation is engaged in the development and marketing of a variety of communications-served data-processing systems. These systems utilize a variety of communications services provided over a large variety of facilities. It has followed the recent developments in public policy with respect to the establishment of alternatives in the provision of telecommunication services and facilities. Of particular interest has been the policy evolution in the matter of domestic satellite communications and its potential impact upon the user community in the matter of so-called "wideband" communications.

From a user viewpoint, one of the major aspects of the establishment of a public policy fostering competitive development[1] of a domestic satellite industry has been in the area of new service concepts in specialized communications. The fundamental nature of a geostationary communication satellite possessing a very large channel capacity has the potential of radically new modes of telecommunications service provision if fully exploited and not merely relegated to the category of another transmission alternative. Indeed, the FCC in the Domestic Satellite proceedings[2] repeatedly recognized this potential, stressing the point of "new and specialized communications services" in several commission orders.[3-5] In support of the FCC deliberations to establish public policy for the authorization of domestic satellite systems, the IBM Company filed comments[6] to encourage policies fostering flexibility of arrangements in the introduction of the benefits of satellite technology to the public at large as well as to the specialized communications requirements of private user entities operating private line networks.

The key elements to the development of such advanced concepts relate to the matter of feasibility of direct access to the satellite repeater channel capacity. Only in this manner can modes of use and services evolve which are truly new and different. The initial introduction of domestic satellites to the provision of telecommunication services will be very much patterned after current offerings provided by the established common carrier community. The essential feature is the "channelization" of the basic communication resource, that is, the satellite, into fundamentally point-to-point channels (to some extent point-to-multipoint) in the manner currently contemplated by the satellite carriers, both the established and the new entries.[7,8]

The "channelization" is based on time-honored techniques of Frequency Division Multiplexing with analog Frequency Modulation and Frequency Division Multiple Access of the satellite repeater resource in a fixed assigned manner. These methods are exceedingly well developed in terrestrial microwave radio relay practice and in the INTELSAT system and should no doubt be cost-effective in the initial introduction of satellite services.

However, it is the matter of new and different "services" that prompted IBM to investigate the potential of this new communications resource. Correspondingly, an experiment in utilization of domestic satellite communications was undertaken to investigate the viability of new service concepts not involving conventional "channelization" concepts and more particularly the ability of this new medium to support those concepts. This paper presents a review of some of the results of that experiment which were reported to the FCC.[9] The following sections present background considerations, the arrangements finally established, and the experimental configuration as finally evolved. The test objectives along with some sample experimental results then are followed by conclusions derivable from the experiment. The referenced report to the commission should be consulted for greater detail in terms of test data, measurements, and the subsequent analyses thereof.

II. Background

As early as 1962, IBM experimented with Telstar to demonstrate the suitability of satellite communication for data communications purposes. In addition, we also have conducted similar tests and demonstrations over INTELSAT facilities through the auspices of COMSAT and the international record carriers. These experiments were restricted to conventional voice-band channels and services.

IBM's Federal System Division long has been active in various areas of the military satellite communications program, particularly in systems engineering studies and modulation/ multiple-access techniques. Of special interest for a long time were the potential advantages of digital transmission in a satellite system environment and the opportunities for efficient multiple-access systems based upon time division (TDMA) of the satellite repeater power and bandwidth. In connection with the Army Satellite Communications Agency (USASATCOMA) program to develop TDMA equipment for the Phase II Defense Satellite Communication System (DSCS), tests were run from the engineering test facility at Ft. Monmouth, N. J., via TACSAT to investigate concepts in computer control of the TDMA pro-

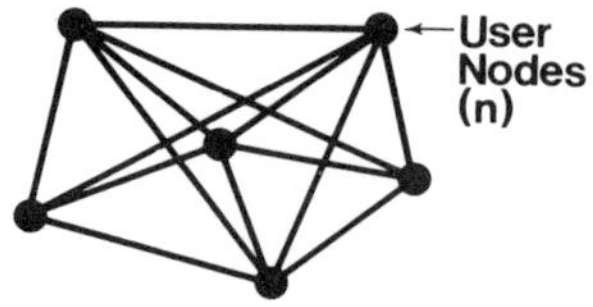

Fig. 1 Potential impact of domestic satellite.

cess.[10] These activities along with the work of others, particularly at COMSAT,[11,12] gradually led to a realization that digital transmission coupled with TDMA methods of sufficient flexibility could permit satellite communication to provide the total communications function, i.e., transmission and switching, required to satisfy all traffic demands between user nodes. Thus the potential impact of domestic satellite communications on private line networks (Fig. 1) provided an impetus to develop the necessary technology.

The utilization of domestic satellites will range from its application to message telephone service by integration into the existing public switched network to direct access to the bulk communication channel capacity resource to accomplish transmission and switching of all traffic in digital form. Technology already greatly has expanded the capacity of satellite systems and can be expected to continue this expansion, thus accenting the practicality of the latter mode of utilization. In between these two extremes would be the provision of channels on a transmission-only basis, ranging from voiceband circuits per the 260 private line service tariffs to digital trunks up to 1 1/2 Mbit/sec (or higher) rates, which, for certain applications, may permit more effective utilization of the system capacity.

Because of the attractiveness of the bulk communication mode of utilization, IBM set out to develop the necessary supporting technology and to carry out experiments to demonstrate that geostationary satellites and associated communication facilities are suitable for "wideband" communications involving all traffic forms, voice, data, image, etc. The experiment objectives have been focused on the use of domestic satellite service in bulk form, i.e., direct access to a whole "transponder" (one rf channel) of a geostationary repeater satellite similar to those to be deployed in U.S. domestic service by 1975, to prove-out IBM-developed supporting technology, and most importantly to verify that the arrangements required are indeed workable. The techniques under consideration are in the following areas generic to satellite communications technology: 1) multiple-access digital communication, 2) network and resource assignment control, and 3) encoding of traffic for digital transmission.

In order to test fully the performance of these technology developments, the actual satellite transmission was embedded in a computer simulation of network traffic flows and operating modes and interconnected with IBM's current voice-band private line network provided as a Common Control Switching Arrangement (CCSA) by the established common carrier community. The scope of the experiment was confined to digital transmission only in burst and continuous modes from 1 1/2 Mbit/sec to 15-30-60 Mbit/sec nominal, with major emphasis placed on efficient handling of voice traffic to provide a technology base for ultimate system architectures integrating all telecommunication traffic forms.

III. Arrangements

With the foregoing introduction and background, the next step was to seek out and consummate the necessary arrangements to undertake the experimental program outlined. It was apparent from the outset that the Telesat Canada space segment for the Canadian domestic satellite system would be most prototypical of an emerging U.S. domestic satellite industry. This space segment is based on the Hughes Aircraft Company (HAC) satellite, designated HS-333, and termed Anik in its Canadian deployment. This basic design[13] would be the space segment for at least two, and possibly three, of the then soon-to-be-operational U.S. satellite entities. Consequently, any assessment on a realistic basis of bulk utilization would be best accomplished via such satellite repeaters.

The initial approach to satisfying the direct rf access from IBM property requirement was to contract with the space segment owner, i.e., Telesat Canada, for occasional use over a finite time period to at least one whole "transponder"(rf channel, to be more precise) and to lease a transportable Earth station for deployment of the Poughkeepsie development laboratory. The governmental agreements, consummated in late 1972,[14] also tended to mitigate against the practicality of such an approach. This approach would have been most useful in exposing the full range of problems a user making direct access to a satellite transponder would have to comprehend. In this time frame, Telesat Canada contracted with two U.S. carrier entities, RCA Globcom and American Satellite, for long-term lease agreements of transponders on a full-time basis with some occasional use provided.

The next-best approach was to select one of the two mentioned entities to provide access to the Telesat space segment through an Earth station leased from them. A specification[15] for the rf ground terminal to be leased was based on minimum requirements to meet reasonable experimental capabilities and maximize possibility of supply, competitive or negotiated. The specification was developed in terms of transmission requirements only for a single-thread receive chain and transmit chain defined to a 70-MHz IF interface (Fig. 2).

RCA Global Communications (RCAG) was selected to provide the satellite access as a tariffed service[16] and to lease an Earth station meeting the minimum specification for an rf terminal to satisfy our objective, thus providing what constituted an equivalent IF channel in a loopback transmission mode accessible only from the RCAG Earth station deployed to IBM property at Poughkeepsie (Fig. 3). The environment provided to a user is thus quite realistic in that such an arrangement has high likelihood of appearing in the U.S. domestic satellite era, particularly in the 4/6-GHz bands, wherein

- **Equivalent IF Channel —**

  $\left(\frac{C}{KT}\right) > +86$ **db-Hz (Threshold)**

- **RF Terminal Transmit Chain —**

  EIRP: +86 dbW
  HS-333 Frequency Plan
  AM-PM: <°3/db at +86 dbW

- **RF Terminal Receive Chain —**

  $\left(\frac{G}{T}\right) > +23.8$ **db/°K**

  5 MHz & 10 MHz Offsets from HS-333 Frequency Plan

- **Common Parameters—**

  Transmission: Amplitude — <±0.5 db
  (±18 MHz) Group Delay — Linear <±0.5ns/MHz
  Parabolic <±0.2ns/MHz^2
  Ripple <±2ns p-p
  Interface: 75 Ω Z_o, 26 db RL, 70 MHz Nominal
  0 dbm to −10 dbm Level

Fig. 2 Transmission service requirements.

Fig. 3 IBM Poughkeepsie Development Laboratory complex.

maximum use of carrier-provided Earth stations appears desirable.

The next important arrangement was that of securing interconnection to the 2-wire telephone plant, since the effectiveness of digitally encoded voice transmission via the satellite link and its long delay and residual impairments is the major focus of the experiment. Appropriate tariff changes to the 260 private line tariff were filed by ATT Long Lines Department to allow unipoint access to IBM's private line network (CCSA) for the duration of the experimental program only.

At this point, all of the key ingredients for utilizing domestic satellite bulk service were at hand. The test program was carried out on an intermittent basis from Aug. 27, 1973 to Feb. 27, 1974, using the configuration to be described.

IV. Configuration

The experiment configuration deployed at the Poughkeepsie Systems Development Division Laboratory (Fig. 3) was made up of the following facilities: 1) one transponder channel on Telesat space segment, RCAG; 2) single-thread rf terminal, RCAG; 3) IF interface facility (trailer), IBM; 4) test bed (engineering models and other utilization equipments), IBM;

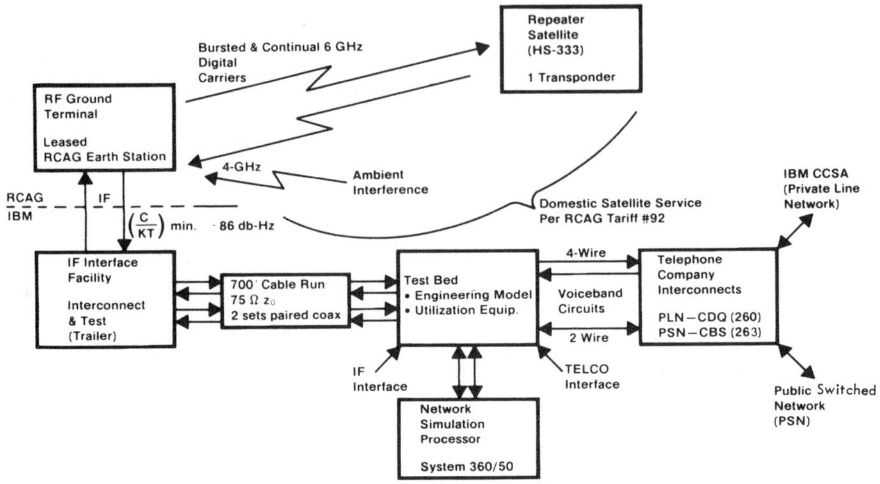

Fig. 4 Experimental configuration.

5) support equipment, IBM; and 6) telephone interconnects, ATT/LL and New York Telephone.

A block diagram of the configuration of the experiment (Fig. 4) shows the relationship and function of each of these facilities. The whole transponder available was one of three on Anik II (really F2 at 109°W long) frequency coordinated by RCA Globcom for the IBM site at which their Earth station was deployed. Limited access also was available to channel 1 of Anik I (i.e., F1 at 114°W long). The bulk of the experimental access of over 100 hr over a 6-month period was confined to the transponders on F2. The RCAG Earth station (Fig. 5) leased to IBM and operated by RCAG (as the licensee) under the tariff[16] satisfied in all respects the minimum requirements established by our specification for the rf ground terminal function. The Earth station as supplied could be described as fairly representative of what might be deployed in actual practice, having no particularly special characteristics or capabilities greatly in excess of our requirement. This consideration is important in assessing the value of our characterization of what amounts to a vendor-supplied (RCAG) equivalent IF channel.

The IF interface facility was housed in an IBM trailer sited at the rf terminal location for purposes of interconnection and test jointly accessible to RCAG and IBM personnel. All characterization of the equivalent IF channel provided by the RCAG service was accomplished at this point with various instrumentation setups primarily organized about an H-P micro-

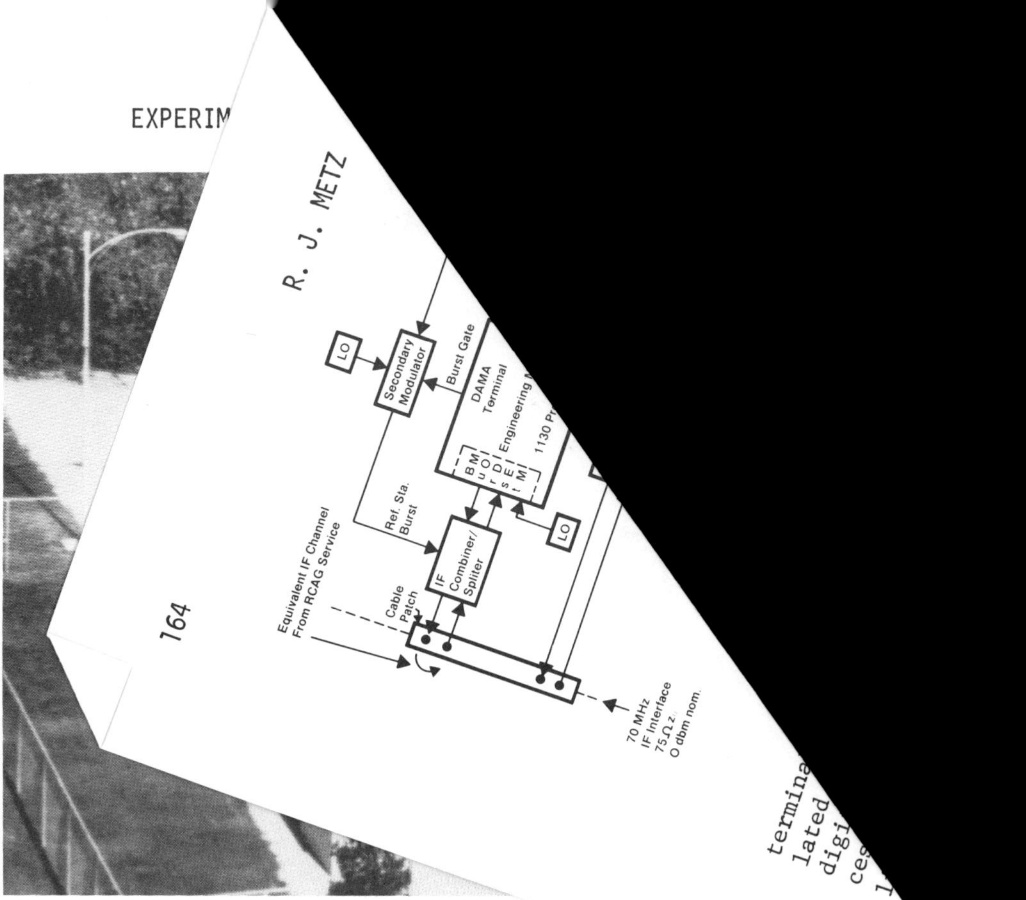

Fig. 5 RCAG Earth station and IBM

wave link analyzer.[17] This instrumentation also allow termination on a real-time basis of specification compliance of the rf terminal function by exercising the use of a test translator capability in RCAG's Earth station.

The other elements of the experiment configuration comprising the test bed, network simulation processor, and the Telephone Company interconnects were located in one of the adjacent buildings in the laboratory complex (Fig. 3) at the end of a 700-ft-long coax cable run. The telephone interconnects allowed both 2-wire and 4-wire access to the IBM CCSA private line network and 2-wire access to the public network. The network simulation processor employed was a dedicated IBM system 360, model 50, with a megabyte of core storage and capability to generate and sustain I/O at 1 1/2 Mbit/sec continuously.

The test bed organization is shown in Fig. 6 and identifies all elements of the experiment wherein specific technology developments are provided as engineering models. The key engineering models were a DAMA (Demand Assign Multiple Access)

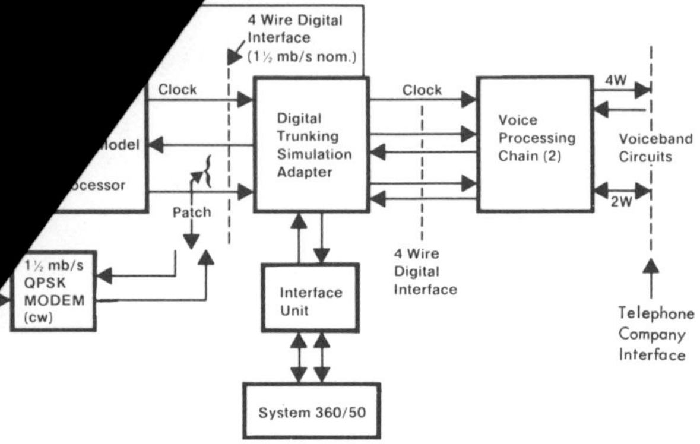

Fig. 6 Test bed organization.

...l operating between the 70-MHz IF interface and a simu-... 1 1/2-Mbit/sec digital trunk carrying all traffic in ...tal form, simulated and/or real, and a pair of voice pro-...ssing chains operating between the trunk multiplex estab-...ished by the digital trunking simulation adapter and telephone interconnects.

The DAMA terminal is based on TDMA, with demand assignment provided by an 1130 processor and its associated software. The distinguishing feature of the TDMA approach adopted in the DAMA terminal engineering model is the use of very long main frame intervals, of the order of 5 msec, and complete software-implemented algorithms for control of the "time" resource of a frame as well as acquisition and maintenance of system synchronization. The burst Modem function was initially supplied by a unit developed at the Poughkeepsie Laboratory and operating coherent QPSK (Quaternary Phase Shift Keyed) at 15 Mbit/sec and 30 Mbit/sec nominal. (When the digital trunk rate was operated at a T1 line rate of 1.544 Mbit/sec, the burst transmission rates were integer-synchronous.) A second unit was developed at IBM-France's La Gaude Laboratory and featured operation at 30 Mbit/sec and 60 Mbit/sec at a constant symbol rate by operating BPSK (Binary Phase Shift Keyed) or QPSK. Both Modems featured different approaches to signal design at the modulator and recovery of carrier and symbol timing references at the demodulator.

To accommodate single-station loopback of a TDMA system, an additional modulator and bit stream source was provided to simulate a reference station burst[12] with a noncoherent IF carrier LO. The 1 1/2 Mbit/sec digital trunk was patchable to a separate continuous mode QPSK Modem used in tests[18] of concatenated satellite-cable TV digital transmission. A special interface unit also was provided to permit the IBM system/360, model 50 network simulation processor to provide continuous full duplex 1 1/2 Mbit/sec bit stream of simulated traffic to the digital trunking simulation adapter.

The voice processing chains incorporate a pair of Deltamod Codecs[19] and associated fully digitally implemented echo-control units. The verification of the performance of the echo control and Codec companding algorithms was considered to be a major objective in assessing this approach to domestic satellite utilization. The arrangement of the voice-processing chains and the telephone interconnects is shown in Fig. 7. A digital delay simulator was provided to permit checkout without the presence of the satellite circuit (two simplex channels) between the two telephone talking paths defined by the interconnects. The 2-wire interconnect to the Centrex serving the Poughkeepsie Laboratory complex provided a worst case for 2-wire plant impedance variation and its consequent potential for echo generation.[20]

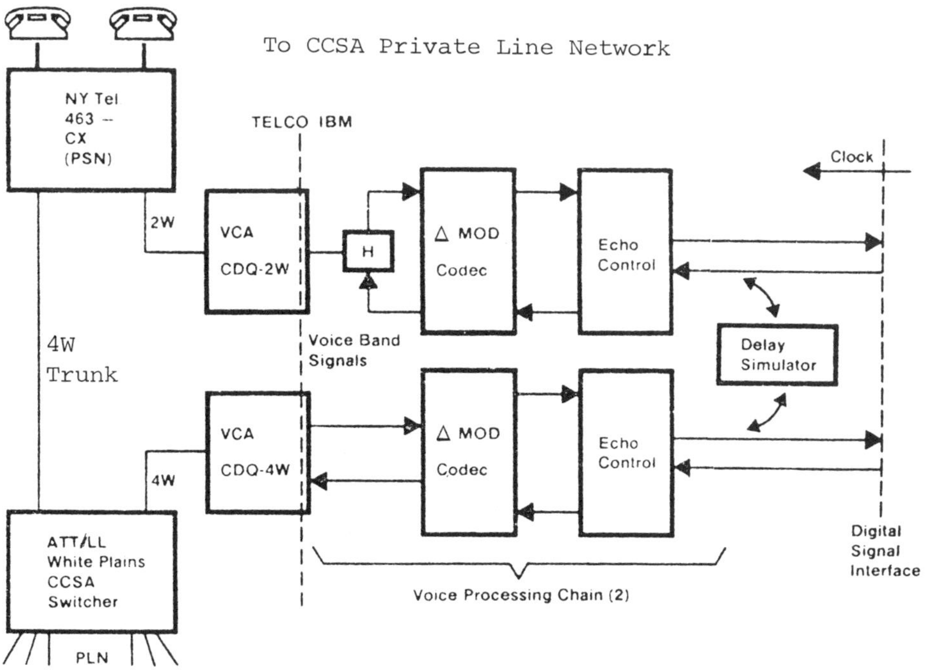

Fig. 7 Telephone interconnect arrangement.

V. Test Objectives

The objectives of the link test program defining the experiment fall into three general areas: 1) verify and characterize utilization of external elements; 2) demonstrate operability with all traffic forms; and 3) evaluate technology developments. Although considerable effort was invested in developing a software-implemented approach to carrying out the TDMA process control algorithms, it was not the purpose of the link test to verify in any way TDMA as a technique, this having been proved conclusively in actual tests and demonstrations since the first true TDMA experiment on Early Bird.[11] In addition, testing of the voice-processing performance in a statistically designed experiment was not planned for, since past work[21] had achieved important conclusions concerning the time delay and its interaction with echo. What was of major importance, however, was user reaction to end-to-end performance through this link inserted into the existing private line network.

Although voice communications received greatest emphasis in the test program, a variety of tests and demonstrations in data and image transfer was undertaken in a mixed traffic environment simulated by the network simulation processor. By utilization with all forms of potential traffic in this manner, a major objective of the program in evaluating technology developments was achieved. The three areas of interest were the performance of the digital echo control implementation, the operability of control algorithms, and the use of a multiplicity of digitally modulated carriers of varying rates.

The test program was organized into four major categories and access to the satellite link scheduled in a manner to accommodate maximum concurrency between measurements and data collection, as well as off-line analysis. Broadly stated, these categories are as follows: 1) IF channel characterization, 2) Modem performance, 3) voice-processing characterization and evaluation, and 4) utilization experiments: voice, data, image.

The IF channel characterization received great emphasis, since this is the "commodity" in which a bulk mode approach to utilizing domestic satellite service deals. The Modem performance tests indirectly comprehend the internal performance and operability of the DAMA terminal hardware, since all of these factors appear in digital transmission performance. In all of these tests, the simulation processor performed as the source and sink of digital traffic bitstreams.

VI. Experiment Results

The experiment results presented here are a few typical samples of the type of data collected and observations made during the program. Reference 9 should be consulted for a more comprehensive view. It should be noted that, in the case of measurements through the equivalent IF channel, no attempt was made to seek out minimum-maximum limits of parameter excursion at this time; rather, the purpose of this data collection was to ascertain which parameters are of primary importance in ultimately correlating performance with the signals and traffic modes used for channel utilization.

A. IF Channel Characterization

This test program was run from the IF interface facility and collected data through the RCAG Earth station on Anik I, rf channel 1, and Anik II, rf channels 1, 2, and 12. The measurements encompassed the following: group delay response, amplitude response, AM-PM conversion coefficient, phase noise characteristics, satellite multipath, noise power spectral distribution, and two-tone intermodulation measurements.

The first three parameters were determined separately for the Earth station by calling for the use of the test translator. However, it should be pointed out that separation and allocation of impairments within the equivalent IF channel can be subject to gross error and so must be done with considerable insight into the exact details of a given configuration. Some sample results obtained are as follows and appear on analysis to indicate that the constituent components of the IF channel are within specification tolerance. Amplitude and group delay responses for channel 2 on Anik II are shown, respectively, by Figs. 8 and 9. These curves have been replotted (and consequently smoothed) from oscilloscope photographs taken from the Hewlett-Packard Microwave Link Analyzer.[17] The derived parameters of interest to high-rate digital transmission waveforms are, of course, the gain slope and the parabolic components of group delay distortion.

The AM-PM conversion properties of the channel were indeed the most troublesome in terms of effect on Modem performance. A typical characteristic for channel 1 of Anik II is shown in Fig. 10 as a function of both relative transponder drive level and Earth station Effective Isotropic Radiated Power. Phase noise characteristics for the same channel were taken from smoothed spectrum analyzer data and are shown in Fig. 11. The ordinate is shown in absolute level with the carrier at 0 dbm. It should be noted that this particular run

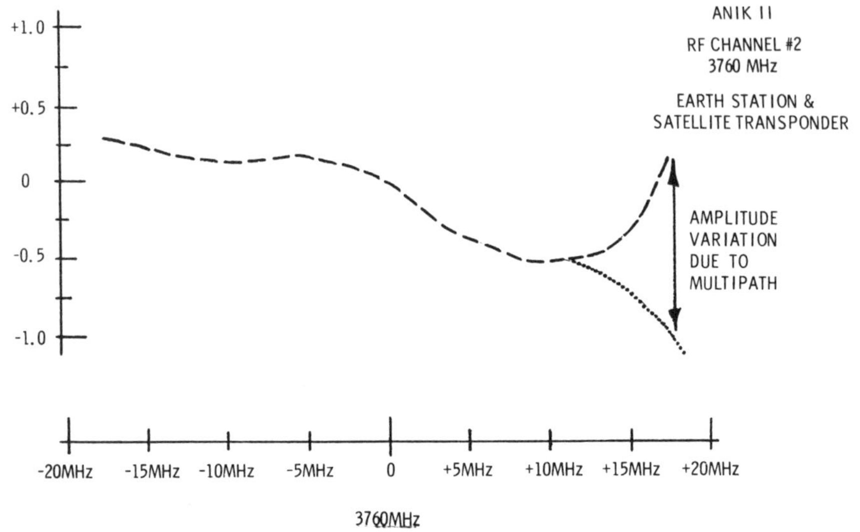

Fig. 8 Amplitude response (typical).

was corrupted by 60-Hz harmonics apparently introduced in the Earth station converters. The requirement specification established by IBM did not limit phase noise at the time the arrangements were being undertaken in order to provide the

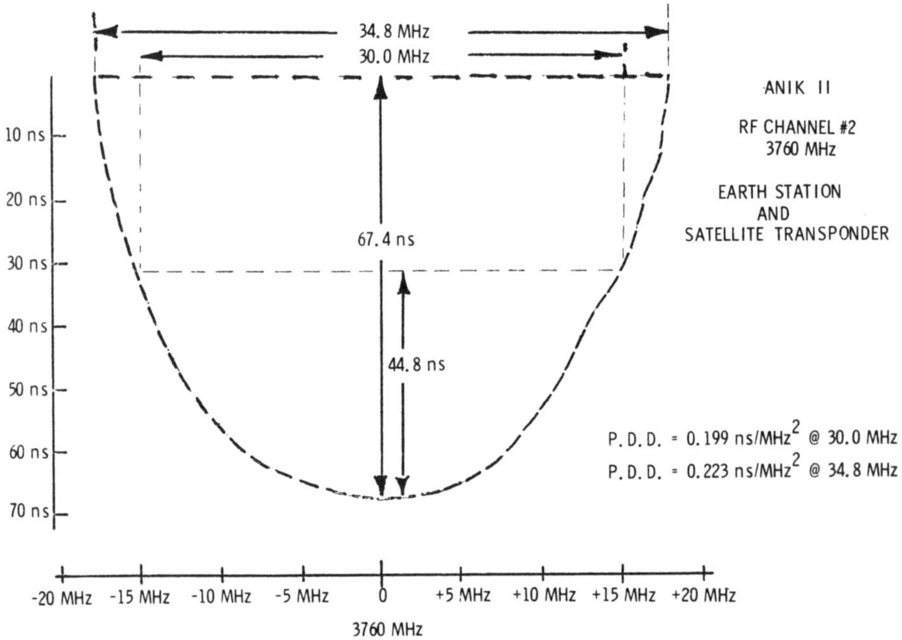

Fig. 9 Group delay characteristic (typical).

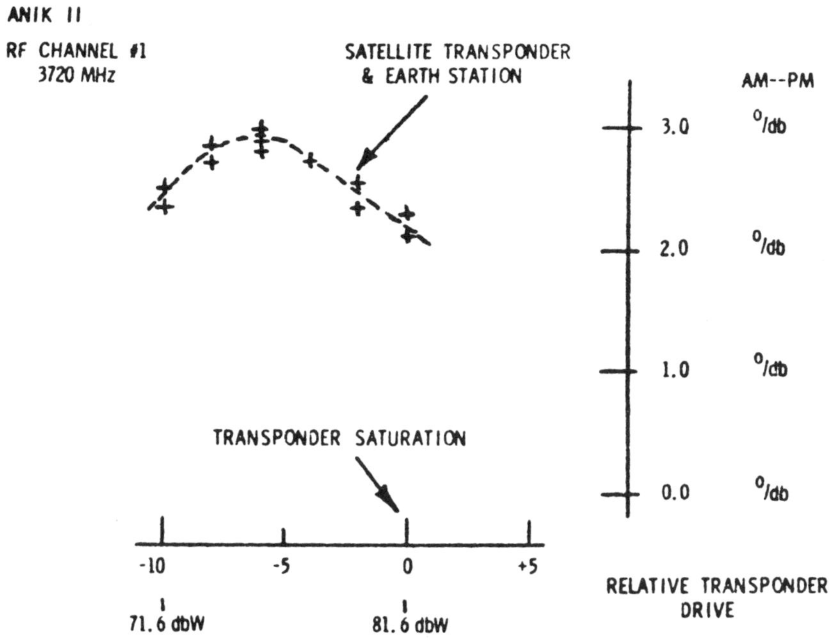

Fig. 10 AM-PM characteristic (typical).

opportunity to experience the full range of such impairments as they might be encountered in real channels.

B. Modem Performance

The usual digital transmission performance was measured as Bit Error Rate (BER) vs C/KT at 15, 30, and 60 Mbit/sec, nominal burst rates for the two different modem approaches tested. It should be noted that the BER is observed at the nominal 1 1/2 Mbit/sec digital trunk rate, representing the digital baseband port of the DAMA terminal. The BER's were taken over very long intervals because the network simulation processor was capable of generating long traffic streams of virtually unlimited complexity. Furthermore, off-line processing of the error syndromes collected allowed correlation with various malfunction modes internal to the DAMA implementation. The measured data are then presented normalized to P_e^B vs E_B/N_O. Figure 12 shows the performance of the Poughkeepsie Laboratory design approach operating in full DAMA configuration for the 15- and 30-Mbit/sec burst rates as compared to theoretical for a White Gaussian Noise perturbed linear channel with no bandlimiting. The 9-db implementation loss incurred was primarily due to the effect of AM-PM, which was

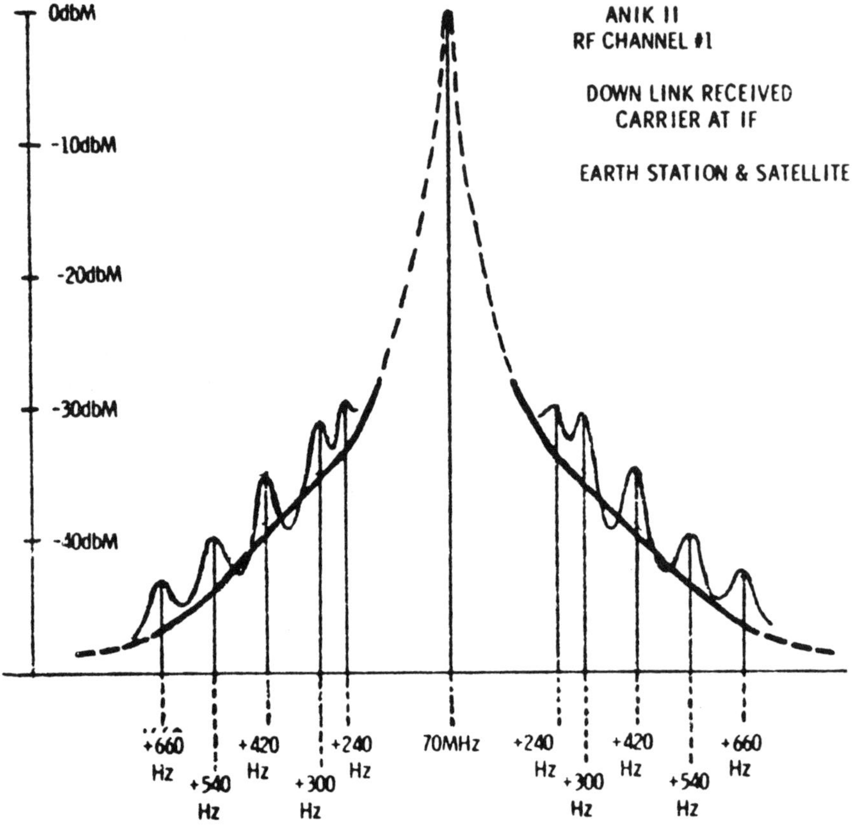

Fig. 11 Residual FM and phase noise characteristics.

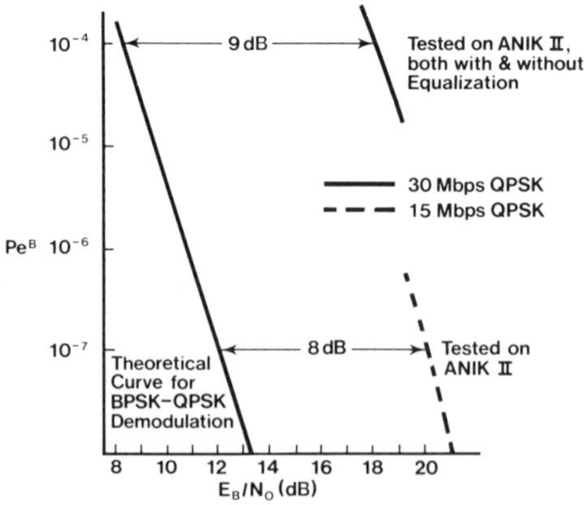

Fig. 12 Poughkeepsie Modem performance.

quite severe for the signal design selected in this instance. Currently contemplated modifications would improve the susceptibility to improvement by the insertion of group delay equalization, retrieving approximately 5 db of the incurred loss.

In the La Gaude Laboratory Modem approach the performance, similarly processed and normalized from the collected data, is shown in Fig. 13. The performance here is for the 30-Mbit/sec mode operating BPSK and clearly shows a 3.7-db improvement with the insertion of optimized delay equalizers. The 5-db loss incurred with equalizers represents a reasonable margin considering bandlimiting and AM-PM conversion effect considering that the symbol rate in this instance is 30 Mbit/sec. At this time, it appears that the order of 3.5 to 4 db loss over theoretical must be allowed for all channel impairments as now determined.

C. Voice-Processing Performance

The Deltamod Codec developed for the experiment was derived from earlier work[20] and featured fully digital-implemented step-size algorithms for companding in the coder feedback path. Key objective design parameters for the Codec were the following: tracking at 1 kHz, +3 to -37 dbm0 < 5% gain linearity; and signal to quantizing noise, +3 to -25 dbm0 > 25 db. With channel measurements conducted from analog input port on one Codec, through the digital satellite link, and out of the analog output port on the other Codec, the following performance was achieved (DAMA link error rate at $< 10^{-4}$): tracking at 1 kHz, +10 to -20 dbm0 < 1% gain linearity; and signal to quantizing noise, +5 to -20 dbm0 > +25 db.

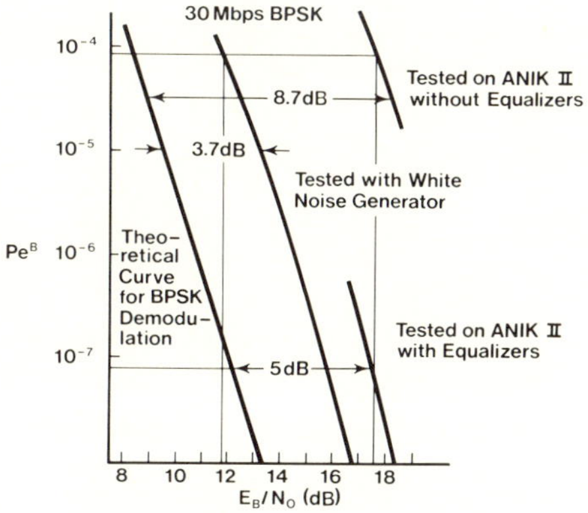

Fig. 13 La Gaude Modem performance.

These two parameters are considered most significant in that they are largely controlling when the Codec is required to handle telephone line data Modem signals in analog form. At this time, it appears that 103-type Modems and the typical range of operating levels encountered are transparently handled through such a digital-link derived "voice" channel.

The preponderant test effort, however, was applied to talker-to-talker communication on IBM's CCSA, with the operating satellite link in tandem. It can be reported that, for most cases with one talker being unaware of the interposition of the satellite path (the originating talker, of course, is aware and thus primed in his expectations), the conversation suffered only initial awkwardness, with a rapid adjustment made in terms of accommodating the delay effect by more care in interruption. It has been observed by several participants that, if anything, a widespread use of satellite connectivity would gradually increase the "politeness" of telephone conversations. For the final design tested of echo control, it was observed on several occasions that a moderate amount of "double talk" is permissible if the two talker volumes are about equal.

Some additional tests over the satellite path were undertaken, substituting a "block-companded PCM" Codec for the Deltamod Codec and assessing voice quality on a subjective basis. It was determined that, for satellite path BER's of 10^{-4} or better, both Codec techniques imparted negligible distortion and modification to "telephone" speech between model 500 handsets.

D. Utilization Experiments

Two areas of considerable interest in this mode of domestic satellite utilization covered the full spectrum of voice-band data transmission by currently employed methods to "wide-band" image transmission. A series of user experiments using remote terminals on internally supplied data processing services (APL and CP/CMS) was routed over the satellite path (usually while other experiments were being carried out) accessed by the CCSA. The signals in these cases were typical 103 Modem voice-band signals routed through the voice processing chains while connecting the terminal to the host CPU. It was concluded that no significant effect was observed on user interaction with the computer service due to the interposition of the satellite path. It should be noted that this experiment configuration embodies all significant elements that would be encountered in a real application environment.

The other user experiment focused on the feasibility of exploiting this mode of domestic satellite transmission for very high-speed document transmission using an experimental scanner available at the Poughkeepsie Development Laboratory. The resolution of the scanner coupled with an auxiliary compression algorithm implemented in the network simulation processor required just 3 sec at the nominal 1 1/2 Mbit/sec digital trunking rate to transmit a complex sample 8 1/2 x 11 document. When translated to the burst transmission rate of 30 Mbit/sec used at the time of the experiment transmission, it can be seen that an interruption of only 150 msec in all other traffic utilizing the satellite repeater could result in a very efficient and low-cost form of remote document reproduction. Refinement in the compression algorithm by hardware implementation rather than software could reduce the "interrupt" to about 50 msec. If done within the confines of a DAMA system such as developed for the experiment on a sufficient low-duty cycle, the impact on the voice traffic in terms of quality or grade of service deterioration would be inconsequential.

VII. Conclusions

On the basis of both subjective and objective evaluation factors of the many facets of the entire scope of the experiment, it appears that the mode of utilization of domestic satellite service investigated shows great promise of allowing a variety of advanced concepts of service for users. Clearly, given the ability to allocate and manipulate the large channel capacity of even one transponder in a current operational space segment to a variety of highly varying and geographically dispersed traffic demands, one should have presaged such a conclusion. But in the real world the proof is in the pudding, and that after all was really what this experiment was all about. There is no one startling breakthrough or technological tour-de-force which lets it all fall in place. Rather, it can be said that the essential robustness of the concept can withstand great abuses of imperfection, of which a technology development program such as this was, as is not unusual, blessed with in abundance, but not to the extent of ever dimming the essential viability of this mode of use. Of course, ultimate deployment of a system "architected" around the key features investigated and/or demonstrated in this experiment depends, as usual, on economics, the need for new and special services, and, to a lesser extent, on related regulatory issues.

References

[1]Hinchman, W. R., "Public Policy and the Domestic Satellite Industry," 1972 International Communications Conference, May 19-21, 1972, Philadelphia, Pa.

[2]"Establishment of Domestic Communication Satellite Facilities by Non-Governmental Entities," Docket 16495, 1966-1972, Federal Communications Commission.

[3]Report and Order 70-306, Docket 16495, March 24, 1970, Federal Communications Commission.

[4]Memorandum Opinion and Order 72-229, Docket 16495, March 17, 1972, Federal Communications Commission.

[5]Second Report and Order 72-531, Docket 16495, June 16, 1972, Federal Communications Commission.

[6]"Comments of IBM Corp. on the Memorandum Opinion and Order Released 17 March 1972," Docket 16495, April 19, 1972, Federal Communications Commission.

[7]Nowakoski, D. B., "The Western Union Integrated Satellite Voice/Data Network," National Telecommunication Conference Proceedings, Nov. 26-28, 1973, Atlanta, Ga.

[8]Puente, J. G., "Designing the American Satellite Corporation System," EASCON Proceedings, Sept. 1973, Washington, D. C.

[9]"Report to the FCC on a Satellite Link Test conducted by IBM at Poughkeepsie, N.Y.," IBM-SDD Report, filed via RCA Globcom (submitted May 15, 1974), Poughkeepsie, N.Y.

[10]Hayase, J. Y. and Goubeaud, G. J., "Computer Controlled Time Division Multiple Access (TDMA) Control System and Link Performance," EASCON Proceedings, Oct. 26-28, 1970, Washington, D. C.

[11]Sekimoto, T. and Puente, J. G., "A Satellite Time Division Multiple Access Experiment," IEEE COMTECH Transactions, Aug. 1968.

[12]Puente, J. G., Schmidt, W. G., and Werth, A. W., "Multiple Access Techniques for Commercial Satellites," IEEE Proceedings, Feb. 1971.

[13]Houterman, M. J., "ANIK Satellite Communications System," International Telemetry Conference Proceedings, Vol. VIII, Oct. 10-12, 1972, Los Angeles, Calif.

[14]"Intergovernmental Agreements between U.S. and Canada on Telecommunication Satellites," U.S. Dept. of State Bulletin, Feb. 5, 1973, pp. 145-147.

[15]"Performance Requirements RF Ground Terminal for a Satellite Communication Experiment," IBM Specification, Feb. 5, 1973 (see Appendix of Ref. 9).

[16]"Developmental Satellite Transmission Service," Tariff 92, filed July 13, 1973, RCA Globcom, Federal Communications Commission.

[17]"MLA Demonstrations" and "MLA Users Manual," HP Publications, 1972.

[18]Blonder, H. H. and Greenhalgh, R. C., "IBM-TelePrompTer Digital Transmission Test-Satellite and CATV," 23rd NCTA, April 22, 1974, Chicago, Ill.

[19]Schindler, H. R., "Delta Modulation," IEEE Spectrum, Oct. 1970.

[20]Gould, R. G. and Helder, G. K., "Transmission Delay and Echo Suppression," IEEE Spectrum, April 1970.

[21]Gould, R. G., "Echo Suppression and Time Delay, A Subjective Evaluation," IEEE COMTECH Transactions, Sept. 1964.

ASTRO-DABS: AN ATC SATELLITE-RELAYED SURVEILLANCE, NAVIGATION, AND DATA LINK CONCEPT

Leonard Schuchman*

The MITRE Corporation, McLean, Va.

Abstract

ASTRO-DABS is a fourth-generation air traffic control (ATC) system, which uses satellite relays for surveillance, navigation, and air/ground data link. It is designed to evolve from the upgraded third-generation ATC ground-based system, which is a discrete address beacon system (DABS) for surveillance and air/ground data link. ASTRO-DABS would coexist with DABS in a noninterfering manner. The paper describes ASTRO-DABS in terms of its system design, component design, communication formats, and design objectives. Particular emphasis is given to illustrating how surveillance information is transmitted in a range-ordered sequential poll that guarantees no garbling at any satellite. Emphasis also is placed on describing the navigation concept, which is designed to give very accurate navigation data at a relatively low cost by use of a ground calibration network and a ground-satellite-aircraft data link.

I. Introduction

ASTRO-DABS[1] originally was conceived, designed, and studied as a satellite discrete address beacon system that could coexist with the terrestial discrete address beacon

---

Presented as Paper 74-440 at the AIAA 5th Communications Satellite Systems Conference, Los Angeles, Calif., April 22-24, 1974. This work was sponsored by the Office of Systems Engineering Management of the Federal Aviation Agency, under the direction of Neal Blake. The author wishes to acknowledge the support and guidance of N. Blake of the FAA. The author also wishes to acknowledge the significant contributions made conceptually and analytically by P. Ebert, B. Elrod, and J. Fee, as well as the analytic contribution made by A. Avant.

*Group Leader, Advanced Communications.

system (DABS), providing very accurate surveillance together with a reliable data link throughout continental United States (CONUS). The concept has been augmented[2] to include navigation (ASTRO-NAV). Although future modifications to the concept will continue to be made, it is the purpose of this paper to describe ASTRO-DABS as it is presently conceived.

The characteristics of the baseline design are summarized in Table 1. ASTRO-DABS has a surveillance satellite-relayed instantaneous capacity of 70,000 active aircraft in the system, which can grow to 140,000 by doubling the transmitting satellite-usable average power, with no additional avionics costs and for a 40% increase in bandwidth. [Two aircraft satellite link modulation techniques have been considered: DPSK (differential phase shift keying) and PAM (pulse amplitude modulation). Although the DPSK system outperforms PAM, PAM

Table 1 ASTRO-DABS: baseline design

Surveillance: data link (DPSK modulation)	
Address population	4,000,000
Instantaneous airborne capacity [a]	70,000
Instantaneous airborne capacity growth capability [b]	140,000
Average interrogation period	5.5 - 6 sec
Three-dimensional position accuracy (anywhere in CONUS) [c]	75 - 150 ft
Number of bits per message per interrogation [d]	
90% of time	50
10% of time	100
Probability of bit error	10^{-6}
Navigation	
Average interrogation period	1 - 2 sec
Position accuracy	100 - 200 ft

[a] This does not include the number of aircraft terrestial DABS would be handling simultaneously. The capacity is reduced to 55,000 if PAM instead of DPSK modulation is used to transmit from the aircraft to the satellite.

[b] The capacity is reduced to 110,000 if PAM modulation is used to transmit from aircraft to satellite.

[c] A two-dimensional version of ASTRO-DABS is described in Ref. 2.

[d] Includes address bits.

reduces the avionics cost.[3] The data link discussed herein assumes a DPSK modulation.] There are two normal surveillance modes of operation: the tracking mode and the acquisition mode. These modes are interleaved, and so aircraft entering the system are acquired with high probability within a 4-6-sec period, and tracking updates occur every 6 sec thereafter. ASTRO-DABS would be designed to provide 75-150-ft (1σ) surveillance positional accuracy for all CONUS. The system will be designed to provide 50-bit messages 90% of the time and 100-bit messages 10% of the time, with a bit error rate of 10^{-6} or better.

ASTRO-DABS is designed to coexist with DABS. No modification to the DABS or air traffic control radar beacon system interrogators has to be made. Its data link is compatible with DABS in that the same messages can be transmitted over either DABS or ASTRO-DABS. Finally, all purchasers of avionics can take advantage of equipment commonality (display, same logic, power supply, and packaging), resulting in cost savings even though ASTRO-DABS and DABS operate in different frequency bands and with different data bit widths. (Aircraft transmit 1-μsec DABS data bits, whereas ASTRO-DABS uses 1.5 μsec/data bit.) ASTRO-NAV, the navigation system of ASTRO-DABS, is estimated to provide a 100-200 ft, three-dimensional position accuracy with an update rate of 1-2 sec. ASTRO-NAV is imbedded in the signal format of ASTRO-DABS. It operates in a noninterfering mode within ASTRO-DABS and over the same frequency channel as the surveillance and data link system, and, although using the same receiver as ASTRO-DABS, it adds a more accurate clock for better timing, a computer, and a display to transform timing pulse data to a useable three-dimensional position display.

II. System Configuration

In this section, the proposed ASTRO-DABS system configuration is given for surveillance, data link, and navigation. The ASTRO-DABS configuration is shown in Fig. 1. Ground-based discrete address beacons (DABS), providing surveillance and data link, are located to cover dense terminal areas. Other airports that are close to busy terminals and within DABS coverage also will be serviced by DABS. The busy terminal control facilities are designed to provide control instructions and assignments to all instrument flight rules (IFR) aircraft and automatic collision avoidance advisories and commands to all visual flight rules (VFR) aircraft. The automated control commands are transmitted via the DABS data link. Backup and special aircraft-to-ground communication would be provided by vhf voice links.

All airspace not surveyed by DABS is covered by the ASTRO-DABS satellite system. Control of this airspace is provided by the 20 enroute centers located throughout CONUS. In airspace covered by ASTRO-DABS, IFR aircraft receive via the ASTRO-DABS data link control instructions and assignments, whereas VFR aircraft receive collision avoidance advisories and commands. Terminals that are not covered by DABS receive terminal surveillance information via land lines from the enroute centers and these terminals for takeoffs and landings.

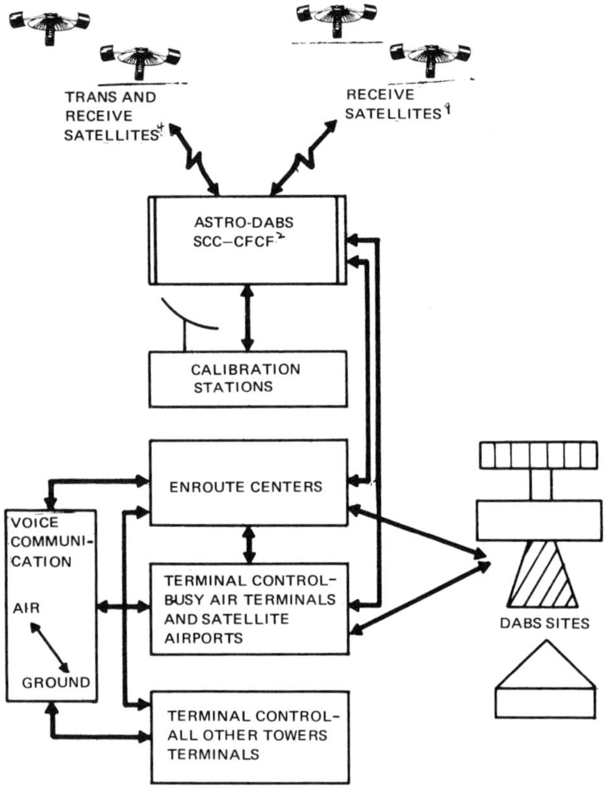

Fig. 1 ASTRO-DABS configuration.

The satellite control center (SCC) handles all communication to and from the satellites. This includes tracking, telemetering, and control of all satellites, as well as surveillance and data communication messages to and from the aircraft. Colocated with the SCC is the central flow control facility (CFCF), which acts as a centralized clearing house for IFR flight plans and near-term (12-hr) airspace planning. The objective is to avoid congestion and delay through a localized matching of demand to capacity. The enroute centers are tied into the ASTRO-DABS satellite control center via C-band satellite data links. There will be two geographically separated SCC's. One backs up the other in case of a catastrophic failure. The enroute centers receive surveillance and data link information from the SCC's. The enroute centers transmit data link messages to the SCC's. In addition, the enroute centers and the SCC are linked via the satellite C-band data distribution system to DABS-controlled terminal areas, so that all control information can flow between CONUS facilities.

In ASTRO-DABS, there are two types of satellites, namely, surveillance and data link (S/D) satellites and surveillance and navigation (S/N) satellites. The S/D satellites transmit data to aircraft and receive from aircraft data link and surveillance information, whereas the S/N satellites transmit only navigation timing pulses and primarily receive surveillance information from aircraft. Calibration stations located throughout CONUS will be used to reduce surveillance and navigation bias errors due to propagation anomalies and satellite position uncertainties. In addition, DABS and ASTRO-DABS surveillance information will be referenced to a common coordinate system by placing ASTRO-DABS calibration stations near DABS sites. It is estimated that on the order of 100 DABS sites would exist in ASTRO-DABS, so that there could be up to 100 calibration stations throughout CONUS.

In ASTRO-NAV, the navigational concept of ASTRO-DABS, the essential components are the satellites, the satellite control center, and the calibration stations. The navigation system would use the same system elements as were used for surveillance and data link. However, no surveillance information derived from ASTRO-DABS is used in ASTRO-NAV. The intent of ASTRO-NAV is to keep the cost of an aircraft navigation computer as low as possible by having a significant amount of computation done on the ground at the SCC. Thus, calibration station information will be gathered at the satellite control center, processed, and data-linked via the satellite relay to all aircraft. This information, together with timing pulses received from all satellites, will be used by an aircraft to determine its three-dimensional position. ASTRO-NAV utilizes timing pulses from all visible satellites. Hence, the ASTRO-DABS S/N satellites also transmit timing pulses to aircraft during the ASTRO-NAV mode of ASTRO-DABS. Aircraft are not required to be ASTRO-NAV-equipped. Thus, aircraft that choose not to use ASTRO-NAV can use the VOR (vhf omnidirectional range) system, which remains as an independent navigation system.

The traditional design of surveillance and navigation systems keeps these systems independent of one another for purposes of reliability. Although navigation and surveillance are not completely independent in ASTRO-DABS, reliability is achieved in several ways. DABS and ASTRO-DABS, which can operate as independent systems, can be utilized to back one another up for the surveillance and data link functions.

ASTRO-NAV and VOR perform similar backup functions for navigation. Within ASTRO-DABS itself, reliability is achieved through redundancy. The satellite constellation is designed so that, if a single satellite fails, there is no loss in data link capacity and only a small loss in surveillance and navigation accuracy. The redundant SCC's provide reliability on the ground, and the avionics is designed so that any single electronic failure will not cause the loss of both surveillance and navigation.

Evolving to ASTRO-DABS would occur in four states, as described in terms of frequency band occupancy as shown in Fig. 2. State one illustrates the ATCRBS (air traffic control radar beacon system) surveillance system as it exists today. DABS service will be initiated in state two. The launching of ASTRO-DABS satellites initiates the operation of state three, whereas state four would occur when the ATCRBS surveillance system is discontinued. In state three, the compatible ATCRBS and DABS coexist with the satellite-relayed ASTRO-DABS, which operates on different frequency channels. (The dashed frequency allocation in the interval from e to f would enable ASTRO-DABS to double its capacity.) Thus, in state three, aircraft not in busy terminal areas will be surveyed by the

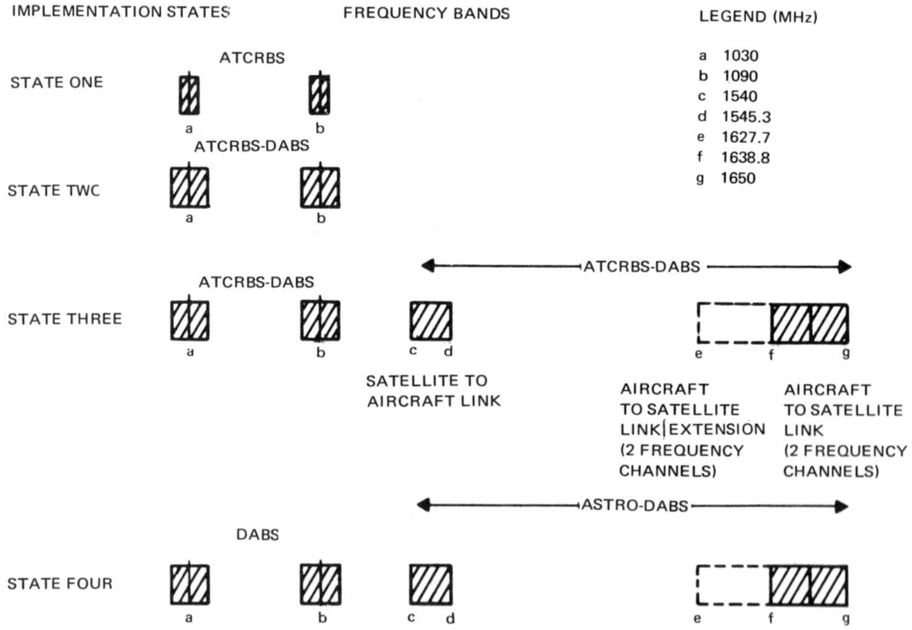

Fig. 2 Evolution to ASTRO-DABS (as seen in terms of frequency band occupancy).

satellites if equipped for ASTRO-DABS, or by ATCRBS if not so equipped. State four will begin when all aircraft are ASTRO-DABS and DABS equipped.

III. Satellite Constellation

The satellite constellation[2,4,5] is given in Fig. 3. There are seven satellites in synchronous equatorial orbit: four surveillance and data link (S/D) satellites and three surveillance and navigation satellites. (Studies to date assumed one-way ranging. The possibility of two-way ranging with its better Geometric Dilution of Precision[6] and/or fewer satellites also is being studied.) An additional six (S/N) satellites are in a geosynchronous orbit. The constellation is designed with one backup S/D satellite and one backup S/N satellite, so that the system performs well even if one satellite fails. (The spare S/D satellite is not shown in Fig. 3.)

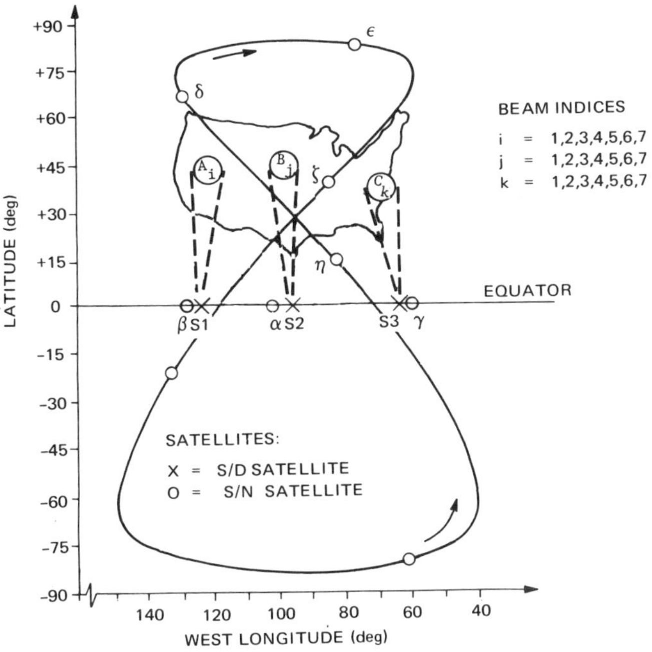

Fig. 3 ASTRO-DABS constellation.

The characteristics of the satellite segment of the system are shown in Fig. 4. ASTRO-DABS aircraft operate on three separate, equal-bandwidth (5 1/3 MHz) channels through the satellite relay. There are two aircraft-to-satellite channels (f_2, f_3) and one satellite-to-aircraft channel. Three large S/D satellites each interrogate and receive responses from one-third of CONUS. At any instant, there are at least six S/N satellites in ASTRO-DABS which receive responses from all CONUS. Thus, every aircraft in CONUS receives interrogations from one satellite and has its reply received by at least seven

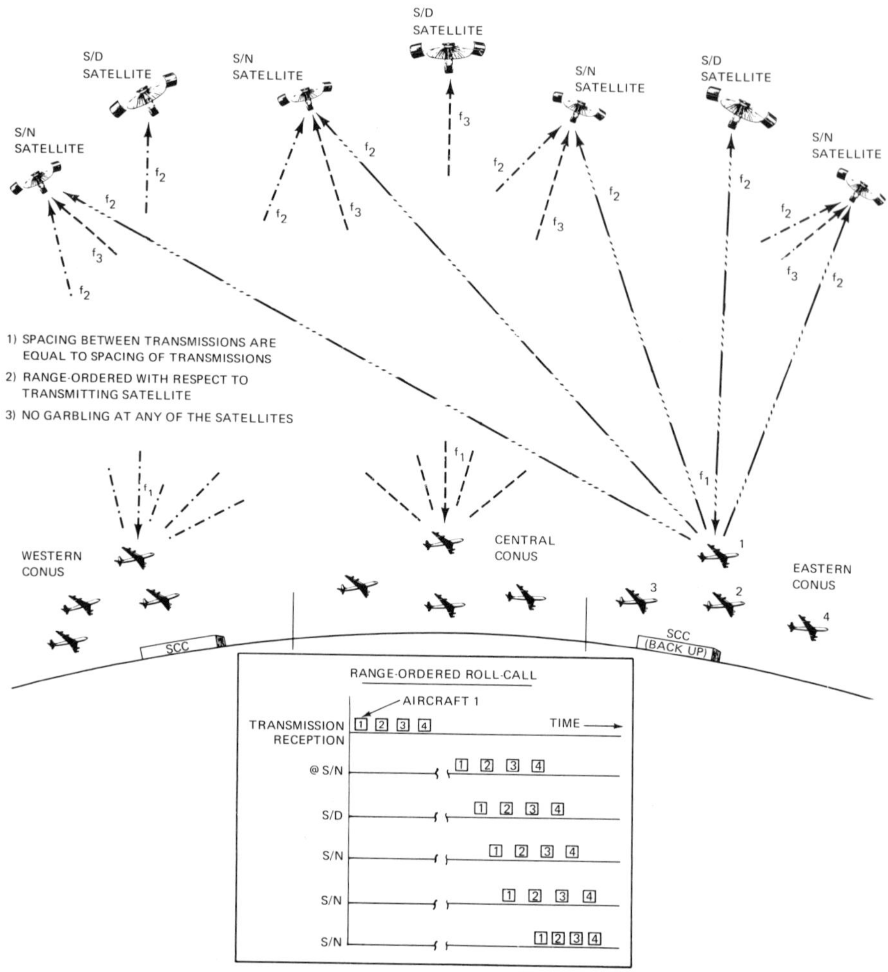

Fig. 4 Satellite-relayed DABS: overview.

satellites. The two aircraft-to-satellite frequency channels prevent aircraft returns from different sections of CONUS from garbling.

Not shown in Fig. 4 are the C-band communication links that connect the satellite control center (SCC) and its backup with all of the satellites.

IV. Surveillance and Data Link

There are four basic surveillance and data link modes in ASTRO-DABS which are described briefly in Table 2. The four

Table 2 Surveillance and data link modes of ASTRO-DABS

Tracking	Range-ordered roll call, frequency, and spatial orthogonality	Average interrogation rate 6 sec. No garbling. 3D-positional accuracy: 100-200 ft
	50-to 100-bit data message on every interrogation	Bit error rate 10^{-6}
Acquisition	21-beam position, 4 (DPSK) acquisition mode addresses	Reduces probability of garbling
	Barker and Neuman-Hofman codes	Reduces probability that garbling results in errors
	Acquisition occurs if correct address detected on only two of seven satellites	Aircraft acquired in 6-sec period
Startup	21-beam position, 1000 (DPSK) startup addresses	Reduce probability of garbling
	Barker and Neuman-Hofman codes	Reduces probability that garbling results in errors
	Aircraft acquired if correct address detected on but two satellites	Startup completed in 2-6 3/4 min (DPSK)
Jamming	Existence of baseline DABS	Reduces problem of jamming
	Jamming but one satellite results in jammer's position being determined within 2 min	Reduces the probability of being jammed and the effect of jamming
	The need to jam several satellites to be effective	
	Multiple beams	
	Successful jamming of all but three of four satellites forces aircraft in jammed areas into a jamming mode	Data link from the ground to the aircraft continues. No data link from the aircraft to the satellite

modes are tracking, acquisition, startup, and jamming. The remainder of this section describes the signal formats used for the tracking and acquisition modes.

Tracking Mode

Each of three S/D satellites located in a synchronous equatorial orbit, as shown in Fig. 5, has responsibility for 1/3 of CONUS. The transmitting satellites use 30-ft-diam antennas, one common frequency channel (at L band), and have beam agility (seven beam positions). Thus, CONUS is divided into 21 areas of coverage, as described in Fig. 5. Satellite transmissions are controlled by the satellite control center, with transmissions from the ground to the satellite made at C-band.

Once acquired, aircraft are interrogated by an L-band transmitting satellite one at a time in a sequential manner. Thus no more than three different aircraft are interrogated at any instant of time in the tracking mode throughout CONUS. Tracking mode interrogations are grouped into approximately 1/8-sec message sets. Tracking mode interrogations can be either 75 or 150 μsec long, and each message set can contain a mix of short and long messages. The complete message set normally will be transmitted in one beam position. The first

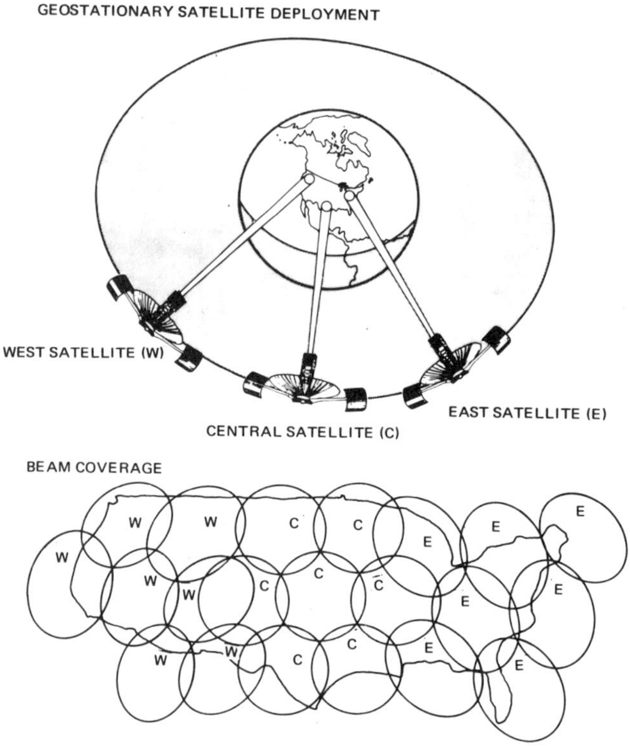

Fig. 5 ASTRO-DABS geostationary satellite deployment and beam coverage.

message of the set, 75 μsec long, is a general synchronizing message (bit, word, mode) to all aircraft within a beam. The remaining messages of the set are individually aircraft-addressed, with gaps left between messages equal to the length of the previously transmitted message. A tracking mode message set can contain up to 832 discretely addressed messages. Data and address bits are 1.5 μsec long, with each bit expanded in bandwidth by a four-chip (3.8 μsec/chip) Barker sequence. This spreading code provides improved timing accuracy and a degree of multipath protection (6 db). The modulation used is differentially coherent phase shift keying (DPSK).

Signal garbling in ASTRO-DABS is eliminated in the tracking mode by the range-ordered polling procedure depicted in Fig. 4. It has been shown[7] that the interrogation period (time between interrogations to the same aircraft) can be bounded between 4.5 and 6.9 sec. (The interrogation procedure and results obtained in Ref. 7 guarantee a sequence of garble-free transmissions from the aircraft to any point in space. Thus the result is independent of the positions of the received satellite positions.) The analysis from which these bounds were derived assumed that the satellite operated at a 50% duty factor, with the time between interrogations being equal to the interrogation time as described in Fig. 4. The interrogation procedure[7] is basically to form a network of hexagonal-cylinder cells (Fig. 6) whose longest dimension d_o is 14 miles; interrogate all aircraft within a given cell in a time-ordered, range-ordered sequence monotonically increasing with respect to the transmitting satellite; leave an appropriate gap and repeat procedure in the nearest occupied cell; sequence through all cells in a given beam in a similar manner; and sequence through all beams, repeating the foregoing procedure until all aircraft in CONUS have been interrogated.

Replies from a given aircraft are received by the transmitting S/D satellite and by at least six S/N satellites. The beam traces from the seven beams of an S/N satellite in an inclined orbit change shape as the satellite's orbit position changes. However, it is believed that East Coast aircraft returns can be kept from garbling West Coast returns by beam separation, without resorting to a third uplink frequency. Two returns of adjacent regions such as from East and Central CONUS within a given beam will not garble, since such returns are on different frequency channels, as illustrated in Fig. 4. The satellite control center knows the order in which aircraft will respond, roughly the time at which they will respond, and which receiver beams (at most three) of each satellite to actuate in order to receive these returns properly. This advantage has been gained by design through the proper ordering of the three S/D satellite interrogation sets. Such knowledge allows the satellite control center to match-filter detect over the 22 address bits of either the 50-bit (75-μsec) or the 100-bit

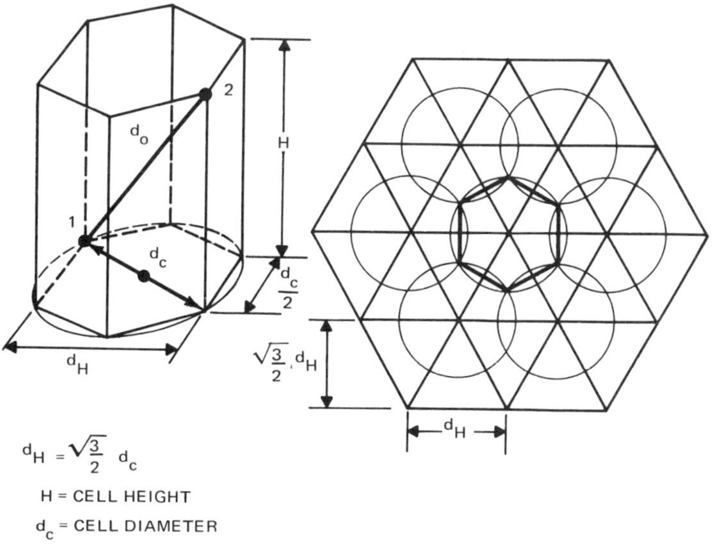

Fig. 6 Interrogation cell configuration.

(150-μsec) transmission to determine range information. This matched filter output also provides bit synchronization, which is used for demodulating the address and data bits of each of the messages. Each bit is 1.5 μsec and, as in the satellite-to-aircraft link, is expanded by a four-chip Barker sequence.

Acquisition Mode†

An aircraft will be acquired by the ASTRO-DABS system while still on the runway. When the equipment is activated on the ground, the aircraft will receive one acquisition interrogation within the next 2 sec and one every 2 sec thereafter until acquired. Acquisition mode transmissions are interleaved with tracking mode transmissions. Nearly all of the time, however, the aircraft will be placed in the tracking mode within a 6-sec period from the time it starts receiving interrogations from the satellite.[9] Acquisition occurs with detection of the correct address on only two of the aircraft-to-SCC links. An indicator light will tell the pilot his aircraft has been acquired. A normal acquisition mode message set is 1/8 sec. Each message set is divided into 1/56-sec intervals. The first message of each interval once again tells aircraft, within a satellite beam, that this subset of the message set is an acquisition mode message set. The second message is 75 μsec long and carries two bits of aircraft acquisition address information and 48 bits of synchronization information. There are no other acquisition transmissions for the remainder of the 4.4-msec period. The reason for the separation of 4.4 msec between acquisition mode addresses is uncertainty in the location of aircraft to be acquired. The 4.4-msec dwell time allows aircraft located anywhere in the beam with a given acquisition address to respond before the next set of addressed aircraft respond. The procedure sequences through acquisition mode addresses in a similar manner and then repeats for each of the other seven beam positions.

Acquisition mode returns contain only the unknown 20 bits of address on the return link, which is the only information required by satellite control to put the aircraft into the tracking mode. An aircraft acquisition message return is 165 3/4 μsec. The first 68 1/4 μsec provides a strong bit-synchronization signal. Such a signal is necessary, since in the acquisition mode system self-interference can be reduced statistically but not completely or absolutely avoided (as it can be for the tracking mode). A 13-chip Barker sequence (3/8 μsec/chip) spreads the signal bandwidth, providing better

†This discussion does not include the impact of ASTRO-NAV on acquisition (see Sec. V).

range information, bit synchronization, and garbling resolution. The Barker sequence, which is used to spread the 14 timing bits of the synchronization preamble, together with a Newman and Hofman code used to code the timing bits, allows resolution to within 1/13 of a transmitted timing bit. The code structure results in low probabilities of false detection and errors due to garbling. (The combined use of the Barker and Newman-Hofman codes is not considered optimum. Preliminary analysis, however, shows that the combined code has good auto- and cross-correlation properties and will work. A study of other synchronization codes will be performed at a later time.) Since the SCC does not know whose return is being demodulated from each of the satellite links, each satellite return must be sufficiently strong to enable this information to be demodulated reliably. Each bit of information in this mode is 4 7/8 μsec long or 5.1 db stronger than a tracking mode data bit. This is sufficient to enable reliable address identification to be made on a bit-by-bit and satellite-by-satellite basis.

Garbling can be resolved only to within 3/8 μsec (Barker code chip width). Thus, it would appear that, if two or more aircraft with the same acquisition address were separated by less than 375 ft and both were ready for takeoff, neither would be acquired by ASTRO-DABS. To resolve this problem, the following technique is used. Upon receipt of the satellite's interrogation with the aircraft's acquisition address, the aircraft delays its response by as little as 3/8 μsec and as much as 3 μsec. There are eight possible delays, in increments of 3/8 μsec, from which the selection is made. The choise is dictated by the decimal value of the first three bits of the aircraft's tracking mode address. Thus, if the first three numbers were 010, the delay used would be 3/4 μsec. If an aircraft is not acquired within 6 sec, the next three bits of the aircraft's tracking mode address are used to determine the delay before transmission. This procedure is repeated until the aircraft is acquired. Two aircraft with different delays and separated by less than 375 ft therefore will be acquired. Since the delay is determined by the tracking mode address, which is unique for each aircraft, the delay selection is guaranteed to separate the acquisition mode replies within a short time.

V. ASTRO-NAV

In ASTRO-DABS, the aircraft surveillance operation alternates between an acquisition mode (for aircraft newly entering the system) and a tracking mode (for aircraft already in the system). Acquisition mode transmissions occur every 2 sec over a 1/8-sec interval that includes considerable time gaps between transmissions. ASTRO-NAV utilizes this gap time for navigation data transmissions plus an additional 1/8 sec, during which navigation timing pulses are transmitted. The interleaved transmission sequence is shown in Fig. 7. In the first 1/8 sec, the three S/D geostationary satellites (S1, S2, S3) transmit acquisition and navigation data sequentially over each of their seven beams, preceded by appropriate preambles. The navigation data pulses (ND) contain satellite ephemerides, local time delays, altitude corrections, and localized navigation algorithm coefficients that can significantly reduce onboard computations. The next 1.8 sec is used to transmit up to seven timing pulse pairs (TP) throughout CONUS (satellites: α, β, γ, δ, ε, ζ, η). [The geostationary satellite pairs (α, S2; β, S1; γ, S3) in Fig. 3 may actually be single satellites with dual antenna functions (multiple-beam and CONUS-coverage).] The navigation preamble pulses (NP) alert aircraft in respective beams to forthcoming navigation data and provide clock references.

Since timing pulse transmissions are CONUS-coverage and utilize the same frequency channel as other transmissions, suitable gaps (guard bands) must be introduced to preclude garbling anywhere in CONUS. Larger guard bands are allotted for timing pulses from the moving satellites to allow for the greater range in propagation times. It has been shown[2] that the time allocations in Fig. 7 are sufficient to preclude garbling in CONUS. The two timing pulses transmitted in time sequence are sent over separate frequency channels. The first pulse is transmitted over the normal satellite-to-aircraft frequency channel and is used by both aircraft and calibration stations. The second timing pulse is transmitted via one of the aircraft-to-satellite surveillance channels and need be used only by the calibration stations. The second transmission will not interfere with aircraft-to-satellite transmissions, since during the navigation mode of ASTRO-DABS there are no aircraft-to-satellite transmissions. The dual frequency transmission would be used by the calibration stations to correct for ionospheric propagation delay. Note that this information would be used for surveillance as well as navigation. (The use of dual frequency transmissions in resolving propagation anomalies is part of a larger analytic study on the impact of the calibration network on surveillance and navigation positional accuracy which is now being conducted.) To minimize chances of multipath problems in ASTRO-NAV, no transmissions are used by aircraft from satellites with local elevation angles less than 19°. Also, moving satellites below the equator do not transmit timing pulses. Hence, in some parts of CONUS only six satellite timing pulses are used, and, at some instances of time, only six satellites will transmit timing pulses. The aircraft

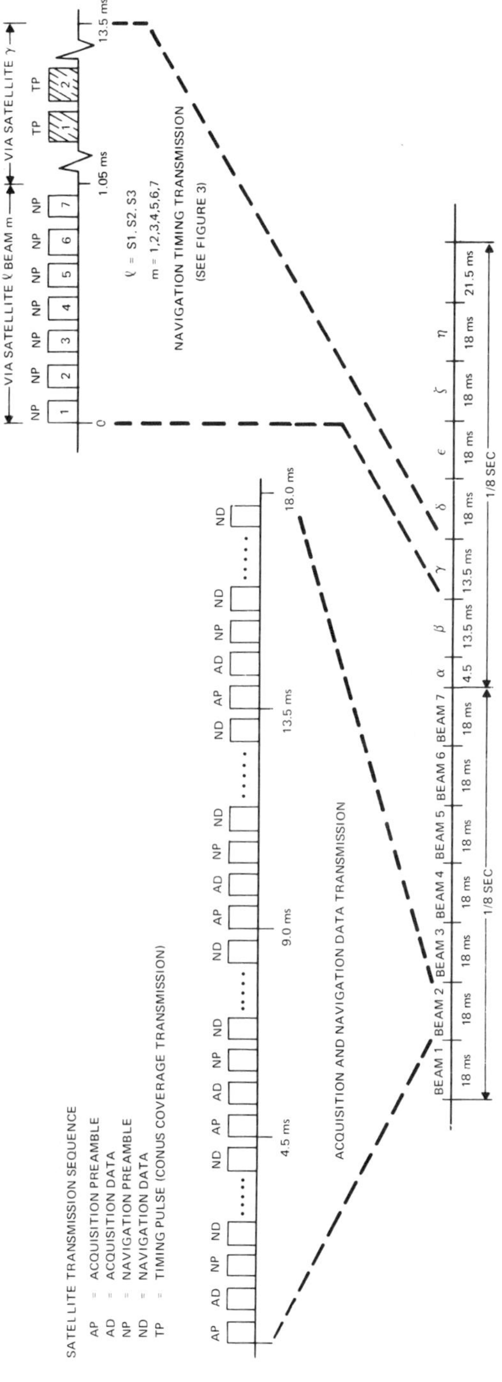

Fig. 7 ASTRO-NAV.

computer is informed as to which transmissions to use as a part of the navigation data sent by the appropriate geostationary satellite.

VI. Performance

In this section, communication, surveillance, and navigation performance are discussed.

Communications

In the previous sections, several different modes of system operation were discussed. It was stated that, for the satellite-to-aircraft link, bit and word synchronization were derived by the use of a preamble to a message set. The aircraft does not know exactly when the preamble is coming, so that it has to make a decision between a message or no message. This implies that the preamble message has to have a large energy-to-noise spectral density ratio. Once a preamble message has been detected, the system has bit and word synchronization. The detection of an addressed message then can be made efficiently on a bit-by-bit basis with relatively little power, using differentially coherent phase shift keyed modulation (DPSK).

In Table 3, the energy budget for the satellite-to-aircraft links is presented. The preamble synchronization pulse arrives with an $E/N_o = 33$ db, strong enough to insure essentially errorless settings for data demodulation. After the 60-μsec synchronization signal is completed, 15 μsec are available for 10 bits of information (to accomplish such things as mode control, etc.). Each of these bits is 1.5 μsec long, so that the E/N_o ratio is reduced to 17.5 db. However, with the accurate timing information available, together with the efficiency of DPSK modulation, an E/N_o of only 11 db is required for a 10^{-6}-bit error rate. Addressed messages within the message set are detected on a bit-by-bit basis, with bit and word synchronization obtained from the preamble and maintained by decoding the sequence of all addressed messages following the preamble. An aircraft responds, however, only to its own detected addressed message. Thus, the E/N_o per bit is again 17.5 db. Finally, the navigation pulses transmitted by S/N satellites are each 180 μsec long, but because of the S/N CONUS coverage antenna an E/N_o of only 25.5 db is provided at the aircraft. This is, however, sufficient to give a thermal noise range measurement accuracy of 14 ft, with 6 db of margin. A nominal 0-db gain for the aircraft antenna is assumed. It has been shown[8] that this value of antenna gain will be exceeded nearly all of the

Table 3 Satellite-to-aircraft energy budget
(500 w peak/250 w average L-band satellite power)

Gain/loss factors	Preamble (60 μsec), db	Data (1.5 μsec), db	Navigation timing pulse (180 μsec), db
Transmitted energy	- 15	- 31.0 db-j	- 20 db-j
Satellite antenna gain	40.5	+ 40.5	+ 27.5
Off-center beam loss	2.0	- 2.0	- 2.0
Path loss	-188.0	-188.0	-188.0
Aircraft antenna gain[a]	0.0	0.0	0.0
Receiver signal energy (E)	-164.5 db-j	-180.5 db-j	-172.5 db-j
Receiver noise power Density (N_o)	-198.0	-198.0 db-j	-198.0 db-j
Signal energy-to-noise spectral density E/N_o[a]	33.0	17.5	25.5
E/N_o required for a DPSK 10^{-6} bit error rate	...	11.0	...
E/N_o required for a satellite-to-aircraft range uncertainty of 14 ft	...	19.5	...
Margin	...	6.5	6.0

[a]Calibration station antenna gain +3 db and an E/N_o = 28.5 db for the navigation timing pulse.

time with respect to the S/D satellites. A similar study soon will be performed for all of the satellites in ASTRO-DABS.

In Table 4, the aircraft-to-satellite tracking mode link energy budget is presented. With respect to the large S/D satellites, timing information is obtained on the first 22 bits transmitted by the aircraft. This results because the ground processor knows the identities of the sequence of returns and can match-filter to each 22-bit address. Therefore, the first 33 μsec of each message can provide bit synchronization. The noise figure of the satellite receiver is 3 db better than the noise figure of the aircraft receiver, so that essentially the same timing performance is obtained on this return link as compared to the forward link. The data still are demodulated on a bit-by-bit 1.5-μsec basis, so that the return link is 3 db better on the larger satellite. The S/N satellites receive 7 db less signal power than the larger satellites, but an E/N_o of 26.9 db still is sufficiently large to provide excellent timing data for both surveillance and the data link. A reduction of 7 db in E/N_o is significant in the data link, but, if

Table 4 Aircraft-to-satellite tracking mode energy budget (500-w peak)

Gain/loss factors	S/D satellite surveillance data, db		S/N satellite surveillance data, db	
	(33 μsec)[a]	(1.5 μsec)[b]	(33 μsec)	(1.5 μsec)
Transmitted energy	- 17.6 db-j	- 31.0 db-j	- 17.6 db-j	- 31.0 db-j
Aircraft antenna gain	0	0	0	0
Path loss	-188.0	-188.0	-188.0	-188.0
Off-center beam loss	- 2.0	- 2.0	- 3.0	- 3.0
Satellite antenna gain	+ 40.5	+ 40.5	+ 34.5	+ 34.5
Received signal energy	-167.1 db-j	-180.5 db-j	-174.1 db-j	-187.5 db-j
Receiver noise power density	-201 db-j	-201 db-j	-201 db-j	-201 db-j
Signal energy-to-noise spectral density E/N_o	33.9	20.5	26.9	13.5
Diversity gain, data 3 satellites	...	...	...	+ 5
E/N_o required for a 10^{-6}-bit error rate	...	11	...	11
E/N_o required for a aircraft to satellite range uncertainty of				
10 ft	22.4	...	...	...
14 ft	...	...	19.0	...
Margin	11.5	9.5	7.4	7.5

[a]68 1/4 μsec for the acquisition and startup modes, and 165 μsec for the jamming mode.

[b]4 7/8 μsec for the acquisition and startup modes. No data link for jamming mode.

the S/D satellite data link were inoperative, the S/N satellite signals could be digitally stored and combined to achieve a 4-5-db gain, which would be sufficient to maintain the data link. An acquisition and startup mode analysis was performed[9] under the assumption of 80,000 aircraft Poisson-distributed throughout CONUS, an average flight time of 50 min, and a peaking factor of 10. It was shown that a very crude, worst-case bound resulted in acquisition within a 6-sec period 90% of the time and that startup is achieved in 3 4/5 min with a probability of at least 0.99.

Multipath and System Interference

Multipath tests have been conducted by Boeing for the FAA and by the Canadian Communication Research Center on the ATS 5 satellite. Balloon experiments have been conducted by NASA, TSC, and ESRO.[10-13] These experiments, although not conclusive, tend to lead to the conclusion that multipath, even over water, is basically diffuse in nature and that the multipath signal becomes depolarized after reflection. Thus it would appear that, for CONUS applications, having lower average reflected power than water, sufficient multipath rejection is afforded by a 6-db coding gain and antenna directivity. Thus we can nearly always operate without being concerned with this phenomenon. If, however, it is determined in the forthcoming ATS F/ATC experiments that this assumption is not valid, the signal bandwidth would have to be widened to obtain greater multipath protection.

Position Determination

From Table 3, it can be seen that thermal noise errors will be in the order of 14 ft for navigation and less than this for surveillance. A significant part of the total error,[14] however, may be due to highly correlated bias-like errors. Ground stations at known locations, which receive satellite timing transmissions, can provide calibration data for reducing these consistent errors. The number and location of calibration stations and their effect on positional accuracy is the subject of a study that we are now pursuing. It is estimated that the equivalent ranging error is 40 ft or less for navigation and a little better for surveillance. The maximum geometrical sensitivity factors over CONUS for the satellite constellation depicted in Fig. 3 are an HDOP (horizontal dilution of precision) of 2.5 and a VDOP (vertical dilution of precision) of 3.8. Thus, 1σ errors in altitude of less than 152 ft and horizontal positional error of less than 100 ft are expected to result from the 40-ft range errors.

VII. Work in Progress

In developing the ASTRO-DABS concept it became clear that the relatively high avionics cost, the vulnerability of the aircraft-to-satellite link to jamming, and the lack of an air-to-air collision avoidance system (CAS) were offsetting the advantages gained from universal coverage and the higher accuracies provided by this satellite ATC concept. Therefore, we have been working on modifications to the concept which will reduce avionics costs, significantly improve the system's performance in a jamming environment, and extend the concept to include an air-to-air CAS. This work will be reported on at a future AIAA Communications Satellite Systems Conference.

VIII. Summary

ASTRO-DABS provides universal surveillance, navigation, and communication via 13 satellites (four large surveillance/ data-link satellites and nine small surveillance/navigation satellites). Its key features are the following. It utilizes a sequential message interrogation scheduling algorithm, together with two aircraft-to-satellite channels on different frequencies, to guarantee a negligible level of mutual interference. Geographic localization, a "burn-through" mode of operation, and integration with a parallel ground system (DABS) in high-density areas provide an important degree of survivability in the event of unintentional or malicious jamming. It is structured to handle 70,000 airborne aircraft with three channels of 5 2/3 MHz bandwidth. This is expandable to 140,000 airborne aircraft with an additional two channels. ASTRO-DABS coexists with DABS and has as a design goal the development of an inexpensive avionics package for the small aircraft user. ASTRO-NAV, the navigation concept of ASTRO-DABS, utilizes a network of calibration stations together with the ASTRO-DABS data link to process a significant amount of navigation information on the ground so that the size and cost of the airborne navigation computer can be minimized. ASTRO-NAV is designed to operate independently of the ASTRO-DABS surveillance interrogation and is immune to uplink L-band jamming.

References

[1]Schuchman, L., "The ASTRO-DABS Concept," MTR-6287, Jan. 1973, MITRE Corp., McLean, Va.

[2]Elrod, B. D. and Fee, J. J., "Preliminary Analysis of the ASTRO-NAV Concept," MTR-6478, Oct. 1973, MITRE Corp., McLean, Va.

[3]Harris, R., "Advanced Air Traffic Management System Summary Report," MTR-6419, Ser. 1, Rev. 1, Jan. 1974, MITRE Corp., McLean, Va.

[4]Elrod, B. D. and Fee, J. J., "Design Considerations for ASTRO-DABS Receiver Satellites," MTR-6453, July 1973, MITRE Corp., McLean, Va.

[5]Fee, J. J., "A Receiver Satellite Constellation for ASTRO-DABS," MTR-6346, March 1973, MITRE Corp., McLean, Va.

[6]Ebert, P. M., "Geometrical Interpretation of GDOP," MTR-6477, July 1973, MITRE Corp., McLean, Va.

[7]Elrod, B. D., "Aircraft Interrogation Scheduling with ASTRO-DABS," MTR-6368, March 1973, MITRE Corp., McLean, Va.

[8]Ekstrom, J. L., "Evaluation of ASQRO-DABS Performance While An Aircraft Turns," MTR-6393, April 1973, MITRE Corp., McLean, Va.

[9]Avant, A. L., "ASTRO-DABS Startup and Acquisition Performance," MTR-6475, July 1973, MITRE Corp., McLean, Va.

[10]"ATS-5 Multipath/Ranging Experimental Program, Interim Results," DG-60149, Dec. 1971, Boeing, Seattle, Wash.

[11]Chinick, J. and Mutt, E., "Spread Spectrum Measurements of Multipath Channel Characterization in a Satellite Mobile Communication System," Astro Panel Session, Nov.-Dec. 1971, Canadian Communication Research Center, Ottawa, Canada.

[12]Wishner, S., Hamby, C., and Reed, D., "Final Report, ESRO ATC Balloon—Aircraft Ranging Data and Voice Experiment," TR X-752-72, 136, May 1972, NASA.

[13]"Final Report, ESRO ATC Balloon—Aircraft Satellite Simulation Experiments, 1971," Vols. I and II, March 1972, NASA.

[14]Fee, J. J., "Techniques for Reducing Ionospheric Propagation Delay Error in Satellite Surveillance and Navigation Systems," MTR-6637, April 1974, MITRE Corp., McLean, Va.

A GROUND SIMULATION OF A SATELLITE NAVIGATION SYSTEM

John J. Courtney Jr.* and Ivan Kadar†

Grumman Aerospace Corporation, Bethpage, N.Y.

Abstract

A system has been suggested for passive position and velocity determination of a user by simultaneously receiving pseudo-noise biphase modulated signals from a constellation of at least four satellites operating at L band. Demonstration flight tests of this system in area navigation and instrument landing system configurations were conducted over a software synchronized ground-based constellation of transmitters. In-flight recorded measurements via a time-varying propagation channel provided navigational solutions for post-flight performance evaluation. The basic navigational and communications systems aspects, test configurations, and test results are presented, verifying the practicality and soundness of the system concepts.

Introduction

A system has been suggested for passive position and velocity determination of a user by simultaneously receiving pseudo-noise biphase modulated code division multiplexed signals from a constellation of at least four satellites operating at L band.[1] The evolution of the system concepts for passive three-dimensional position and velocity determination of a user has gone through a number of proposals: 1) synchronous satellite constellations using five satellites in an X configuration; 2) four synchronous satellites in a "rotating-Y" constellation and a worldwide 20-satellite eggbeater

---

Presented as Paper 74-426 at the AIAA 5th Communications Satellite Systems Conference, Los Angeles, Calif., April 22-24, 1974.

*Manager, Navigation Systems.

†Staff Technical Specialist.

configuration using pseudo-range solutions; and 3) a constellation of medium-altitude satellites resulting in the Global Positioning System (GPS) (NAVSTAR) system utilizing pseudo-ranging algorithms.[2-7]

Analytical studies and computer simulations have demonstrated over the past few years that a system of satellites utilizing L-band carrier transmission with pseudo-noise modulation would provide a degree of navigation precision and multipath rejection not available from any present system. The predicted system accuracy has been substantiated by simulation and by a variety of laboratory correlation receiver breadboard tests.[8] Past studies also have identified a number of system error sources which could affect the system accuracy, such as satellite ephemeris, ionospheric and tropospheric delay uncertainty, signal multipath, receiver bias uncertainty, and noise. This paper highlights the results of a test program of 32 flights aimed primarily at validating, in typical field and flight environments, the analytical estimates of the performance of two four-channel correlation receivers constituting the "user" segment of the system.[9] To achieve this goal, a simulated satellite constellation geometry with typical receiver power profiles was established at White Sands Missile Range (WSMR). Ephemeris and ionosphere errors were not included in the test configuration. The flight profiles flown over the simulated satellite cluster of transmitters represented compressed time flights at representative satellite-to-user look angles. The test system accuracy, including user dynamics in the presence of multipath, was to be demonstrated both in area navigation and instrument landing system (ILS) flight configurations.

The integrity of the tests was to be established by calibration and accurate survey of the location of the transmitting antennas and ground systems and by comparison of the derived trajectories with WSMR-provided best-estimate trajectories (BET's). The WSMR best-estimate trajectory is a position and velocity solution obtained by combining data from optical and rf trackers, and inertial system data. Two correlation receivers were developed, differing in the design details, but both capable of performing the measurements of pseudo-range and range-rate relative to each of the transmitters necessary for the unique determination of user's position and velocity coordinates. In particular, it was important to verify the predicted (both theoretical and laboratory measured) receiver performance parameters in an operational environment with dynamics, low signal levels, and in the presence of multipath.

Review of Passive User Navigation

There are three major approaches to deriving three-dimensional passive position location information from a navigation satellite system: range differencing, pseudo-ranging, and interferometry. The interferometer principle represents an approach differing from the main theme of this paper and will not be considered here, but it has been treated in extant literature.[10]

In range differencing, the user measures the difference in the time of arrival of signals broadcast from a network of time-synchronized satellites. With the satellite ephemerides and system time available to the user, simultaneous measurements from four satellites are necessary to solve for the three-dimensional position and velocity coordinates. Since the user's clock is not synchronized with respect to the satellite's clock, there is an apparent bias between the user and satellite clocks. The clock bias measurement is eliminated in the range-differencing solution, resulting in the intersection of three hyperbolic surfaces which locates position. However, a fifth measurement (i.e., using five satellites) would allow for the solution of bias and bias rate.

In pseudo-ranging, the user also measures the time of arrival of signals simultaneously received from time-synchronized satellites. By performing four simultaneous measurements, the user establishes four independent pseudo-ranges that differ from the geometrical ranges by the difference between the satellite oscillator phase and the user oscillator phase (expressed in range units). The user, knowing the satellite ephemerides, can solve four equations in four unknowns for his position in three dimensions and the common bias of the range measurements. A similar computation algorithm enables the user to determine his three-dimensional velocity coordinates based upon doppler (pseudo-doppler) measurements derived from the navigation signal. Thus, pseudo-ranging is equivalent to range differencing utilizing five satellites. The pseudo-ranging technique, therefore, is preferred in systems having accurate clocks and requiring high-accuracy navigation. This technique was used in the field test.

Measurements

The measurements in this system implementation are performed by a four-channel correlation receiver operating at

L band, receiving signals phase-shift keyed by unique pseudorandom sequences, providing code division multiplexing of signals. Other implementations may use tone ranging with narrow-band phase modulation (PM) or double side-band amplitude modulated (DSBAM) signals utilizing time or frequency division for channel multiplexing.[11,12] The measurement errors introduced into the system are a function of the signal design, the receiver implementation, and the time-varying channel characteristics described by the relative system geometry and by the properties of the propagation medium. The errors affecting range and range-rate measurements include ionospheric and tropospheric uncertainties, multipath, receiver noise, receiver precision, receiver dynamics, and time-varying biases introduced by the receiver. Other system error sources include calibration station location, range measurements, satellite clock drift, and calibration station clock drift.

Only the user-to-satellite line-of-sight components of these errors contribute to "system" errors, with the major contribution being the satellite position errors, ionospheric uncertainties, and residual uncompensated biases. The system implementation, in part consisting of ground-based receivers at known locations and a master calibration station, provides for calibration of many of the error sources associated with the satellites, atmosphere, and clocks, resulting in the computation of error-correction parameters, which are sent to the user over the satellite-to-user ranging link so that he may improve the accuracy of his position and velocity.

Measurement Device

It is important at this point to highlight typical implementation and operation of the measurement device (correlation receiver) used in the tests, since it is its unique implementation that either introduces or eliminates measurement errors that directly affect the accuracy of the navigation solution. However, before one can discuss the receiver implementation, one has to define the signal structure, which in conjunction with the line-of-sight (LOS) propagation channel characteristics between transmitter and receiver influences the receiver design. The transmitted signal is a distinctive 25-stage, M-sequence code (truncated to 2^{25}-255 bits) that is combined with a common eight-stage, M-sequence code to biphase modulate an L-band carrier at a code rate of 10 Mbit/sec. The eight-stage, common M-sequence is used as a preamble and is power-boosted to assure sufficient received power required for acquisition. These parameters of the signal, along with the cross-correlation properties of the code and received power levels, set the lower limits on thermal, quantization, multipath, and dynamic error performance of the receiver (see discussion of error sources).

The generic correlation receiver block diagram employing delay lock and carrier tracking loops is shown in Fig. 1 (which may represent one of the "channels" of a four-channel receiver). The delay lock loop tracks the incoming code with respect to its local replica, thereby providing relative estimates of time of arrival (TOA), whereas the carrier loop tracks the reconstructed rf carrier, enabling relative estimation of doppler (range rate). It is this combination of the design of the delay lock and carrier tracking loops that affects the multipath error behavior of the receiver.[13, 14]

In the signal acquisition process, a synchronization signal is received (derived either from the boosted preample using a passive matched filter or search algorithm, or from a separate acquisition signal in the satellite system), initiating a search in the code loop. When the received and replica codes are running in approximate synchronism, there is sufficient reconstructed carrier available for the carrier loop to

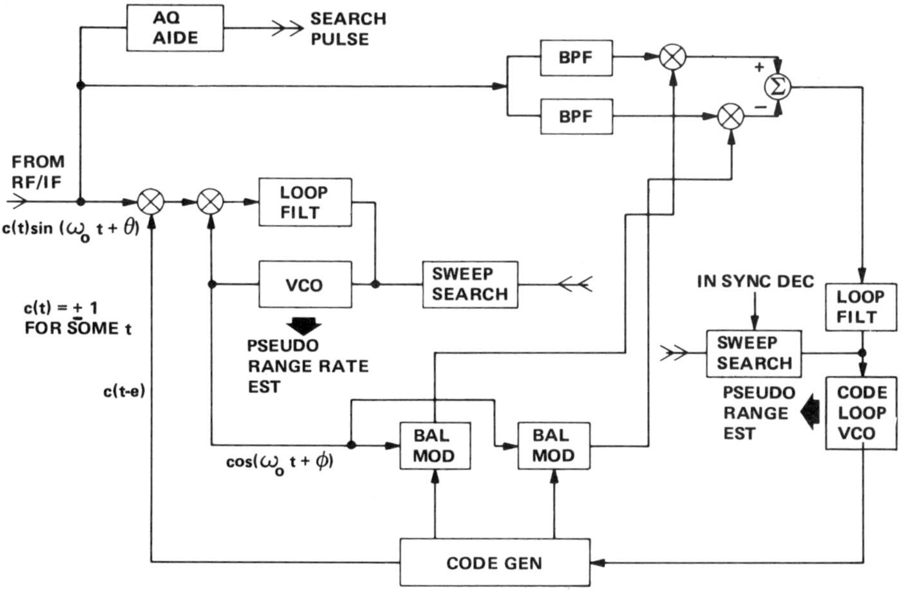

Fig. 1 Generic correlation receiver.

commence search and acquire the carrier, at which time the code loop pulls in and the code search is disabled. Therefore, the carrier tracking loop acquires first, followed by the code loop. This requires that the carrier loop noise bandwidth be wider than the code loop (providing faster acquisition times), which in turn forces the carrier tracking loop to follow all signal dynamic changes and at the same time be narrow enough to satisfy constraints of the code repetition period governing the interference rejection criterion and finally meet the thermal noise range-rate error requirements subject to receiver input power constraints and receiver noise figure. The code loop, on the other hand, has to be narrow enough to provide tracking of the signal and to meet thermal noise range error requirements at the specified sensitivity level.

The loop designs are complicated further by specifying the dynamic input changes they have to follow without losing lock at a given signal level (with specified duration of dynamics) with either a specified or zero steady-state error. This could require the loops to be of third order to track acceleration and to maintain lock under acceleration rate of the LOS dynamics introduced by a maneuvering aircraft. If the loop design is analog, the resultant error will grow with time for a constant dynamic input (i.e., acceleration rate). Digital (discrete) loop implementations, however, using recursive digital filters and shift register memories, replace the analog nonideal components (integrators) and have zero growth of the steady-state error for constant input dynamics. Furthermore, the digital processing lends itself to time-shared processor implementations reducing complexity and size, and it is easily interfaced with digital logic used for range and range-rate tallies.

Another aspect of the receiver implementations is the use of automatic gain control (AGC) to keep the loop noise bandwidths independent of input signal level vs the use of adaptive bandwidth approach. These features (along with the previously discussed features) were tested, and the receiver with AGC was seen to provide some discrimination against scintillating multipath, whereas the adaptive bandwith receiver could tolerate increased LOS dynamics with strong signal levels.[9] Equally important design features include schemes for calibrating the adjacent channels for relative time delay, by self-test or software methods, and control of asymmetries of the received and locally generated signals prior to range processing to reduce measurement bias errors, which were found to be very significant. The receivers utilized on the test program embodied most of the aforementioned concepts and design features by having both analog, Hazeltine Corporation (HC), and digital, Magnavox Research Laboratories (MRL), implementations available.

Theory

To determine his position and velocity, a user cross-correlates pseudo-random (PRN) codes (locally generated in the receiver) with the code-division-multiplexed navigation signals simultaneously emitted at the same carrier frequency by the satellites. The relative phase or equivalent time displacement between his local code and the incoming code is measured, allowing determination of the relative range. Simultaneous with range measurements, the user is measuring relative range rate in each channel from the doppler on the reconstructed rf carrier frequency. This information is sufficient for the calculation of user position and velocity, given the satellites' (R_i) ephemerides and the time of day of the satellites (GMT). The system provides these data as each satellite transmits its own ephemerides and timing errors as part of the navigation signal.

For this discussion, assume that Earth-centered relative (ECR) coordinates will be the basic computational reference system (see Fig. 2).[4,9] The pseudo-range from the ith satellite to the user is (i.e., true range plus bias)

$$\rho_i = \sqrt{(\bar{R}_i - \bar{R}_\mu) \cdot (\bar{R}_i - \bar{R}_\mu)} + b_\mu + b_i \qquad (1a)$$

or

$$\rho_i = \bar{\epsilon}_i \cdot (\bar{R}_i - \bar{R}_\mu) + b_\mu + b_i \qquad i = 1, \ldots, 4, \ldots, N \quad (1b)$$

$$\bar{\epsilon}_i = (\bar{R}_i - \bar{R}_\mu)/(\rho_i - b_i - b_\mu) \qquad (1c)$$

where b_u is the user oscillator phase bias (feet) to be determined by user, and b_i is satellite oscillator phase bias (feet) transmitted to user. In the same fashion, the pseudo-range rate from the ith satellite to the user is

$$\dot{\rho}_i = \bar{\epsilon}_i \cdot (\dot{\bar{R}}_i - \dot{\bar{R}}_\mu) + \dot{b}_\mu + \dot{b}_i \qquad i = 1, \ldots, 4, \ldots, N \quad (2)$$

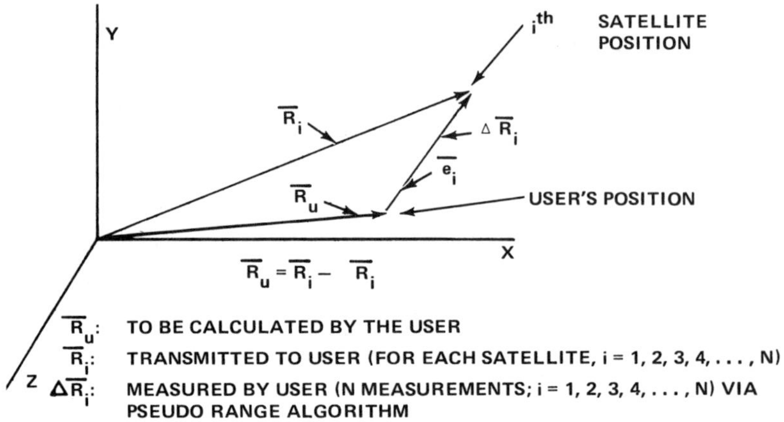

Fig. 2 ECR coordinates.

where b_u is the user oscillator frequency bias (feet per second) to be determined by user, and b_i is satellite oscillator frequency bias (feet per second) transmitted to user. Rearranging the preceding equations,

$$\bar{\epsilon}_i \circ \bar{R}_\mu - b_\mu = \bar{\epsilon}_i \cdot \bar{R}_i + b_i - \rho_i \tag{3}$$

$$\bar{\epsilon}_i \circ \dot{\bar{R}}_\mu - \dot{b}_\mu = \bar{\epsilon}_i \cdot \dot{\bar{R}}_i + \dot{b}_i - \dot{\rho}_i \tag{4}$$

The matrix equation for the four satellites becomes

$$\underset{(4 \times 4)}{\begin{bmatrix} \bar{\epsilon}_1^T & -1 \\ \bar{\epsilon}_2^T & -1 \\ \bar{\epsilon}_3^T & -1 \\ \bar{\epsilon}_4^T & -1 \end{bmatrix}} \underset{(4 \times 1)}{\begin{bmatrix} R_\mu \\ b_\mu \end{bmatrix}} = \underset{(4 \times 16)}{\begin{bmatrix} \bar{\epsilon}_1^T \; 1 & 0000 & 0000 & 0000 \\ 0000 & \bar{\epsilon}_2^T \; 1 & 0000 & 0000 \\ 0000 & 0000 & \bar{\epsilon}_3^T \; 1 & 0000 \\ 0000 & 0000 & 0000 & \bar{\epsilon}_4^T \; 1 \end{bmatrix}} \underset{(16 \times 1)}{\begin{bmatrix} R_1 \\ b_1 \\ R_2 \\ b_2 \\ R_3 \\ b_3 \\ R_4 \\ b_4 \end{bmatrix}} - \underset{(4 \times 1)}{\begin{bmatrix} \rho_1 \\ \rho_2 \\ \rho_3 \\ \rho_4 \end{bmatrix}} \tag{5}$$

$$G_\mu \overline{X}_\mu = B_\mu \overline{X}_s - \overline{\rho} \tag{6}$$

where the direction cosine matrices G_u and B_u are known to sufficient accuracy. The system measurements become

$$\overline{\rho} = B_\mu \overline{X}_s - G_\mu \overline{X}_\mu \tag{7}$$

$$\overline{\dot{\rho}} = B_\mu \overline{\dot{X}} - G_\mu \overline{\dot{X}}_\mu \tag{8}$$

with the resulting measurement error equations

$$\delta\overline{\rho} = B_\mu\, \delta\overline{X}_s - G_\mu\, \delta\overline{X}_\mu + \overline{d} + \overline{n}_1 \tag{9}$$

$$\delta\overline{\dot{\rho}} = B_\mu\, \delta\overline{\dot{X}}_s - G_\mu\, \delta\overline{\dot{X}}_\mu + \overline{\dot{d}} + \overline{n}_2 \tag{10}$$

where $B_\mu \delta\overline{X}_s$ and $B_\mu \delta\overline{\dot{X}}_s$ are the position and velocity ± clock errors due to satellite orbit determination and satellite oscillator errors in the line-of-sight directions; $G_\mu \delta\overline{X}_\mu$ and $G_\mu \delta\overline{\dot{X}}_\mu$ are the projection of the error vector due to user position and velocity error together with user oscillator errors in the line-of-sight directions; $\overline{d}$ and $\overline{\dot{d}}$ are the measurement error due to environment (ionosphere signal delay correction, multipath delay, troposphere signal delay correction, and vehicle dynamics); and n_1 and n_2 are the measurement error due to uncorrelated receiver noise.

In reference to the Earth-centered coordinates given in Fig. 2, the user, by performing four simultaneous pseudo-range ρ_i and pseudo-range-rate $\dot{\rho}_i$ measurements, can compute his three-dimensional position and velocity coordinates by the solution of Eqs. (1) and (2) for $\overline{R}_u$ and $\overline{\dot{R}}_u$, respectively.

Position Solution

Starting with Eq. (1), one can write, defining $b = b_u + b_i$,

$$\bar{R}_i \cdot \bar{R}_\mu = -\frac{(\rho_i - b)^2}{2} + \frac{\bar{R}_i \cdot \bar{R}_i + \bar{R}_\mu \cdot \bar{R}_\mu}{2} \tag{11}$$

$$\bar{f} = \mathrm{col}\left[R_{\mu x}, R_{\mu y}, R_{\mu z}, b\right]$$

Subtracting pairs of equations for i = 1, 2, 3, 4, one can define a vector f in terms of the unknowns as

$$D\bar{f} = C \tag{12}$$

where

$$D = \begin{bmatrix} (\bar{R}_2-\bar{R}_1)^T(\rho_1-\rho_2) \\ (\bar{R}_3-\bar{R}_2)^T(\rho_2-\rho_3) \\ (\bar{R}_4-\bar{R}_3)^T(\rho_3-\rho_4) \end{bmatrix}, \quad C = \frac{1}{2}\begin{bmatrix} \rho_1^2-\rho_2^2+\bar{R}_2\cdot\bar{R}_2-\bar{R}_1\cdot\bar{R}_1 \\ \rho_2^2-\rho_3^2+\bar{R}_3\cdot\bar{R}_3-\bar{R}_2\cdot\bar{R}_2 \\ \rho_3^2-\rho_4^2+\bar{R}_4\cdot\bar{R}_4-\bar{R}_3\cdot\bar{R}_3 \end{bmatrix}$$

Equation (12) is solved satisfying the constraint [from Eq. (11) by selecting i = 4 for reference, i.e., the fourth transmitter]

$$\bar{R}_4 \cdot \bar{R}_\mu = -\frac{(\rho_4 - b)^2}{2} + \frac{\bar{R}_4 \cdot \bar{R}_4 + \bar{R}_\mu \cdot \bar{R}_\mu}{2} \tag{13}$$

This solution, satisfying the constraint, results in a quadratic. The appropriate roots are selected based upon physical realizability.[9]

Velocity Solution

The velocity solution is obtained from Eq. (2) containing a set of four linear equations in four unknowns:

$$\bar{\epsilon}_i \cdot (\dot{\bar{R}}_i - \dot{\bar{R}}_\mu) = -\dot{\rho} + \dot{b} \tag{14}$$

Let

$$\bar{\dot{f}} = \text{col}\left[\dot{R}_{\mu x}, \dot{R}_{\mu y}, \dot{R}_{\mu z}, b\right] \tag{15}$$

where

$$M\bar{\dot{f}} = I\dot{\rho}_i + A \tag{16}$$

$$M = \begin{bmatrix} \epsilon_{11} & \epsilon_{12} & \epsilon_{13} & \epsilon_{14} \\ \epsilon_{21} & \epsilon_{22} & \epsilon_{23} & \epsilon_{24} \\ \epsilon_{31} & \epsilon_{32} & \epsilon_{33} & \epsilon_{34} \\ \epsilon_{41} & \epsilon_{42} & \epsilon_{43} & \epsilon_{44} \end{bmatrix}, \qquad A = \begin{bmatrix} \bar{\epsilon}_1^T \cdot \bar{\dot{R}}_1 \\ \bar{\epsilon}_2^T \cdot \bar{\dot{R}}_2 \\ \bar{\epsilon}_3^T \cdot \bar{\dot{R}}_3 \\ \bar{\epsilon}_4^T \cdot \bar{\dot{R}}_4 \end{bmatrix}$$

$$\dot{f} = M^{-1}\left[I\dot{\rho}_i + A\right] \tag{17}$$

provided that M is nonsingular.

Static Geometric Dilution of Precision

The apparent time of arrival and apparent doppler measurements are corrupted by errors caused both by the propagation medium and by the receiver error sources, which are a strong function of the signal design. Starting with noisy measurements, the solution algorithms for position and velocity in three dimensions (based upon pseudo-range and pseudo-doppler) will result in noisy position and velocity solutions. The position and velocity solutions in three dimensions obtained directly by linear transformations of pseudo-range and pseudo-range-rate measurements, respectively, are called the "static" position and velocity solutions.[9]

The sensitivity of the position solution to unity uncorrelated measurement errors as a function of the linear transformation (a function of the relative position of the user with respect to the satellites) is a measure of the reduction (or dilution) of the system precision. The covariance of the transformation to unity measurement error variances is called the geometric dilution of precision (GDOP), which is a measure of the degree to which measurement errors are amplified by system geometry effects to produce position errors. The sensitivity equations for position follow from linearizing Eq. (1) with $\delta \bar{R}_i = \bar{0}$ as

$$\bar{\epsilon}_i \cdot \delta \bar{R}_\mu + \delta b = \delta \rho_i \qquad i = 1,2,3,4 \tag{18}$$

It is easy to show from Eq. (1) by covariance operations with $E\{\delta\bar{\rho}_i\,\delta\bar{\rho}_i^T\} = I$ (identity matrix) that the static GDOP's are given by[9]

$$\text{GDOP}\begin{bmatrix} X \\ Y \\ Z \\ b \end{bmatrix} = \left\{ \text{diagonal} \left[G^{-1}(G^{-1})^T \right] \right\}^{1/2} \tag{19}$$

where G is a (4 x 4) direction cosine matrix.

Experiment Program Implementation

The principal elements of the field test navigation system consisted of the ground transmitters, calibration receiver segment, and the airborne receiver (user) segment. Both the ground and airborne segments utilized four-channel correlation receivers to extract apparent time of arrival and doppler history of the incoming signals. Each ground transmitter generates a navigation signal consisting of a PRN code modulated on an L-band rf carrier, as discussed previously. The test program was designed with certain constraints with respect to the rotating-Y satellite system concept in order to keep the overall execution time of the test program within reasonable bounds. The prime areas where constraints were applied are the following:

1) The signal did not contain "satellite" or time information but consisted only of the PRN-modulated range code signal.

2) Hardware synchronization of the transmitters was not attempted. Instead, a time history of the transmitted signals was recorded, and synchronization of the data was accomplished postflight via software.

3) The satellite environment was simulated by scaling the distance down and inverting the system. The test range space available at WSMR and the flight envelope of the test aircraft limited realistic simulation and data collection regions to narrow corridors approximately 6 naut miles long by 2 naut miles wide for area navigation tests. The selected configuration is given in Figs. 3 and 4. The static GDOP's for the area navigation geometry at 30,000 ft above ground level (AGL) are shown in Fig. 5.

The ILS site configuration was developed with the following concepts in mind: 1) minimize altitude error at flare, 2) minimize sink rate error between flare and touchdown, and 3) maintain reasonably good horizontal accuracy.

It can be shown that three ground transmitters forming an approximately symmetrical Y about the aircraft flare point would provide minimum horizontal GDOP, and a fourth transmitter located directly above the flare point would provide the required Z-axis accuracy.[8,9] In practice, the fourth transmitter was 1000 ft horizontally from the flare point in order to keep the balloon tethering cable clear of the aircraft flight path zone, as shown in the site configuration (Fig. 6). This did not affect the Z-axis GDOP appreciably (see Fig. 7).

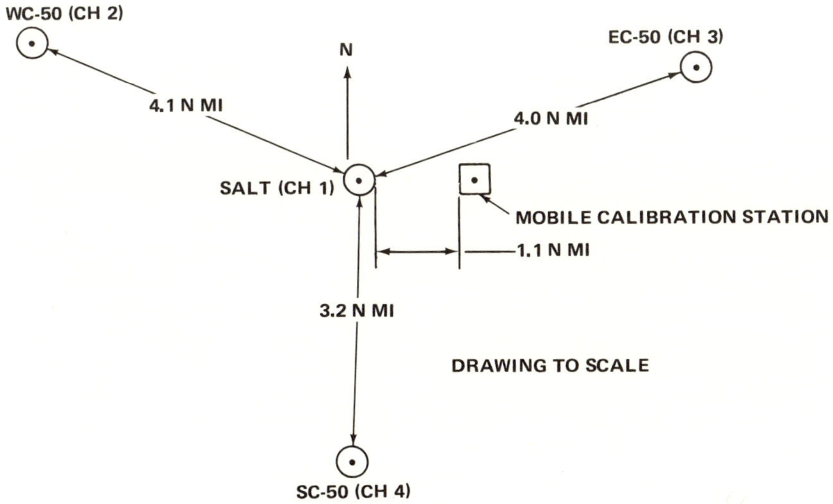

Fig. 3 Sites selected to simulate a rotating Y satellite configuration.

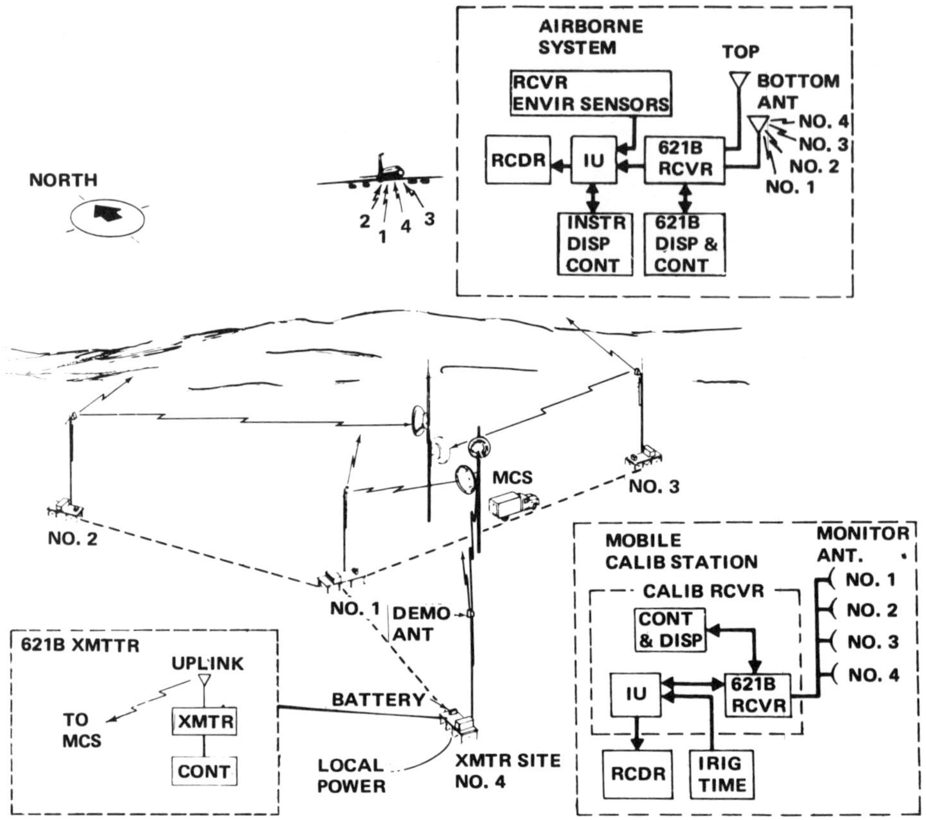

Fig. 4 System demonstration configuration.

Software Synchronization

In a satellite system, the signals emanating from each transmitter would be synchronized approximately in both code phase (time) and frequency to a reference oscillator in a master ground station. Synchronization errors with respect to the reference oscillator would be measured and made available to the system user, so that the user would be, in effect, viewing "perfectly" synchronized transmitters. In the field test program at WSMR, this same general concept was implemented but, of course, mechanized quite differently than in a satellite system. Since there was no requirement to compute the navigation solution by the user in real time, a signal synchronization scheme using postflight data reduction software was implemented. The scheme basically consists of measuring the code phase and oscillator frequency of each transmitter with respect to a reference oscillator at a

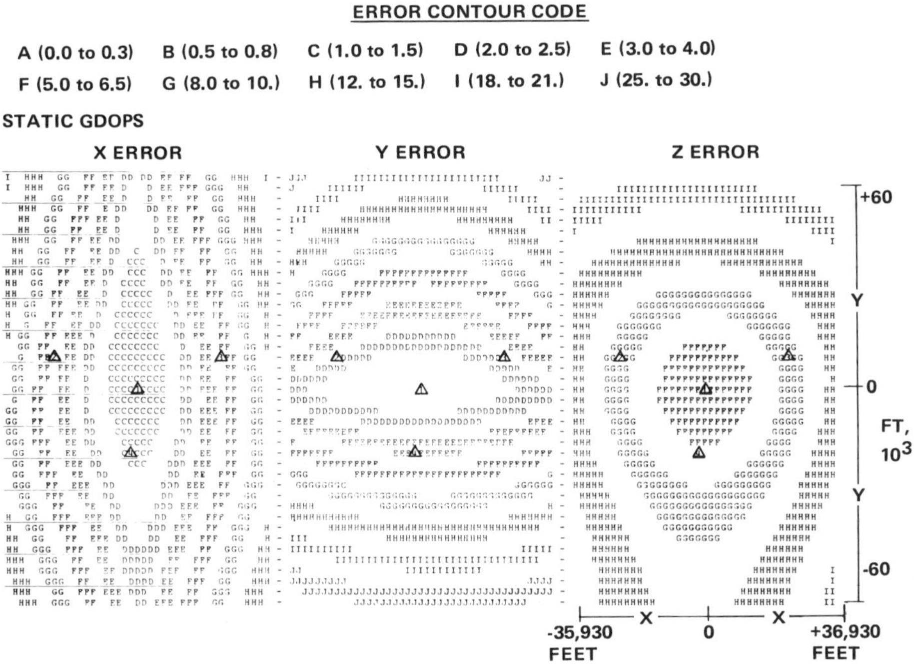

Fig. 5 Static (static solution) geometric dilution of precision contours for area navigation test geometry. Contours shown in X-Y plane at 30,000 ft above ground level.

mobile calibration station (MCS). Since the four transmitters are not even approximately synchronized to a reference oscillator (unlike the satellite case), the user does not require a synchronization error term for each signal as in the operational case; instead, the entire value of the code phase and frequency offsets from the reference oscillator must be provided. The offset information is recorded at the MCS and made available to the postflight data reduction process, in conjunction with the signal history of pseudo-range-rate recorded in flight. The total pseudo-range and pseudo-doppler from the ground and airborne components is thus formed (adjusted to common sample intervals based upon WSMR time) and used to compute user position and velocity in a convenient coordinate frame for WSMR tests.

Site Measurements

Three basic kinds of site measurements were performed: 1) determination of the specific location of the transmitting

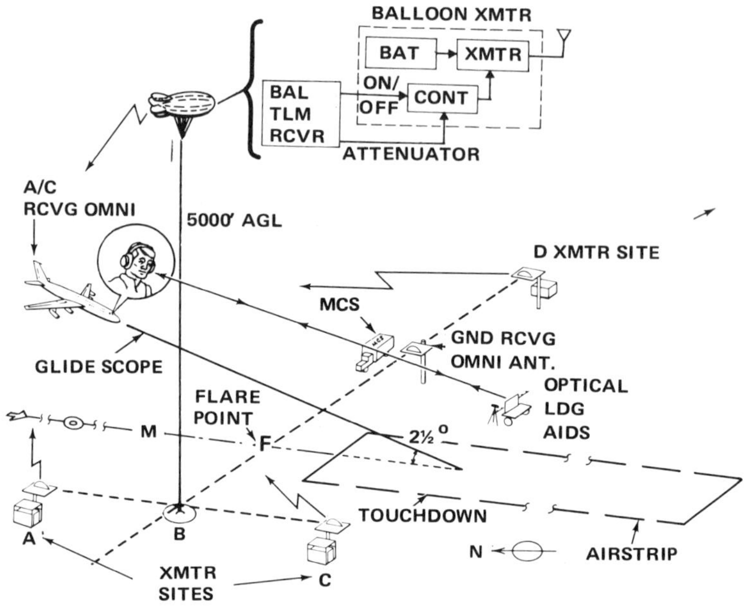

Fig. 6 System demonstration configuration: ILS.

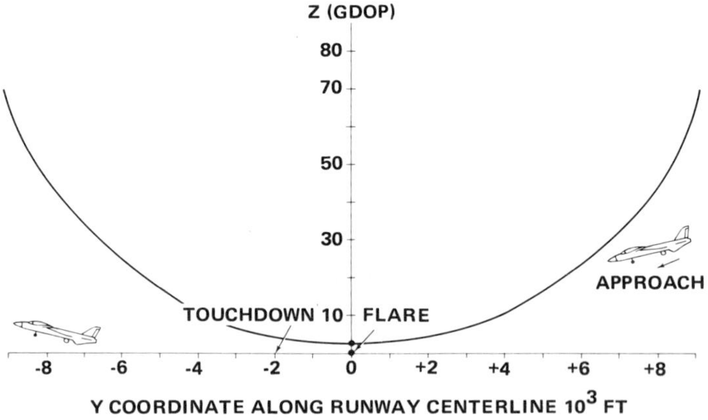

Fig. 7 GDOP contour as a function of position along ILS Y axis.

antennas; 2) signal path time-of-travel measurements; and 3) measurements that insured the integrity of various components. Two of these measurements were necessary elements in the formation of the navigation solution, and the other was necessary to understand certain aspects of the ground system operational characteristics (i.e., error sources).

Frequency Sources

All frequency sources (i.e., transmitters and receivers) utilized high-quality quartz crystal oscillators. The frequency stabilities of the oscillators were in the range of one part in 10^8 to one part in 10^{10}/day.

Flight Profiles

The area navigation test utilized four basic flight profiles for a majority of the tests. The flight paths are summarized below:

1) Flight path A: the A flight path, simulating a rotating 30° Y satellite configuration as seen from the equator, covers a ground track over Salt Site with a heading of 0° or 180° true north, at an altitude of 34,000 ft Mean Sea Level (MSL) straight and level (20-mile path).

2) Flight path B: the B flight path, simulating a rotating 30° Y satellite configuration as seen from the continental United States, covers a ground track east of the transmitter Y configuration.

3) Flight path C: the C flight path covers the same ground track as that of the A flight path but at an altitude of 22,000 ft MSL. The 20-mile flight path has the same terminal points as the A flight path. This flight path represents the GDOP and power profile of an "eggbeater" satellite configuration.

4) Flight path D: the D flight path covers the same ground tract as that of the A flight path but at an altitude of 7000 ft MSL. The 20-mile flight path has the same terminal points and headings as the A flight path. This flight path objective is to exercise dynamics of the system in R, $\dot{R}$, $\ddot{R}$, and $\dddot{R}$. One version of the D flight path is to fly over one channel at 300 ft AGL on the A ground track to measure $\dddot{R}$ lock-up capabilities. Figure 8 shows typical flight profiles flown in area navigation tests.

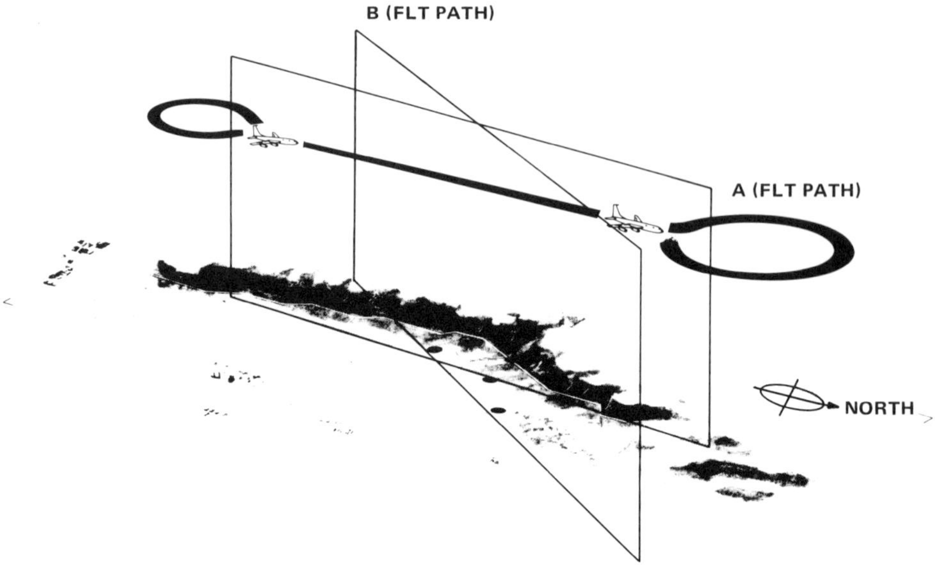

Fig. 8 Typical flight profiles centered at Salt.

The ILS flights were simulated all-weather landing missions with various categories of approach (II-III); see Fig. 6.

Flight Equipment

The airborne equipment for this test consisted of the MRL receiver, HC receiver, Grumman antenna installations, Grumman data interface system, and an instrumentation recording system. The equipment was installed in a NC-135A aircraft. The antenna system was designed to be able to select top or bottom antennas (ILS or area navigation test, respectively) and to insert a fixed amount of attenuation in series with the signal for tests of various received power profiles.

Ground Equipment

The ground equipment can be considered in two groups, one that is used to generate the experiment program data consisting of transmitters, antennas, calibration receivers, and instrumentation, and the other used to evaluate the navigation solution data consisting of WSMR instrumentation hardware.

Transmitters

Four transmitters were provided, one of which has the capability of operating on batteries while carried aloft by a balloon. All transmitters were designed to radiate spread spectrum signals to simulate the type of signal which would be obtained from satellites, as described previously.

Antennas

A primary goal in the system simulation was to obtain a signal profile (i.e., strength and polarization) at the aircraft which would be similar to that received from a satellite. To achieve this goal, the system was configured with a transmitter uplink array antenna with a pattern shaped to give a constant power level for constant-altitude flight paths, a hemispherical coverage omniantenna mounted within a radome at the aircraft and at the balloon transmitter, and parabolic dish antennas on the ground to route the signal from the transmitters to the calibration receiver. The antennas were right-hand circularly polarized (RHCP), except for the parabolic dish, which was linearly polarized. A typical smoothed three-dimensional radiation pattern of the uplink antenna is shown in Fig. 9. Although the pattern shown is geometrically correct, minor lobes and irregularities have been smoothed for clarity. This is especially true for the lower hemisphere.

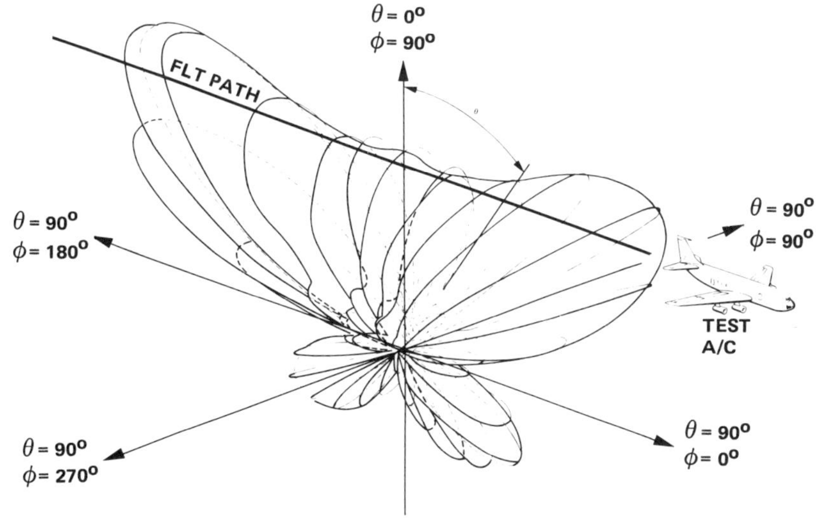

Fig. 9 Typical uplink antenna radiation pattern.

Data Processing

The data collection, reduction, and analysis processes of the flight tests separated into seven distinct parts. Three independent instrumentation systems collect time-tagged data for each of the three receivers and direct it to separate magnetic tapes; the engineering units/validation program (Euval) simultaneously reads the three raw data tapes, reformats the tapes for computer compatibility, validates the time data, and writes a single, chronologically ordered output tape; the navigation and calibration program (Navcal) forms the total user pseudo-range quantities from ground and airborne measurements, performs the deterministic navigation solution, and supports, in part, the field data analysis effort by estimating system random errors in the navigation solution; the filter program (Filter) implemented the eleven-state and two-state clock bias filters; the trajectory merge program (Merge) laid the foundation for complete data analysis by computing system vs reference trajectory residual defined as navigation solution minus WSMR reference (that is, estimates of navigation solution errors) and plotting them vs time and other system parameters on a per-run basis; the trajectory residual characterization program (Inquire) statistically analyzed short time series of Merge residuals (data slices within a run) for bias, random, and periodic content; the cross-plot analysis program (Corlate) supports the overall data analysis function by plotting the Inquire error statistics against system state parameters taken over the ensemble of runs and flights. Thus, the data system allowed investigations of data from raw receiver measurements to plots that display system error quantities as a function of system state.

Two-State Filter

The method of passive-user, four-coordinate radio navigation used in the experimental test program employs the pseudo-range formulation. The pseudo-ranging equations contain information about the clock bias (i.e., the fourth coordinate) between the user reference oscillator and the overall system time reference. The two-state clock bias filter takes advantage of the predictable behavior of clock bias and clock bias rate to provide smoothing of these quantities.[8,9] If the two-state filter could provide exact estimates of the states, the position solution algorithm would become a four, true-range least-squares solution. However, the two-state filter can provide only best, and not exact, estimates of clock bias and bias rate. Therefore, the solution can only approach the four, true-range least-squares solution.

The state transition matrix is given by

$$\phi = \begin{bmatrix} 1 & \Delta t \\ 0 & 1 \end{bmatrix}$$

where Δt is the elapsed time between samples. The plant noise matrix models clock bias acceleration (oscillator drift rate) as a zero mean normal random variable with variance $\sigma_{\ddot{B}}^2$. The filter is initialized by the deterministic navigation solution and subsequently accepts inputs of 1) the predicted pseudo-range and pseudo-doppler measurement errors; 2) the predicted bias acceleration variance; and 3) the measured static biases B_c and $\dot{B}_c$.

Once the filtered values of the biases are available, this added information is applied to the coordinate trajectory. This is accomplished by forming four true ranges and range rates:

$$\left.\begin{aligned} r_{fi} &= R_i - B_f \\ \dot{r}_{fi} &= \dot{R}_i - \dot{B}_f \end{aligned}\right\} \quad i = 1, 2, 3, 4 \qquad (20)$$

where the f subscript indicates the filtered value. These four true measurements are used in a least-squares solution for the coordinate position and velocity. The least-squares solution is realized by the pseudo-inverse technique. The resulting decrease in random errors in the user trajectory is characterized by the "dynamic geometric dilution of precision" ($GDOP_D$) corresponding to perfect estimates of bias and bias rate, given by

$$GDOP_D \left\{\text{col } (x, y, z)\right\} = \sqrt{\text{diag } \left[G^{*T} G^*\right]^{-1}} \qquad (21)$$

where G^* is a (3 x 4) direction cosine matrix. Comparing the GDOP's in Fig. 10 with the static GDOP's in Fig. 5, we see that the contours become almost spherical as a result of an ideal two-state filter. The X and Y GDOP gradients become much smaller in the dynamic case, and the Z GDOP improves by a factor of 5 from the static to the dynamic case.

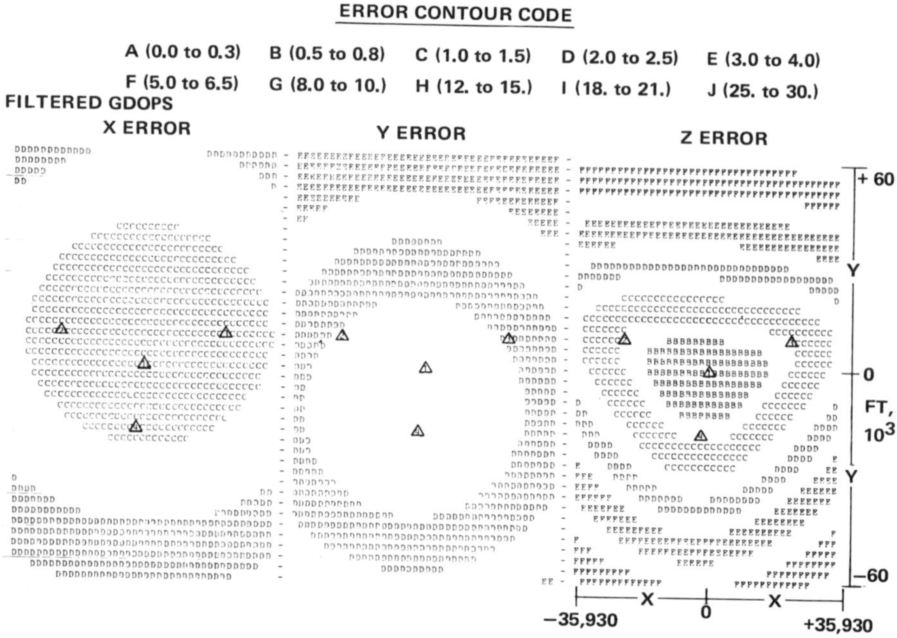

Fig. 10 Dynamic (filtered solution) geometric dilution of precision contours for the area navigation test geometry. Contours shown in X-Y plane at 30,000 ft above ground level.

Eleven-State Filter

An 11-state Kalman filter also was implemented to smooth the trajectory data. The state vector for this filter consists of position, velocity, acceleration, and clock states x, y, z, $\dot{x}$, $\dot{y}$, $\dot{z}$, $\ddot{x}$, $\ddot{y}$, $\ddot{z}$, B, $\dot{B}$. The filter is mechanized along classical Kalman filter guidelines. The effect of filtering the data is shown in Fig. 11. This figure contains a data run from area navigation flight 19, which has been processed by the 2- and 11-state filters and also contains an unfiltered version for comparison. The plots show the residual between the various filtered versions of test system trajectory and the WSMR reference. The upper row of plots shows the X-coordinate residual. There is practically no difference between the unfiltered data and those processed through a two-state filter for both the X- and Y-coordinate data for the region of the flight path considered here. The 11-state filter, however, reduces the X and Y data noise by approximately an order of magnitude. The Y-coordinate data are now shown here, since they are very similar to the X data.

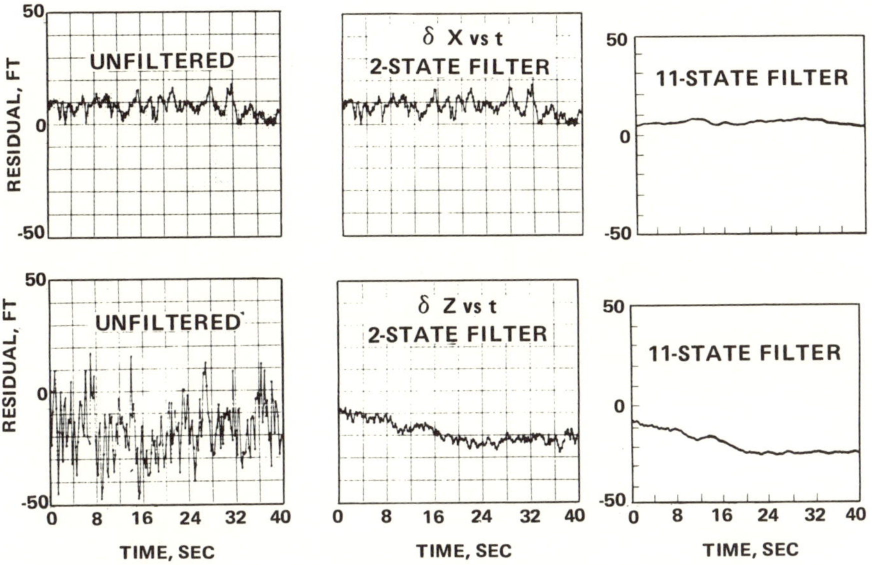

Fig. 11 Comparison of unfiltered, two-state filtered, and 11-state filtered data: flight 19, run 6.

The Z-coordinate data, on the other hand, have a large reduction, about an order of magnitude, in noise content resulting from the two-state filter, as shown by the bottom row of plots in Fig. 11. The 11-state filter improves the two-state Z data by approximately a factor of 3. The improvement of the 11-state filter data over the two-state bias filter is a result of modeling the additional position, velocity, and acceleration states, and also is influenced by the selection of Q values, a measure of uncertainty in the estimates of the state dynamics. The state dynamics are known to be relatively small during the field tests. The area navigation flights were essentially straight and level runs that follow closely the mathematical model of rectilinear motion with constant velocity. The trajectory dynamics during the ILS flights were similar to those experienced during area navigation. The improvement in the 11-state filtered ILS trajectories over other versions is substantial, since the ILS system configuration has good geometries only for small region centered around the flare point on the flight path. In this case, the 11-state filter is necessary to remove the trajectory data noise effectively. A quantitative measure of the effectiveness of these two filters is provided in Fig. 12. This figure presents the trajectory noise (standard deviation) in each coordinate for an area navigation flight path as a function of the degree of filtering. The lines connecting the

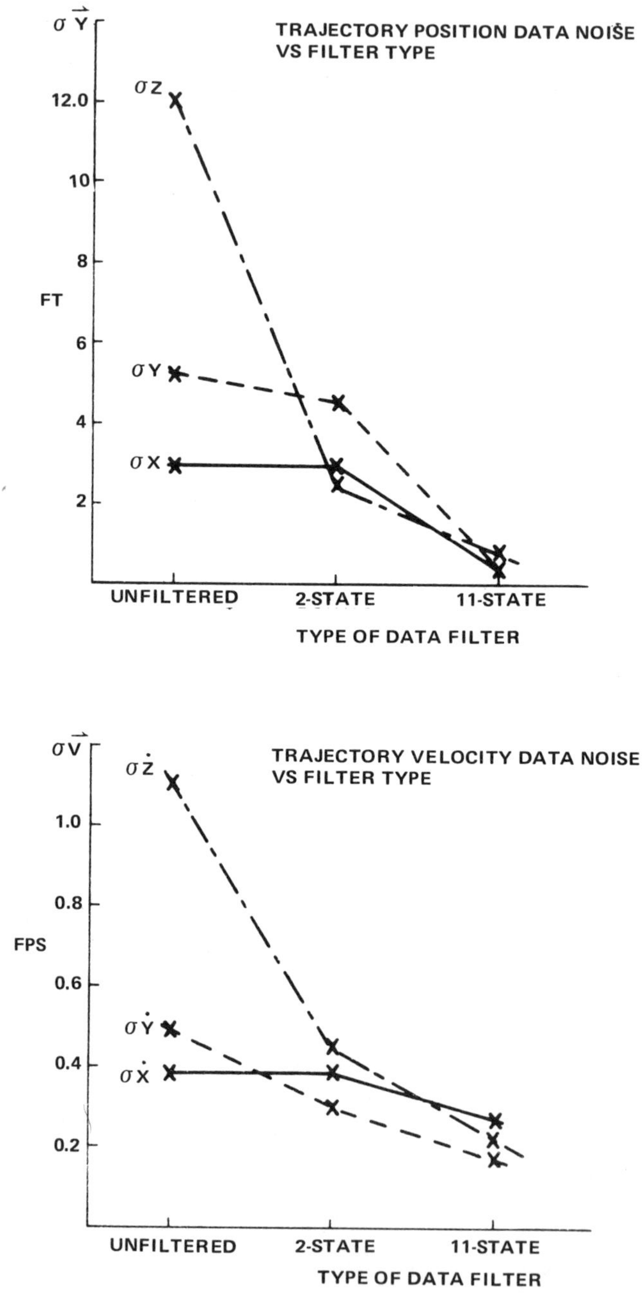

Fig. 12 Improvement in trajectory data smoothing resulting from two- and 11-state filters.

sigmas for each filter type are shown for clarity in determining the improvement trends and should not be used to imply the performance of potential in between (e.g., six-state) filters. Although the reduction of noise in navigation solution data is real, as shown in Fig. 12, there are certain physical realizability parameters absent from the two-state filter formulation which make the comparison of the two filters on noise reduction alone somewhat misleading. Since the two-state filter does not model the equation of motion, the measurement noise via the geometry of transformation can cause trajectory perturbations physically impossible for the aircraft and dynamics of motion. This is the basic weakness of the application of the two-state filter, which explains why reduction of noise on the trajectory data is not the fundamental driving force in the use of the 11-state filters.

Error Analysis

In order to present a meaningful discussion of the results of the field test program, we must first define the errors whose evaluation will provide the measure of system performance. In the terminology of this paper, the set of user's vector position and velocity data, which is the desired system output, is referred to as the "solution domain." Any deviation of the output from an "ideal reference" vector position and velocity represents the navigation errors. However, an ideal (i.e., perfect) reference is a physical impossibility. Therefore, we can produce only an estimate of the navigational error. The precision of the error estimate is a function of the precision of the reference trajectory.

The disturbances creating the navigational errors are assumed to be present at the input system. These disturbances are created by the imperfect measurement process or measurement errors. The set of measurements is referred to as the "measurement domain." The two domains can be related to each other by a linear geometric transformation, which is an implicit function of time due to user's motion with respect to the transmitters. This transformation also applies to the errors in the data sets of the two domains and allows identification and modeling of errors in whichever domain it is most appropriate or useful to do so. A block diagram highlighting the error analysis flow by identifying the major steps in the analysis is given in Fig. 13. Figure 14 represents a system functional flow diagram indicating the source and type of errors associated with the test system. The error sources shown in Fig. 14 have been classified into three broad categories: random, steady-state (bias), and periodic errors.

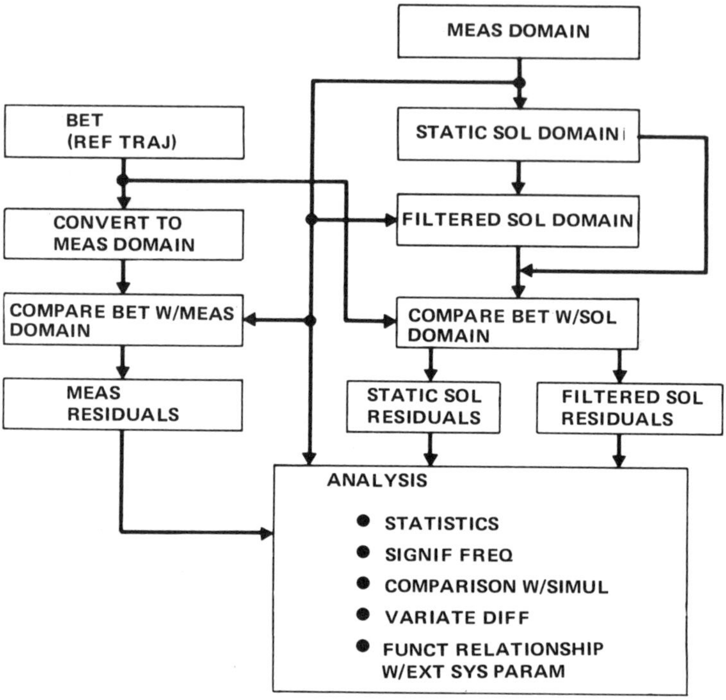

Fig. 13 Error analysis approach.

Some of the error sources are well known, for example, receiver thermal noise and atmospheric refraction correction residuals. There are other error sources that became evident during the data analysis, such as sensitivity of the receivers to the transmitter spectrum shape and apparent signal delay in some of the antennas. The error-producing mechanisms and the contribution of each error source on the navigation solution accuracy were investigated.

Table 1 gives a summary of the observed and the theoretically predicted random errors. Of the sources of random errors, the effects of thermal noise, ground multipath, and (in the doppler measurement) oscillator phase noise and tropospheric doppler are most prevalent. The effect of the multipath, for example, is most prevalent on the northern part of the range.

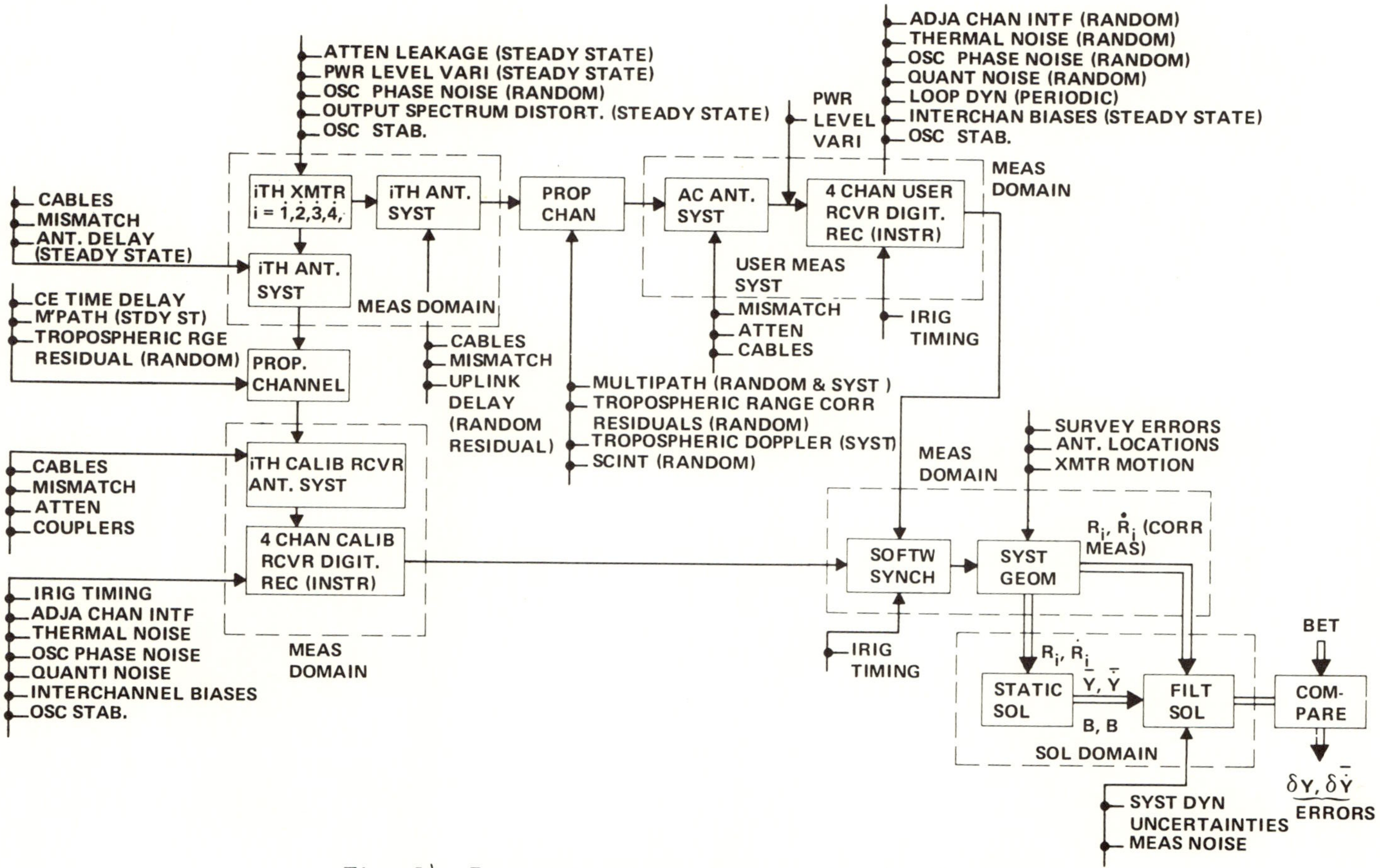

Fig. 14 Representative error sources in field test.

Some Atypical Sources of Error

Ground Calibration System Errors. The calibration link is defined as the signal path from the transmitters to the calibration receiver located at the MCS site. The effective signal path length, expressed as a propagation time in nanoseconds, had to be measured precisely to obtain the necessary constants used in the calculation of the navigation solution. The radiated signal path portion of the calibration link was measured with the ground components of the test system stabilized by two cesium-beam frequency standards. These measure path lengths were combined with the rf cable path length measurements to establish the total calibration link delay quantities used in the navigation algorithm. Cable propagation time delays were measured by a substitution technique in the rf link and also with a time domain reflectometer. The magnitude of the corresponding error sources is listed in Table 1.

Uplink Antenna Delay Errors. The ground transmitting antenna arrays caused a distortion of the transmitted signals as a function of the antenna look angles. The distorted signal envelopes are perceived by the correlation receiver as apparent signal delays.[14] Typical measured data are presented in Fig. 15. The delays are presented in terms of the measured bias-corrected numbers relative to an arbitrary constant.

The effect of the antenna delay error processed by the airborne receiver on the navigation solution is shown in Figs. 16 and 17. Figure 16 shows the time history of the residual error obtained by differencing the navigation solution with WSMR BET ($z = z - z_{BET}$). The navigation solution in this figure is computed ignoring any antenna delay error effects. Thus the antenna error is present in the solution

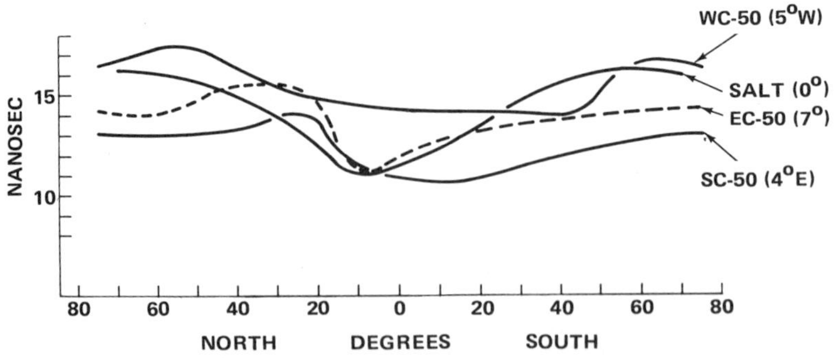

Fig. 15 Antenna delays for 30K-ft altitude.

Table 1 System random error summary

	Hazeltine		MRL		Remarks
	σ_R	$\sigma_{\dot{R}}$	σ_R	$\sigma_{\dot{R}}$	
A) Aircraft receiver					
1) Thermal noise	1.5	0.05	0.6	0.02	At -123 dbm,
2) Oscillator noise	0.02	0.03	0.02	0.03	low vibration
3) Quantization	0.43	0.08	0.145	0.008	
4) Time delay mismatch	0.61	0	0.2	0	
5) Dynamics					
A) Clock loop	1.3	N.A.	1.3	N.A.	ILS only
B) Doppler measurement	N.A.	0	N.A.	0	
6) Adjacent channel interference	0.05	0	0.02	0	ILS
B) Uplink					
1) Multipath					
A) Ground	3.5	0.1	2.5	0.1	Average all flts
B) Aircraft	0.7	0.07	0.7	0.07	
2) Troposphere	1	0.1	1	0.1	Maximum
3) Scintillation	0.2	0.05	0.2	0.05	
C) Transmitter oscillator	0.4	0.1	N.A.	N.A.	Maximum
D) Cal receiver					
1) Thermal noise	1.5	0.04	1.5	N.A.	At -120 dbm
2) Oscillator noise	0.001	0.003	0.001	N.A.	
3) Quantization	0.43	0.008	0.43	N.A.	
4) Time delay mismatch	0.61	0	0.61	N.A.	
5) Adjacent channel interference	0	0	0	N.A.	
E) Cal link propagation					
1) Fading stability	0.06	0	0.06	N.A.	
2) Refraction	0.005	0	0.005	N.A.	
3) Scintillation	0.2	0	0.2	N.A.	
Total, rss	4.7	0.2	3.2	0.17	

and causes a distortion of the z-coordinate position. Figure 17, on the other hand, has the antenna delay error effect modeled in the solution.

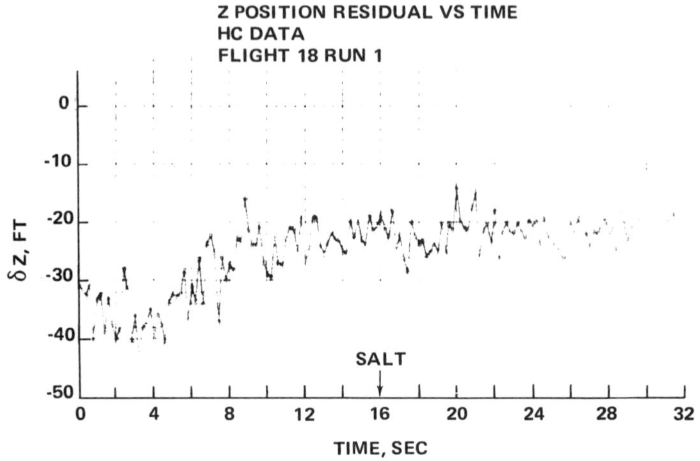

Fig. 16 No corrections for observed antenna delays (case 1).

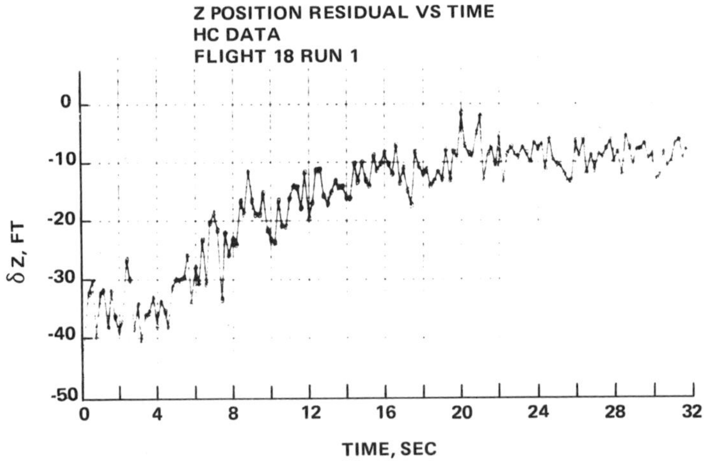

Fig. 17 Antenna delays modeled.

Propagation Channel Characteristics

A propagation channel (atmospheric, i.e., ionospheric, tropospheric, and multipath) can be characterized by a time-varying impulse response that is a function of space and time. This allows the description of the output of the channel (received signal) for a given input (transmitted signal). The time variation can be an implicit property of the medium or can be induced by relative motion between the receiver and the transmitter (explicit time variation). In a satellite system, for example, spatial and temporal variations of the electron density cause random scintillations, causing changes in group delay, whereas relative motion of the satellites with respect to the user creates ionospheric (and also tropospheric) doppler shifts due to the time rate of change of excess group delay. Furthermore, dispersion in the ionospheric causes differences between phase and group velocities affecting range and range-rate measurements. Precise models of those phenomena are extremely important to remove bias terms in measurements and to drive the residuals to zero.

During the field tests (in the absence of ionosphere), the primary propagation effects were tropospheric, multipath, and relative dynamics. The channel impulse response in the field test data analysis was described by ray theoretic models. For example, the two-ray specular (two-component multipath) description of the channel over a smooth surface was found to be in excellent agreement with measurements (including depolarization effects and the Rayleigh criterion) described by differential delay and differential doppler (fade frequency) models.[9,11,13] In addition, signal dynamic effects created by relative motion along the LOS between transmitter and receiver also were included in the channel description, affecting receiver tracking performance.[11] Figure 18 depicts the salient area navigation channel characteristics for straight

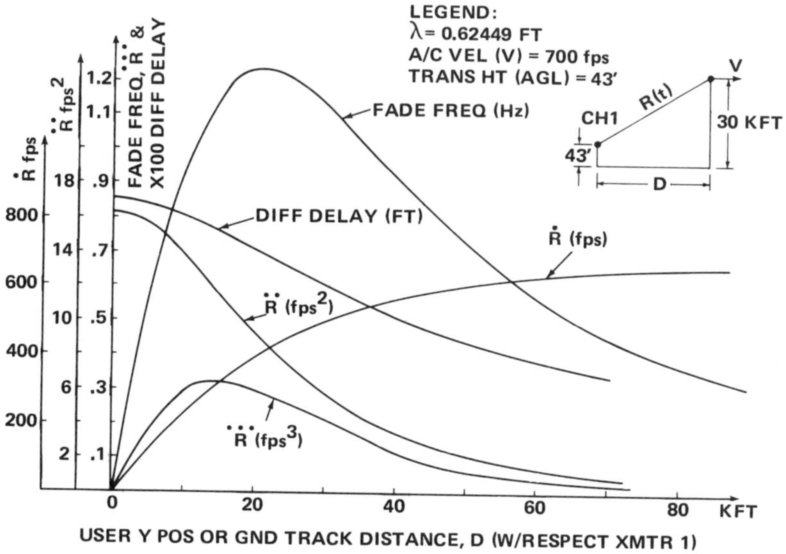

Fig. 18 Area navigation channel characteristics.

and level flights over and with respect to the center transmitter. The effects of relative dynamics are shown in Fig. 19, illustrating computer simulation results for maximum tracking errors on ILS, which were small enough to be negligible in comparison with other sources of error.

Multipath

Multipath can cause both bias and random errors in the measurement of pseudo-range and pseudo-doppler, depending on the properties of the propagation medium and on system geometry. The multipath errors in the system demonstration (area navigation) and ILS tests could be attributed to errors in ground-to-aircraft links (both diffuse and specular), error due to aircraft structure (predominantly diffuse), and errors on the calibration links (specular).

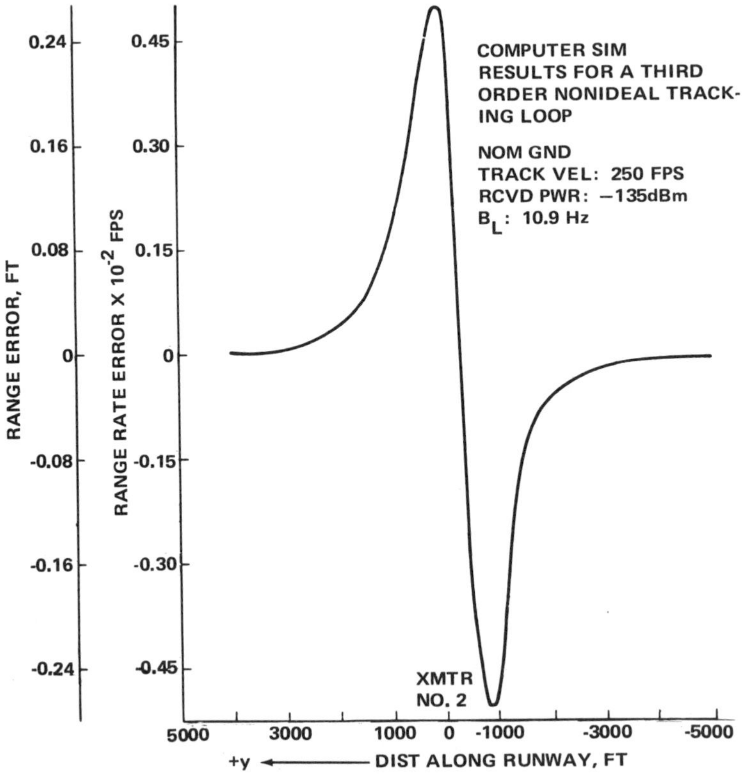

Fig. 19 Worst-case tracking error from relative dynamics on ILS.

Range Error. The multipath range bias error characteristics of the correlation receiver in terms of the received signal characteristics are divided into two regions.[13,14] The error characteristics in region I in the absence of band limiting can be described by

$$\Delta R_m = \begin{cases} d\left[R_1/(1+R_1)\right], & d < (T/2)(1+R_1) \\ \left[(3T/2)-d\right]\Big/\left[R_1/(2-R_1)\right], & (T/2)(1+R_1) \le d < 3T/2 \\ 0, & d \ge 3T/2 \end{cases} \qquad (22)$$

where $R_1 = R\cos[(2\pi/\lambda)d-\phi]$, considering ϕ taking on values 0 or π only, for this representation. The region II error characteristics are described in Refs. 13 and 14 (also see the Appendix). R is the magnitude of the effective voltage reflection coefficient for either vertical or horizontal polarization; λ is the wavelength (feet); T is the chip time of the code expressed in range units ($T \simeq 100$ ft); d is the differential delay (feet); $(2\pi/\lambda)d$ is the multipath relative phase (radians); and ϕ is the phase angle (radians) for either vertical or horizontal polarization. The specular multipath contribution (in region I) to range error was found to be approximately 2 ft. The region II errors (which are always positive) were not achieved during the area navigation tests (see the Appendix).

Range-Rate Error. The changing system geometry causes a beat between the direct and specularly reflecting signals, called the differential doppler or fade frequency. The rate of change of phase between the direct and interfering signals can be shown to be

$$\dot{\alpha} = \Delta\omega R\,[R+\cos(\Delta\omega t+\phi)]/[1+R^2+2R\cos(\Delta\omega t+\phi)] \qquad (23)$$

where $\Delta\omega = (2\pi/\lambda)\dot{d}$ is the differential delay or fade frequency, and $\dot{d}$ is the rate of change of differential delay. Equation (23) can be expanded in a Fourier series of the form

$$\dot{\alpha} = \Delta\omega \sum_{i=1}^{n} (-1)^{n+1} R^n \cos \left[n\,(\Delta\omega t + \phi) \right] \tag{24}$$

In terms of the carrier tracking loop parameters, the velocity error is

$$\Delta \dot{R}_u = \frac{\lambda}{2\pi} (\Delta\omega) \sum_{i=1}^{\infty} (-1)^{n+1} R^n P_{B_L} (\Delta\omega) \cos \left[n(\Delta\omega t + \phi) \right]$$

$$P_{B_L}(x) = \begin{cases} 1 & |x| \le B_L \\ 0 & \text{otherwise} \end{cases} \tag{25}$$

where B_L is the one-sided loop noise bandwidth of the carrier tracking loop. Note that Eq. (25) shows the effects of band-limiting on velocity error; i.e., if $n\Delta\omega > 2\pi B_L$, there is no error in velocity due to multipath. The multipath contribution to range-rate error is significant. It was found to be on the order of 0.6 fps at the worst flight/transmitter geometries (see Fig. 20).

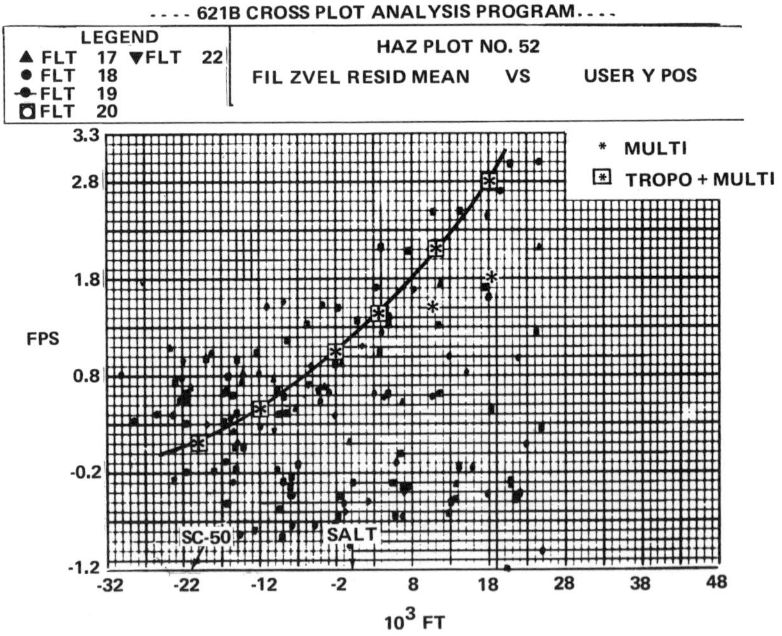

Fig. 20 Predictable effects of SC-50.

Tropospheric and Multipath Doppler Measurement Biases. The doppler measurements are obtained in the receivers by averaging (differencing) the doppler over a 100-msec counting interval. The result is the average range-rate over the sample interval. This is equivalent to deriving range-rate by differencing two range samples per time interval. Therefore, the tropospheric correction for doppler is the difference between two tropospheric corrections:

$$\Delta \dot{R}_{i_j} = (\Delta R_i - \Delta R_j)/\Delta t_{i_j} \qquad (26)$$

The contribution of Eq. (26) becomes significant at low ground elevation angles, where the tropospheric corrections exhibit the steepest slope, with ground track distance measured from a ground transmitter.

The multipath contribution to doppler measurements exhibits a monotonic functional behavior with distance similar to the tropospheric doppler contribution but with larger magnitudes. In addition, the mechanism causing the multipath doppler bias is different. Multipath doppler is caused by the rate of change of differential delay due to the aircraft's motion, which creates a beat between the direct and specularly reflected signals. The resultant composite signal phase varies with time. The rate of change of phase induces an error in the carrier tracking loop, creating a doppler bias in the measurements, given in Eq. (25); see Fig. 20.

Some Multipath Errors in Data. There was a shift found in time bias error between a northerly run and a southerly run on the same area navigation flight path which was representative of multipath presence. Multipath effects could be seen in the data primarily on channels 1 and 4. The multipath signals reflecting from the terrain are affecting the aircraft data either at low elevation angles when far away from the transmitting site or at very high elevation angles encountered while passing over a transmitter. The unfiltered navigation solution is used in the analysis of multipath, since filtering can mask certain of these effects. The interchannel residual between the navigation solution and the WSMR BET was formed without, however, averaging the data. A spectrum analysis was conducted on blocks of data to determine the presence of statistically significant frequency components. The spectral lines were calculated for each sample; hence the units are cycles per sample (with data sampling rate of 5 samples/sec).

Figure 21 shows that the spectral lines of R4-1 data show a gradual rolloff of frequencies from 1.25 Hz at 15,000 ft north of Salt (channel 1) to approximately 0.90 Hz at 25,000 ft north of Salt. A comparison with a simulated differential doppler (fade frequency) between station SC-50 and the aircraft (user) position shown in Fig. 18 clearly identified the "upper region significant" frequencies as specular ground-to-aircraft multipath from station SC-50 (channel 4), which is located approximately 20,000 ft south of Salt (35,000 - 45,000 ft from the user). Figure 18 shows the theoretical multipath fade frequency as a function of user distance from SC-50, channel 4. An examination of the plot of range 1 residual (navigation solution range 1, WSMR range 1) vs range 4 (Fig. 22) at approximately the same user Y location (since the plot against range

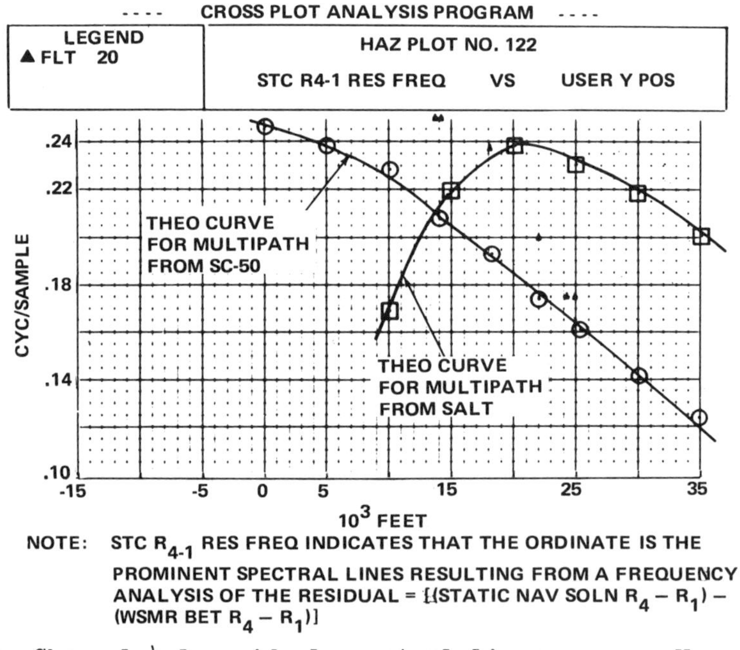

Fig. 21 Channel 4-1 residual spectral lines vs user Y position.

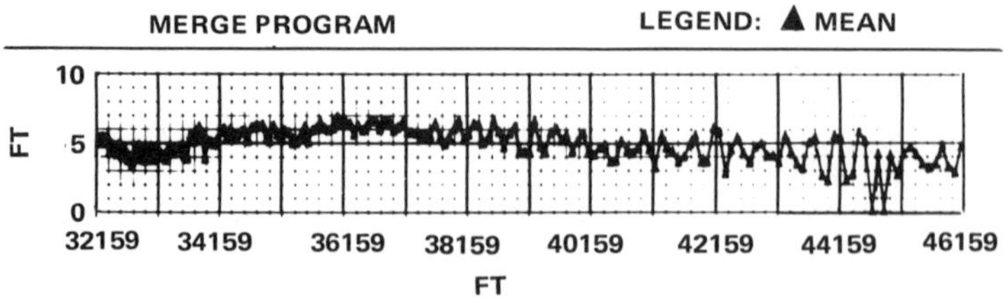

Fig. 22 Range 1 residual vs range 4 HC: flight 20, run 02.

4 is almost the same as plot against Y) shows a strong cyclic variation. The presence of specular multipath at distances of greater than 35,000 ft from SC-50 is expected, since the aircraft antenna is essentially linearly polarized on the horizon and cannot reject the ground-reflected signals.

Multipath Environment During ILS. The fade frequencies have been evaluated for the ILS flight geometry, and Δf ranges from approximately 0.1 to 0.6 Hz for transmitters 1, 2, and 3 and from approximately 1 to 48 Hz for transmitter 4, which is located on a tethered balloon. The amount of multipath rejection provided in a particular channel is a function of the relative geometry, antenna patterns, polarization effects, and tracking loop noise bandwidths. Both the HC and MRL receiver tracking loop bandwidths are wider than a fraction of a Hertz, and therefore the multipath scintillations from transmitters 1, 2, 3, and at times from transmitter 4 should be present in the data from both receivers.

The HC receiver carrier tracking loop noise bandwidth varies with received power. At a received signal strength in the balloon channel of -120 to -130 db-m, the HC carrier tracking loop single-sided noise bandwidths are, respectively, 109 and 34.5 Hz, which will pass the balloon transmitter-caused multipath at an average fade frequency of 24 Hz. At the same time, the MRL receiver is not expected to respond to balloon-caused multipath, since the MRL carrier tracking loop noise bandwidth is approximately 10 Hz (single-sided) and is not a function of power level.

During the ILS flights, a cyclic variation of the Z-position residual highly correlated with the variation of the received power from the balloon transmitter (channel 4) as a function of user Y-position was observed from the HC receiver. A good example of this phenomenon is shown in run 6 of flight 25. A plot of the static Z-position residual and the received power in channel 4 (balloon channel) as a function of user Y position (Fig. 23) shows marked similarity to the multipath error and power plots of ground tests.[9,13] As expected, this occurs only in the HC receiver data due to the adaptive increase in loop noise bandwidth at this signal level. On the same run, the channel 1 Z-position residual of both the HC and MRL receivers shows a slow cyclic variation at a rate of approximately 0.25 Hz, as expected. It is important to observe that the balloon channel multipath scintillation rate at times is very much above the data sampling rate of 5 samples/sec, representing a 2.5-Hz bandwidth system. Therefore, the highest frequency in the recorded data cannot exceed 2.5 Hz. The approximate frequency of Z-velocity difference be-

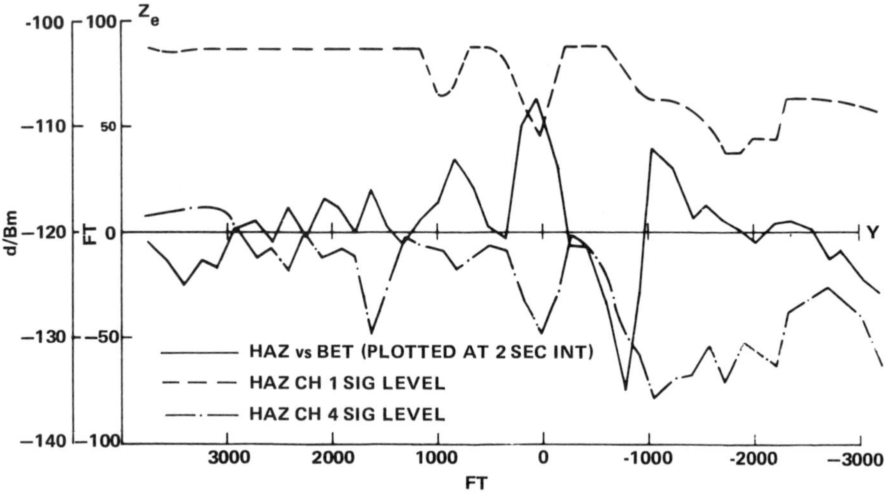

Fig. 23 Aircraft position along runway centerline (Y position).

tween MRL and HC is in the range of 0.5-1 Hz, although the scintillations could have been as high as 48 Hz.

Summary of Bias and Periodic Errors

Table 2 gives a summary of the magnitude of the observed and theoretically predicted systematic errors. The most significant steady-state errors are the interchannel biases, cesium time delay measurements, uplink antenna apparent signal delays, and transmitter attenuator leakage. These four sources of systematic errors contribute 18 ft of the total 19.3 ft of the root sum squared (rss) errors.

Views of the System Performance

Having arrived at an understanding of the data behavior, it is possible to look at the user's navigation performance from different viewpoints. Since the overall objective of this test program was to verify through field experiments the theoretical understanding of the four-channel correlation receiver behavior and to assess the potential performance that might be expected of this type of equipment, it is useful to examine the impact of the field test error mechanisms on the user navigation performance. The field test error mechanisms may be divided into two broad categories. The experiment program was a simulation of the major features of a satellite system configuration. As such, it suffers the problems of all simulations, namely, the inability to represent perfectly the actual system. Error mechanisms such as the ground calibration link measurements, the uplink antenna delay, and multipath interference from ground reflections caused by the proximity of the transmitting antennas to the ground are unique to

Table 2 System steady-state and periodic errors

	Hazeltine		MRL		Remarks
	ΔR	$\Delta \dot{R}$	ΔR	$\Delta \dot{R}$	
A) Aircraft receiver					
1) Interchannel biases					
A) Power level variation	12	N.A.	0.5	N.A.	Maximum
B) Spectral distortion	0.5	N.A.	0.5	N.A.	Estimated minimum
C) Other	1	N.A.	N.A.	N.A.	Calib cycle
2) Dynamics	0	0	0	0.4	Maximum
3) Clock error	0.1	0	0.1	0	
B) Uplink					
1) Multipath	2	0.6	2	0.6	Maximum
2) Antenna delay	4[a]	N.A.	4[a]	N.A.	Over Salt
3) Troposphere	0	0.3	0	0.3	Not modeled
C) Transmitter					
1) Spectral distortion	4	N.A.	1	N.A.	Maximum
2) Attenuator leakage	5/0.5	N.A.	5/0.5	N.A.	First number of ΔR refers to flights 1-15, second to all other flights
3) Cable length uncertainty	2	N.A.	2	N.A.	
4) Mismatch	0.1	N.A.	0.1	N.A.	
D) Cal receiver					
1) Interchannel biases					
A) Spectral distortion	0.5	N.A.	0.5	N.A.	Estimated minimum
B) Ant interaction	1	N.A.	1	N.A.	
2) Cable length uncertainty	2	N.A.	2	N.A.	
3) Power combiner	0.5	N.A.	0.5	N.A.	
4) Mismatch	0.1	N.A.	0.1	N.A.	
E) Cal-link propagation					
1) Refraction	3	0	3	0	
2) Cesium time delay	12	0	12	0	
3) Multipath	3	0	3	0	Slow variation (not calibrated)
Total, rss	19.5	0.67	14.7	0.77	

[a] Applicable to flights other than 17-20 where it is modeled.

the WSMR tests. From an overall system evaluation viewpoint, these are shortcomings of the simulation.

Another category of error mechanisms is those indigenous to the receiver equipment. These include items influencing the pseudo-range and pseudo-range-rate measurement accuracies such as tracking loop implementation, basic oscillator characteristics and frequency management within the receiver, data output techniques, automatic calibration of receiver path delays, and basic path delay stability. An assessment of these types of error mechanisms was the primary purpose of the program, since this equipment represents some of the current concepts for highly accurate correlation receivers. Some typical data results showing Merge program output residual of unfiltered Z-position and Z-velocity data for area navigation flight 19, run 5 are given in Figs. 24 and 25. The abscissa represents elapsed time along the flight path. At approximately 20 sec into the flight, the aircraft was over the center transmitter (channel 1 or Salt). The Z-velocity residual fluctuations in Fig. 25 were attributable primarily to fluctuations in the reference trajectory.[9] Filtered navigation solution data from the same flight family were illustrated in Fig. 11 during the 11-state filter discussion. The improvement gained by filtering is quite explicit in comparing the figures. Figure 26 is a time history of the Z-position residual for ILS flight 25, run 5. Each of the major divisions on the abscissa represents 4 sec, during which time the aircraft travels approximately 1300 ft. Flare occurs at 20 sec. It should be noted that the error at that point is only a few feet. These test system trajectory data have been processed with the 11-state Kalman filter. The primary cause of the differences between the HC and MRL data is a result of multipath interference.

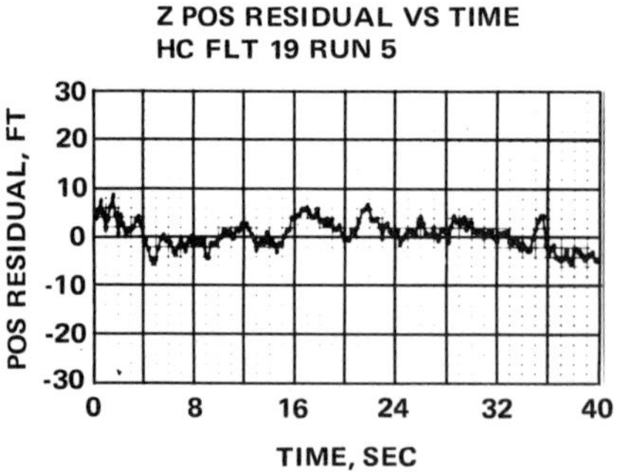

Fig. 24 Merge outputs, Z-position residuals.

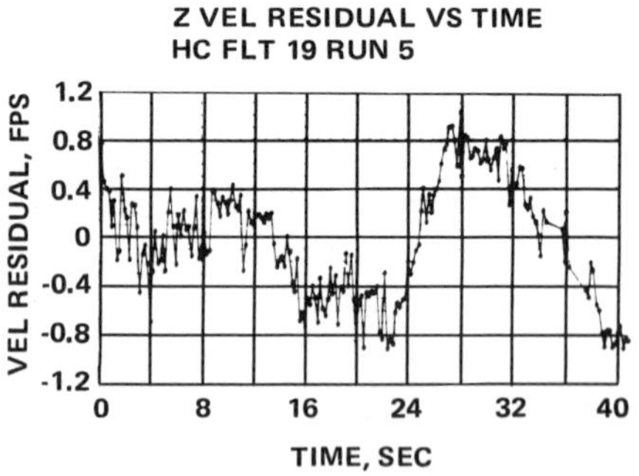

Fig. 25 Merge program outputs, Z-velocity residuals.

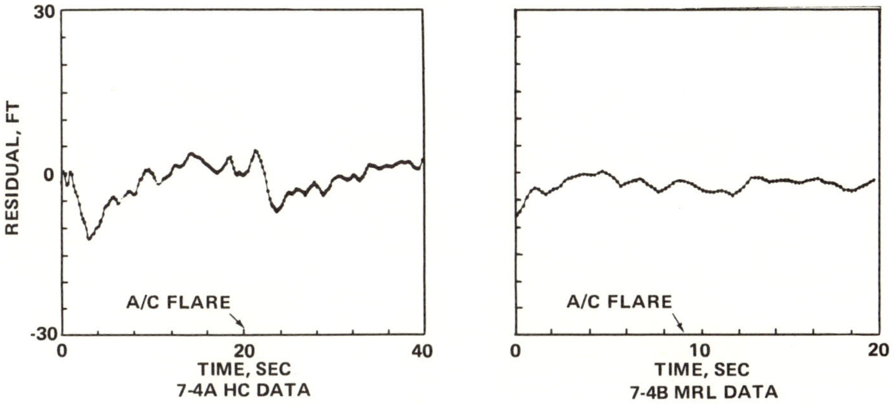

Fig. 26 Z-position residual for flight 25, run 05.

Conclusions

The test program successfully met the principal objective of evaluating a four-channel receiver in typical flight and field environments, and confidence in the system concept was verified through the actual tests performance. Analysis of the error components indicates that a small average range measurement difference of about 2 ft still exists between the WSMR best estimate of trajectory (BET) and the test trajectory. The range-rate measurement errors were less than 1 fps.

Several observations can be made about the system behavior. Perhaps the most important of these is that ranging bias errors (steady-state errors) are extremely difficult to measure and control. Although almost all of the error mechanisms have been determined through analysis of the data, the small (2-ft) steady-state differences referred to above still remain unexplained. But perhaps of more importance, the known error mechanisms point up the extreme care that must be exercised during design and manufacturing to control signal delay variability to less than 1 nanosec throughout the system.

The ILS testing conclusively demonstrated the suitability of the system in an aircraft landing environment. The mean error for vertical position was about 1.5 ft and for vertical velocity less than 0.5 fps. These are representative of absolute errors obtainable with a filtered solution. This quality of system performance is more than adequate for landing, and the test results indicate that the signal environment (large changes in power and multipath effects) during this phase of aircraft operation does not appreciably degrade the navigation accuracy.

Appendix

The multipath error performance of the generic correlation receiver (see Fig. 2) employing coherent delay lock and carrier tracking loops can be described by the zero crossings (i.e., zeros) of the composite delay lock discriminator characteristics given by[13, 14]

$$D_{comp}(\tau) = [R(\tau + T/2) - R(\tau - T/2)] \cos \phi_{comp} + \alpha [R(\tau + T/2 - d) - R(\tau - T/2 - d)] \cos(\phi_m - \phi_{comp}) \tag{A1}$$

where $R(\tau) = 1 - |\tau|/T$ and

$$\phi_{comp} = \begin{cases} \tan^{-1} \{\alpha R(\tau - d) \sin \phi_m / [R(\tau) + \alpha R(\tau - d) \cos \phi_m]\}, & |\tau| < T \\ \phi_m, & |\tau| > T \end{cases}$$

The composite delay lock discriminator characteristic is a function of the multipath relative amplitude α and phase ϕ, the multipath relative phase $\phi_m = (2\pi/\lambda)d$, and the composite signal phase (i.e., direct and multipath) which the carrier loop tracks ϕ_{comp}, the code chip time T, and the differential delay d. It is the tracking of the composite phase that results in the two-region breakdown of the multipath errors. The multipath error envelope corresponding to unity relative multipath amplitude as a function of differential delay in the absence of carrier tracking loop threshold (i.e., carrier power) considerations is shown in Fig. 27, depicting the two regions of multipath.

Region II multipath was observed during ILS, possibly at one time instant (see Fig. 23), showing deep received power nulls, and in one test configuration during separate ground multipath tests.[13] The predominant multipath performance, both during in-flight and during the separate ground multipath test, can be characterized adequately by the region I model [see Eq. (22)], which is shown to be justified by actual measured data, summarized in Figs. 28 and 29, representing all of the data taken during the ground tests.[13]

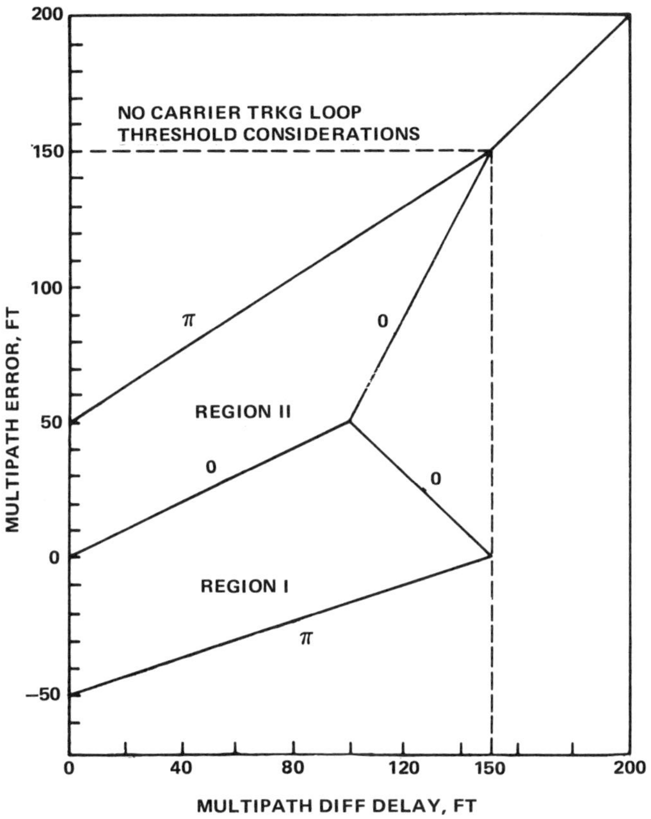

Fig. 27 Multipath envelope for coherent delay lock loop with unity relative multipath amplitude and relative multipath phase, 0 or π rad.

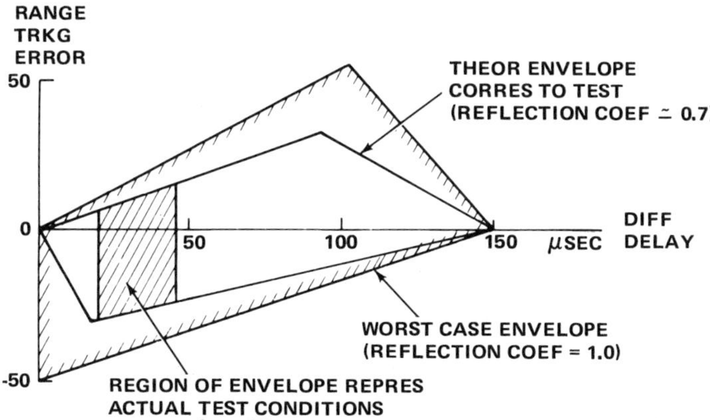

Fig. 28 Theoretical envelope of range tracking error due to multipath during ground tests.

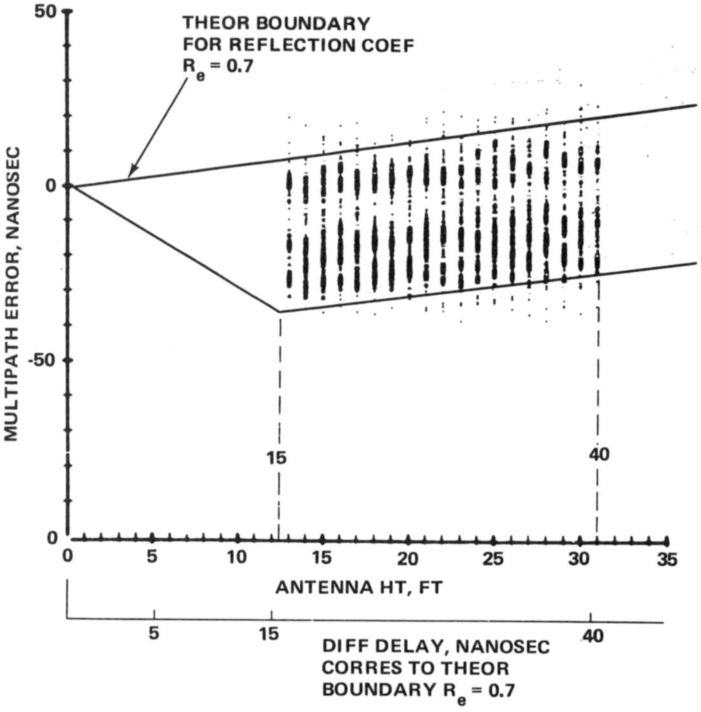

Fig. 29 Multipath measurement results (ground tests).

References

[1] Woodford, J.B., Melton, W.C., and Dutcher, R.L., "Satellite Systems for Navigation Using 24-Hour Orbits," EASCON'69 Conference Record, 1969, pp. 184-189.

[2] Diamond, P.M., "Satellite Systems for Integrated Communications, Navigation, and Identification," AIAA Progress in Astronautics and Aeronautics-Communication Satellites for the 70's: Systems, Vol. 26, edited by N.E. Feldman and C.M. Kelly, The MIT Press, Cambridge, Mass., 1971, pp. 437-456.

[3] Otten, D.D., Craigie, J.H., Garabedian, A., and Morrison, D.D., "Satellites for Domestic Air Traffic Control," AIAA Progress in Astronautics and Aeronautics-Communication Satellites for the 70's: Systems, Vol. 26, edited by N.E. Feldman and C.M. Kelly, The MIT Press, Cambridge, Mass., 1971, pp. 209-249.

[4] Bogen, A.H., "System 621B Geometric Performance for a 20 Satellite Global Deployment," TR-0172(2521-01)-1, Nov. 5, 1971, Aerospace Corp.

[5]Miller, B., "USAF Studies Navsat Proposal," Aviation Week and Space Technology, May 8, 1972, pp. 50-57.

[6]Klass, P.J., "Plans for Defense Navset Readied," Aviation Week and Space Technology, Aug. 20, 1973, pp. 65-69.

[7]Klass, P.J., "Compromise Reached on Navsat," Aviation Week and Space Technology, Nov. 26, 1973, pp. 46-51.

[8]Courtney, J.J., Jr., Laho, R., Kadar, I., et al., "System 621B User Equipment Definition and Experiment Program, Phase I," SAMSO TR70-440, Final Report, Vols. I and IV, Dec. 1970, Grumman Aerospace Corp., Bethpage, N.Y.

[9]Courtney, J.J., Jr., Laho, R., Kadar, I., et al., "System 621B User Equipment Definition and Experiments Program," SAMSO TR-73-65, Final Report, Vols. I-III, April 1973, Grumman Aerospace Corp., Bethpage, N.Y.

[10]Thoma, G.R. and Berkowitz, R.S., "A Hybrid Navigation Concept Using a Spinning Satelliteborne Interferometer and Self-Contained Equipment," IEEE Transactions on Aerospace and Electronic Systems, July 1972, pp. 528-537.

[11]Kadar, I., "General Signal Design and Systems Considerations for Navigational Satellites and Links," unpublished TR, 1969, revised April 1972, Grumman Aerospace Corp., Bethpage, N.Y.

[12]Kadar, I., "An Analysis of Helicopter Rotor Modulation Interference," IEEE Transactions on Aerospace and Electronic Systems, Vol. AES-9, No. 3, May 1973, pp. 434-441.

[13]Kadar, I. and Neglia, M., "Results of a LOS Ground Multipath Propagation Test on Time-of-Arrival Measurements Using Correlation Reception," to be published.

[14]Kadar, I., "Performance of Correlation Received in an Interference Environment," Conference Record, National Aerospace and Electronics Conference, June 1975, Dayton, Ohio.

CHAPTER V—COMMUNICATIONS SATELLITE SYSTEMS

Communications satellite technology has reached maturity and is now able to provide operational systems that have demonstrated high availability and superior performance. The INTELSAT system has become the principal means of international voice communication and of course the only means for transoceanic television broadcast and wideband digital service. The increased demand has required INTELSAT to develop improved satellites with expanded capability. The current generation of INTELSAT IV's are being supplemented by INTELSAT IV-A's, which promise greater capacity, principally through frequency reuse. The paper by Dicks and Brown provides the reader with an insight into the multiple performance tradeoffs associated with the overall design. At present, however, INTELSAT is no longer the only operational system. Several domestic systems are now operational, or about to become operational, and the Defense Satellite communications System (DSCS) has demonstrated a full capability.

Three U.S. domestic companies are currently providing service to users, although Western Union is the only company actually to have operating satellites in orbit. The American Satellite Corporation is leasing transponders from Western Union, and RCA GLOBCOM is leasing service from the Canadian TELESAT System and Western Union. The first COMSAT satellite is scheduled for launch early in 1976, and it will be utilized by A.T.&T. and G.T.&E. Earth terminals. RCA also will launch its own spacecraft in late 1975 or early 1976. American Satellite does not have definite plans for launching its own satellite. Both American Satellite Corporation and CML plan to develop a multipoint satcom service. This service, would utilize many small satellite terminals located at the users' facilities and would bypass dependence on A.T.&T. For local distribution of service.

Foreign domestic satellite service also is developing rapidly. Most of these systems, however, are (or will be) nationally owned and operated. The TELESAT system is already available for service in the Canadian region. The MOLNIYA system is primarily a domestic system (U.S.S.R.), but, because of a highly elliptical orbit, it has a usable pass over the northern hemisphere. The MOLNIYA system, together with INTELSAT, will soon provide the operational links for the Washington-Moscow hotline service. Other countries (e.g., Japan) and regions (e.g., Europe) are developing their own capabilities and likely will have operating systems in the 1980 time frame. Other countries such as Algeria and Brazil are availing themselves of the transponder leasing service offered by INTELSAT. The interruptible service that Algeria is obtaining will give it an excellent system at a very reasonable annual cost. A description of this system is contained in the paper by Bairi. Indonesia has awarded contracts to Hughes Aircraft for spacecraft to provide domestic satellite service to that region of the Pacific.

The MARISAT system will provide service to commercial ships at sea, using L-Band frequencies, and to U.S. Navy ships at VHF/UHF (200-400 MHz). European Space Research Corporation (ESRO) is pursuing a similar venture to be called European Maritime Communications Satellites (MAROTS). MAROTS and other programs sponsored by ESRO are described in the paper by Collette and Stockwell. ESRO is also a partner with Canada and the U.S. (COMSAT) in the Aerosat program, which will provide improved communications and surveillance for transoceanic air-traffic control. The Federal Aviation Agency (FAA) will be the U.S. operator of the Aerosat system.

The original communications satellite programs were sponsored by the DoD, but, as NASA evolved, much of the R&D in this new field was transferred to NASA. However, the DoD has continued to develop communications satellite programs to satisfy "unique and vital" DoD requirements. The major operational DoD satellite communications system is the Defense Satellite Communications System (DSCS). In 1974, the "Phase II" DSCS satellites were declared operational in the Pacific and Atlantic regions of the world. The DSCS is described in detail in the paper by LaVean.

The DoD employed R&D assets to provide an Interim Operational Capability (IOC) for mobile users, primarily ships. This capability lasted for nearly three years (1970-1973), using the TACSAT and LES-6 spacecraft. This service is to be restored, using a leased spacecraft (MARISAT) in early 1976. By 1978 the FLEETSAT/AFSATCOM system will provide increased capability for this class of user. This capability is described in the paper by Wardle. For the more distant future, DoD is looking to survivable satellite techniques which will enhance the command and control capability of the strategic forces.

INTELSAT IV-A SATELLITE TRANSMISSION DESIGN

Jack L. Dicks* and Martin P. Brown Jr.†

Communications Satellite Corporation,
Washington, D.C.

Abstract

The Intelsat IV-A satellite is expected to be ready for launch in 1975. This 20-transponder satellite will achieve expanded capacity over the Intelsat IV through frequency reuse by spatial separation of its East and West antenna beams. Compared to the Intelsat IV, transmission planning is more complex because of the effects of cochannel interference, as well as adjustments made in communication performance for the normally encountered signal impairments: Earth station rf out-of-band emission, satellite group delay and dual path distortion, adjacent transponder intermodulation interference, and crosstalk. This paper discusses preliminary Intelsat IV-A transmission design in consideration of these impairments.

Introduction

Because of the rapid international telecommunications traffic growth in the Atlantic Ocean region, a new generation Intelsat satellite series will be required to replace the existing Intelsat IV system in the 1976 time frame. As is true

---

Presented as Paper 74-474 at the AIAA 5th Communications Satellite Systems Conference, Los Angeles, Calif., April 22-24, 1974. This paper is based upon work performed at Comsat under the sponsorship of the International Telecommunications Satellite Organization (Intelsat). Views expressed are not necessarily those of Intelsat.

*Director, Communications Transmission Engineering, International Systems Division.

†Senior Communications Engineer, Communications Transmission Engineering, International Systems Division.

with most technology, the new series, designated Intelsat IV-A, has retained many of the proven characteristics of its predecessor while introducing new concepts as a step toward achieving expanded capacity with the same size launch vehicle: the Atlas Centaur.

Spatial separation of East and West antenna beams will allow the same frequencies to be used in both the East and West hemispheres. As illustrated in Figs. 1 and 2, frequency reuse will enable the Intelsat IV-A to almost double the 500-MHz bandwidth that had been available to the Intelsat IV. Although the fundamental transponder configuration found in the Intelsat IV will be preserved, there have been many changes necessary in individual subsystems to make a suitable tradeoff between weight, transmission performance, and new technology. This paper focuses on the transmission performance characteristics of the Intelsat IV-A and how they differ from the Intelsat IV.

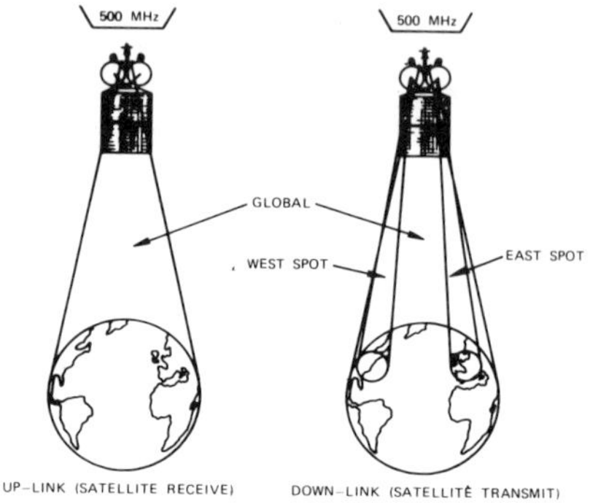

Fig. 1 Intelsat IV antenna coverage capability.

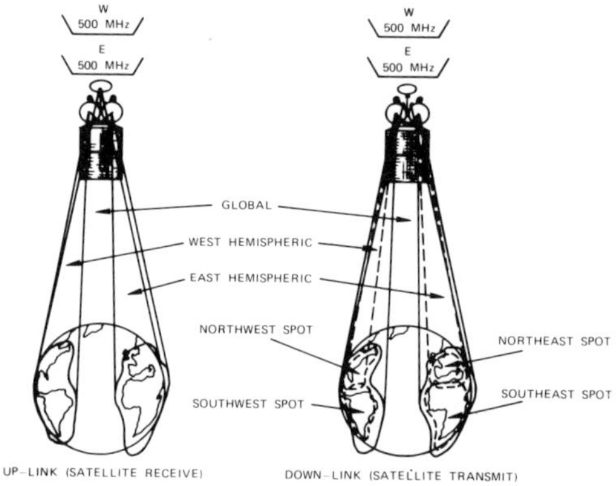

Fig. 2 Intelsat IV-A antenna coverage capability.

Satellite Communications Configuration

Traffic Distribution Patterns

The cross-connectivity requirements for the Intelsat IV-A satellite initially were based on meeting the traffic distribution pattern of the Atlantic Ocean region. Later, in the planning stages, the usefulness of this satellite in the other operating regions, i.e., the Pacific and Indian Oceans, also was considered. Consequently, the satellite as now configured is expected to operate in the Atlantic region between year-end 1975 and year-end 1980 and can handle traffic in the other regions in later years.

Figure 3 shows the basic traffic distribution pattern in the Atlantic Ocean region. It should be recognized, however, that, although most of the traffic can be represented in this manner, there are many small links that do not fit this pattern and therefore require the use of a global beam capability.

Antenna Patterns and Transponder Connectivity

Based on the data in Figs. 1– 3, it is apparent that reuse of the frequency in the East and West hemispheres provides the maximum spatial separation and at the same time encompasses most of the land mass in both the Atlantic and Pacific Ocean areas. Initially, it was considered undesirable to include a global beam capability, since this allows a frequency to be used only once. Geographic locations requiring service which tended to fall generally between the East and West beams therefore were to be serviced by "secondary coverage" beams. These would be either separate offset antenna

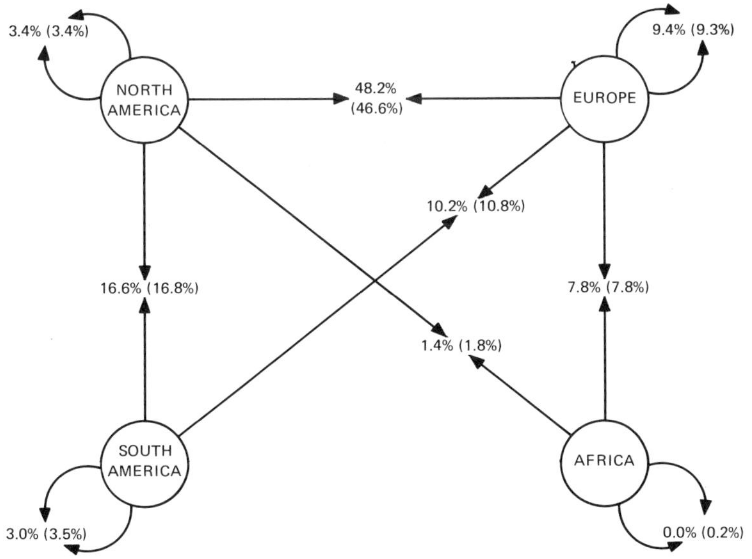

Fig. 3 Typical Atlantic primary traffic distribution for 1977 and (1978).

feeds or carefully defined areas on the slope of the main beam with specifically defined parameters (e.g., antenna gain, satellite G/T, and saturation flux density). Examples of such coverage requirements are Canary Island, Ascension Island, and Senegal in the Atlantic and New Zealand and Indonesia in the Pacific.

When this satellite later was evaluated for use in the Indian Ocean region, it was considered necessary to incorporate global beams, since large areas of land mass with fairly large traffic requirements fall between the two main beams. In operational use, it is expected that the secondary coverage capability will be utilized where possible because it allows greater overall channel capacity to be realized than the use of global beam transponders. The resulting antenna pattern is depicted in Fig. 2.

In terms of providing flexible connectivity between transponders to meet the traffic patterns of Fig. 3, transponders were made available for operation in either the hemispheric or spot beam modes. Careful consideration of all operating modes in each ocean area resulted in the switching connections shown in Fig. 4.

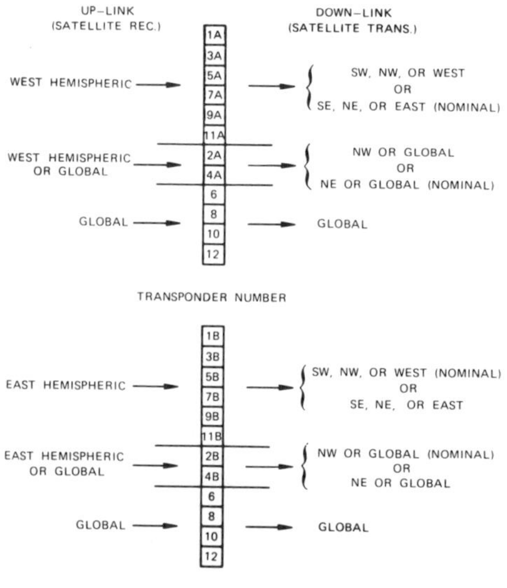

Fig. 4 Switching connections available on Intelsat IV-A.

Referring to the block diagram in Fig. 5, note that three different switch designations have been applied. The purpose of each of these switches for selecting the desired antenna mode is as follows:

1) C switch: These are connected to the input of each transponder and provide the facility for cotransponder pairs to be interchanged between East or West downlink beams.

2) S and T switches: The output of each transponder is connected to either an S or T switch. For even-numbered transponders 2 and 4, the S switch is used to select between a North spot beam or the global mode of operation. On odd-numbered transponders 1-11,

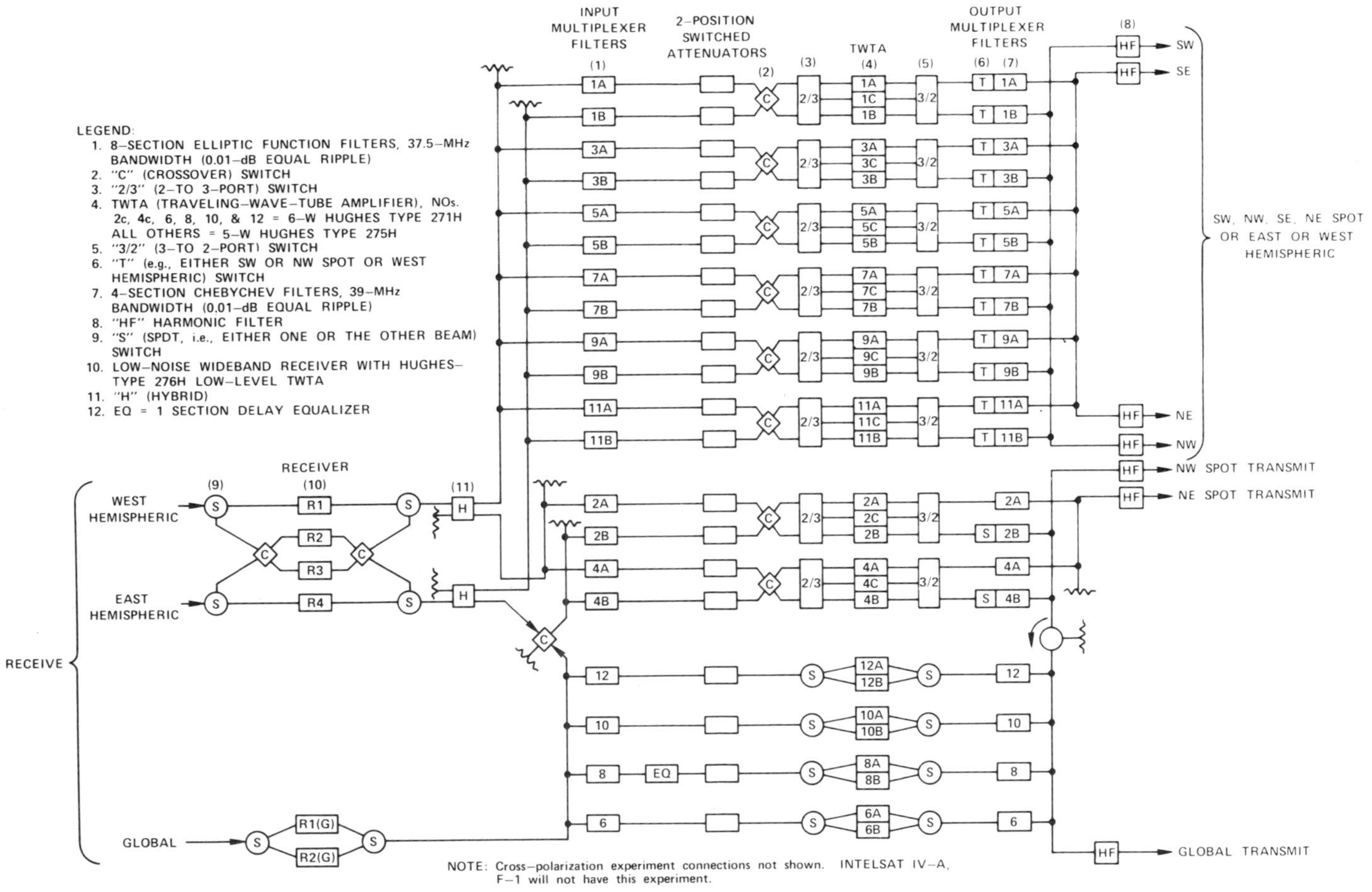

Fig. 5 Intelsat IV-A block diagram.

the T switch connects transponders to either North or South spot beams and also permits two spot beams to be combined into a composite hemispheric beam by paralleling the appropriate North and South antenna beams.

The T switch plays an important part in Intelsat IV-A system development. In the early years of the satellite, i.e., before saturation, combining the output of a transponder into a hemispheric mode will mean that many Earth stations transmitting a single global carrier will be required to add only one more transmit chain if they transmit to both the East and West hemispheres, thus minimizing their transition costs. Under these circumstances, maximum use is expected of the T switch connection (hemispheric). As saturation is reached, more spot beam operation (with its greater EIRP capability) will develop. There will be circumstances, however, where a transponder switched to the South mode may be only partially filled, whereas those in the North are fully saturated. Switching this unfilled transponder to the T mode still will allow it to fulfill its South traffic requirements while allowing the unfilled portion also to be utilized for more North beam traffic.

Satellite Transmission Characteristics

One of the main underlying features in the development of the Intelsat IV-A satellite was to attempt to minimize its impact on Earth stations as it is introduced into service, while at the same time obtaining the maximum channel capacity available. Accordingly, the transmission characteristics were selected to reflect closely those of the Intelsat IV satellite within, of course, any constraints dictated by actual Intelsat IV-A development.

Transmission Link Parameters

Since the global beam transponders in the Intelsat IV-A satellite are similar in output power to those of Intelsat IV, the satellite receive G/T and saturation flux densities also are similar. For the Intelsat IV, the G/T is specified as -18.6 db/°K, and the single-carrier saturation flux densities (ψs) are specified as -73.7 db-w/m^2 (for single-carrier operation) and -68.7 db-w/m^2 (for multiple-carrier operation). In the Intelsat IV-A, the saturation flux densities are specified as -75.0 db-w/m^2 and -67.5 db-w/m^2, respectively. The minor differences between the two specifications result from compensation necessary to mate with values selected for Intelsat IV-A hemispheric antenna gain.

On the Intelsat IV, all transponders are connected to the global receive antenna and therefore have the same satellite G/T for both spot and global beam modes. In the Intelsat IV-A, the receive antenna for the spot/hemispheric transponders is a hemispheric beam antenna, one each for the West and East hemispheres. Since the coverage area has been decreased, the antenna has an increased gain of about 6 db over the global beam, with a specified G/T (at beam edge) of -11.6 db/°K. This improved sensitivity will provide sufficient uplink carrier-to-noise performance while employing the same saturation flux densities of -75.0 db-w/m^2 for single-carrier saturation and -67.5 db-w/m^2 for multicarrier operation. Based on these specifications, typical link parameters were developed as shown in Table 1.

With the exception of the effects of cochannel interference, transponder optimization on the Intelsat IV-A satellite is carried out in the same manner as the Intelsat IV by using a tradeoff between down-path noise and transponder intermodulation as a function of transponder input backoff.[1]

Table 1 Typical Intelsat IV-A link parameters

Item	Spot Beam		Hemispheric Beam		Global Beam
	Single Carrier	Multi Carrier	Single Carrier	Multi Carrier	Multi Carrier
EIRP at beam edge, db-w	29.0	29.0	26.0	26.0	22.0
$Backoff_{in}$, db	...	-12.0	...	-11.0	-10.0
$Backoff_{out}$, db	...	- 7.0	...	- 6.0	- 4.2
Usable bandwidth, MHz	36.0	32.4	36.0	32.4	32.4
(36 MHz - 10% guardband for multiple carriers), db-Hz	75.5	75.1	75.5	75.1	75.1
Saturation flux density, db-w/m^2	-75.0	-67.5	-75.0	-67.5	-67.5
Satellite G/T, db/°K	-11.6	-11.6	-11.6	-11.6	-17.6
$C/N_{up\text{-}path}$, db	29.0	24.9	29.0	25.9	20.9
$C/N_{intermod}$, db	...	23.5	...	21.8	18.9
$C/N_{down\text{-}path}$, db (includes 0.6 db geo. adv.)	26.7	20.1	23.7	18.1	15.9
C/N_{total}	24.7	17.6	22.6	16.1	13.3
Loss due to co-channel interference, C/I ~ 24 db	~ 1.5	~ 0.8	~ 1.5	~ 0.8	...
$C/N_{available}$, db	23.2	16.8	21.1	15.3	13.3
Typical number of channels per transponder	1300	700	1100	600	450

Beam Isolation

Since the same frequencies are used in both the East and West beams, the required degree of isolation needed to keep cochannel interference within acceptable levels is quite important. The present specification requires 27 db of isolation between one beam and its interfering sidelobe component in the other beam. This number applies to both the uplink and the downlink. Since the spectrum of an FM signal has a nonuniform distribution, it cannot be treated as thermal noise. As a result, two carriers located at the same frequency would cause greater mutual interference than they would if they were offset in frequency. Based on an analysis of a number of frequency plans, it is estimated that, with 27 db of isolation and considering the types and sizes of carriers involved, the interference per carrier can be kept in the region of 1000 to 1500 pw0p (pico-watts psophometrically weighted, measured at a point of zero reference level). A more detailed description of this analysis is presented in a later section on cochannel interference.

Impact on Earth Stations

Transmission design for the Intelsat IV-A system has been developed toward minimizing the transition of Earth stations presently configured for Intelsat IV operation. Accordingly, carrier sizes are not expected to change when the new system is implemented, with the possible exception of the addition of a few new carriers to the existing list found in Ref. 2.

Although the Intelsat IV-A will have a 6-db better G/T than the Intelsat IV, the maximum Earth station power levels presently used as guidelines in Ref. 2 will remain almost the same. There are several reasons:

1) The gain stability of the Intelsat IV-A has been specified to be ±2 db over its operating lifetime. Toward the end of satellite design life, Earth station power could be required to make up for a difference in sensitivity.

2) Cochannel interference will be a factor that must be compensated for by power adjustment. Small carrier sizes paired with large carriers may need from 1 to 2 db more power to compensate for this impairment than if it were not present.

3) As many as six global beam transponders could be placed in service, particularly in the Pacific and Indian Ocean regions. Since the global beam on the Intelsat IV-A is essentially the same as the Intelsat IV, carriers in these

transponders would need the same power levels as found in the present system.

Other areas that may impact Earth stations are summarized below:

1) With the modified filter specifications on the Intelsat IV-A, transmission impairments such as crosstalk and group delay will require equalization at the Earth stations for a few large carrier sizes.

2) Earth station rf out-of-band emission could be a potential problem in the Intelsat IV-A system and could have a direct impact on Earth station high-power amplifier (HPA) size and operation.

3) The concept of frequency reuse will mean that more smaller carriers are required to be transmitted from most Earth stations because of the physical separation of geographic locations by spot and hemispheric beams. This will mean the addition of one or two more up-converter chains.

4) All existing Intelsat IV EIRP monitoring facilities in the Atlantic Ocean region are located above the equator. Two new monitoring Earth stations will be required to control carriers that can be viewed only in the SW and SE spot beams.

Transmission Impairments

With respect to transmission planning, the Intelsat IV-A satellite will be different from the Intelsat IV, not only because of the introduction of cochannel interference but also because of modifications in filter requirements. Each of the following Intelsat IV-A impairments therefore will be discussed separately: 1) cochannel interference (due to beam isolation requirements); 2) Earth station rf out-of-band emission; 3) satellite group delay and dual path distortion; 4) adjacent transponder interference (due to satellite TWT intermodulation); and 5) satellite and Earth station crosstalk.

Cochannel Interference

Considering the wide geographic dispersal of Earth stations accessing Intelsat satellites and variations in the electromagnetic environment surrounding those stations, careful frequency coordination is necessary. Extensive use of multidestinational carriers (used in order to minimize the number of carriers in the system) dictates that a single carrier frequency be cleared with as many as 12 to 20 different Earth station locations. Since there is also a wide range in carrier sizes, Intelsat IV-A frequency plans show carriers with large variations in capacity and power density located through the satellite band. In a typical frequency plan, maximum interference is expected to occur in the carrier with the smallest power density from the carrier with the largest power density if they happen to be cochannel (or nearly cochannel). The overall problem reduces to a search for a plan in which a certain amount of flexibility in distributing the capacity, power density, and frequency locations is retained within the constraints of the available isolation between the two cochannel beams.

Another important aspect when setting up such a frequency plan is that the interference between the two systems should remain within certain specified bounds, irrespective of the loading of the carriers (i.e., whether the carrier is carrying the full "busy hour load" for its nominal capacity or if it has a light load with an energy dispersal signal applied). When carriers are located nearly cochannel, the interference on one carrier actually increases with decreasing load on the other carrier.

The method used to develop frequency plans is to lay out manually all of the carriers assigned to each of the cochannel transponders in a manner so as to maximize carrier frequency offsets. In practice, each transponder is initially brought to its optimum operating condition and then adjusted for cochannel interference. A program is presently under development to automate this evaluation fully by employing computerized techniques. Figure 6 shows a typical representation of how the interference between carriers can vary as a function of the carrier offset.[3]

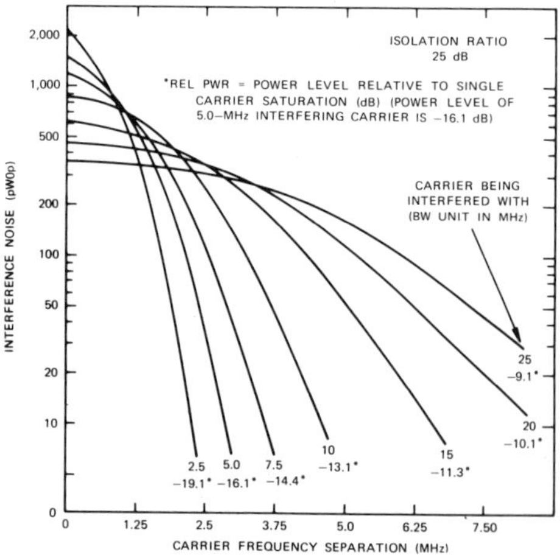

Fig. 6 Cochannel interference noise (effect of 5-MHz bandwidth carrier on other carriers).

Earth Station Radio-Frequency (rf) Out-of-Band Emission

Out of the total per channel system noise budget, 500 pwOp has been allocated for Earth station rf out-of-band emission. (The phrase "rf out-of-band emission" may require clarification. In this paper, it refers to the unwanted radiation generated by the intermodulation of carriers within an Earth station's wide-band HPA. When transmitted, these intermodulation products may fall within the satellite frequency band or outside of the satellite band. Radio-frequency out-of-band emission therefore is meant to indicate rf emission that is out of the wanted carrier's band.) There is a tradeoff between usable transmitter output power and the amount of emission noise the satellite transmission system can tolerate. When a high-power, wide-band traveling-wave tube (TWT) is operated at saturation, its full output power can be realized; however, it also will produce severe intermodulation emission with multiple-carrier transmission. In the Intelsat IV system, it was determined that an output backoff of about 7 db could be employed and still allow the system to be operated within a 500-pwOp allowance.[4] For a 12-kw TWT, this would mean a loss of 9.6 kw, or a usable output power of 2.4 kw just to compensate for the effects of out-of-band emission.

The determination of an output backoff of approximately 7 db as a criterion was in itself a compromise. Depending on the carrier arrangement, it was determined that an emission product of 26 db-w/4 kHz could be produced by a transmitter operating with a 7-db output backoff. With the particular link budget of carriers adjusted in the Intelsat system, this could equate to as much as 4000 pwOp of emission noise, the amount related to the carrier size affected. Therefore, even with a 7-db output backoff, there would be no guarantee that 500 pwOp could be met in the system noise budget unless another alternative were available. In this respect, it was noted that, if the affected carrier were to increase its uplink power level, the amount of emission noise could be reduced back down to 500 pwOp through an increase in carrier power. By using this give-and-take method, it was felt that a 7-db output backoff could be applied, and at the same time emission noise could be kept within the system noise budget of 500 pwOp.

The tradeoff between transmitter output power and uplink adjustment has worked favorably in the Intelsat IV system because there has been sufficient power margin in the satellite transponders to allow the uplink of some carriers to be increased, followed by a reoptimization of the remaining carriers for the best transponder intermodulation structure and carrier performance. However, as the transponders become more heavily loaded, there is less tolerance for any increase in uplink power without upsetting the delicate power balance of the other carriers in that transponder. The point here is

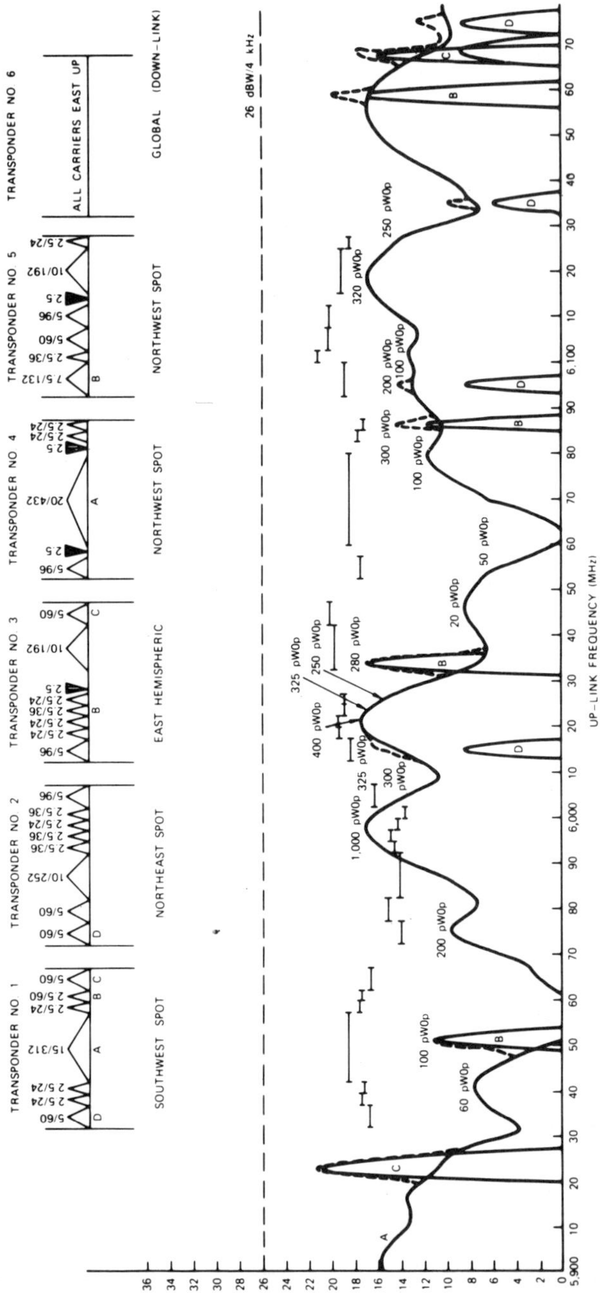

Fig. 7 Intelsat IV-A Earth station rf out-of-band emission environment (1976 Atlantic primary).

(Figure 7 is continued on opposite page.)

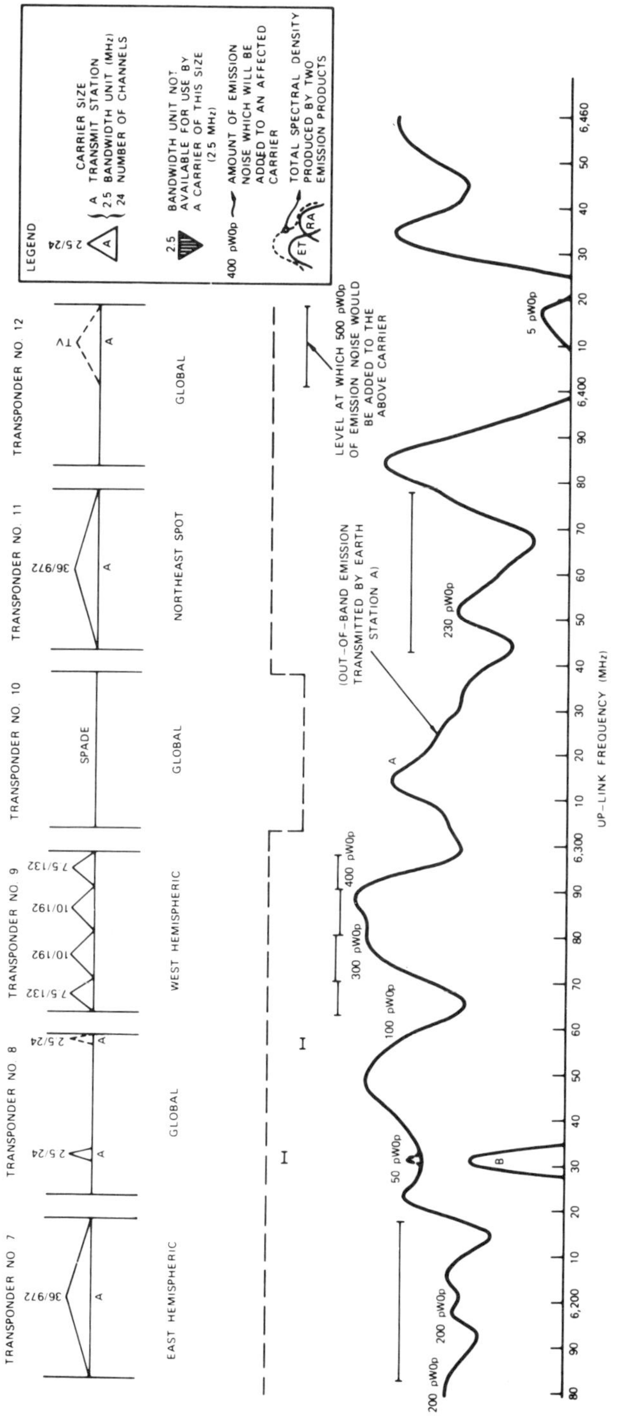

Fig. 7 (Cont.) Intelsat IV-A Earth station rf out-of-band emission environment (1976 Atlantic primary).

that in a fully loaded system it will be necessary to limit Earth station out-of-band emission to 500 pwOp through some means other than uplink power adjustment. This can be accomplished by methods such as careful frequency planning, larger TWT's, or multiplexed klystrons, but not through readjustment of uplink power.

A typical out-of-band emission environment expected for the West uplink of the Intelsat IV-A system in 1976 is shown in Fig. 7. In general, most of the products are below the Intelsat IV criterion of 26 db-w/4 kHz by about 8 db. This is because of the large number of small carrier sizes transmitted from each Earth station, which tends to spread the total emission spectrum out over the 500-MHz band instead of concentrating them in peaks. The Intelsat IV-A system will not be as flexible for uplink power adjustment to overcome the effects of out-of-band emission as the Intelsat IV system. As has been discussed in the section on cochannel interference, an intricate balance in power levels will be needed just to insure that cochannel interference is kept to a minimum. To introduce another factor, such as out-of-band emission, into the juggling act may be difficult to implement, especially if contingency carriers are transmitted on an intermittent basis.

Satellite Group Delay and Dual Path

Since the Intelsat IV-A system will be configured primarily on an FDM-FM-FDMA network, it was felt that the input and output filter specifications could be modified with respect to the Intelsat IV in order to permit a weight reduction. Intelsat IV filters and group delay equalizers were designed with the expectation of extensive use of high-capacity, low-modulation index FM carriers. Because of a somewhat different course in system development, Intelsat IV-A transponders will employ many relatively small carriers and be able to accept up to 4-phase, 60-Mbit/sec TDMA transmission without requiring group delay equalizers (with the exception of transponder 8, which includes a one-section equalizer for TDMA field trial tests).

With this in mind, the Intelsat IV-A will be configured with an eight-section pseudoelliptic[5] function Invar filter in place of the ten-section Chebyshev input filter used on the Intelsat IV. The output filter will be a four-section Chebyshev, whereas a six-section Chebyshev had been chosen for the Intelsat IV. In addition, only transponder 8 on the Intelsat IV-A will retain a group delay equalizer. A graphical comparison of Intelsat IV and Intelsat IV-A filter specifications is shown in Figs. 8 and 9.

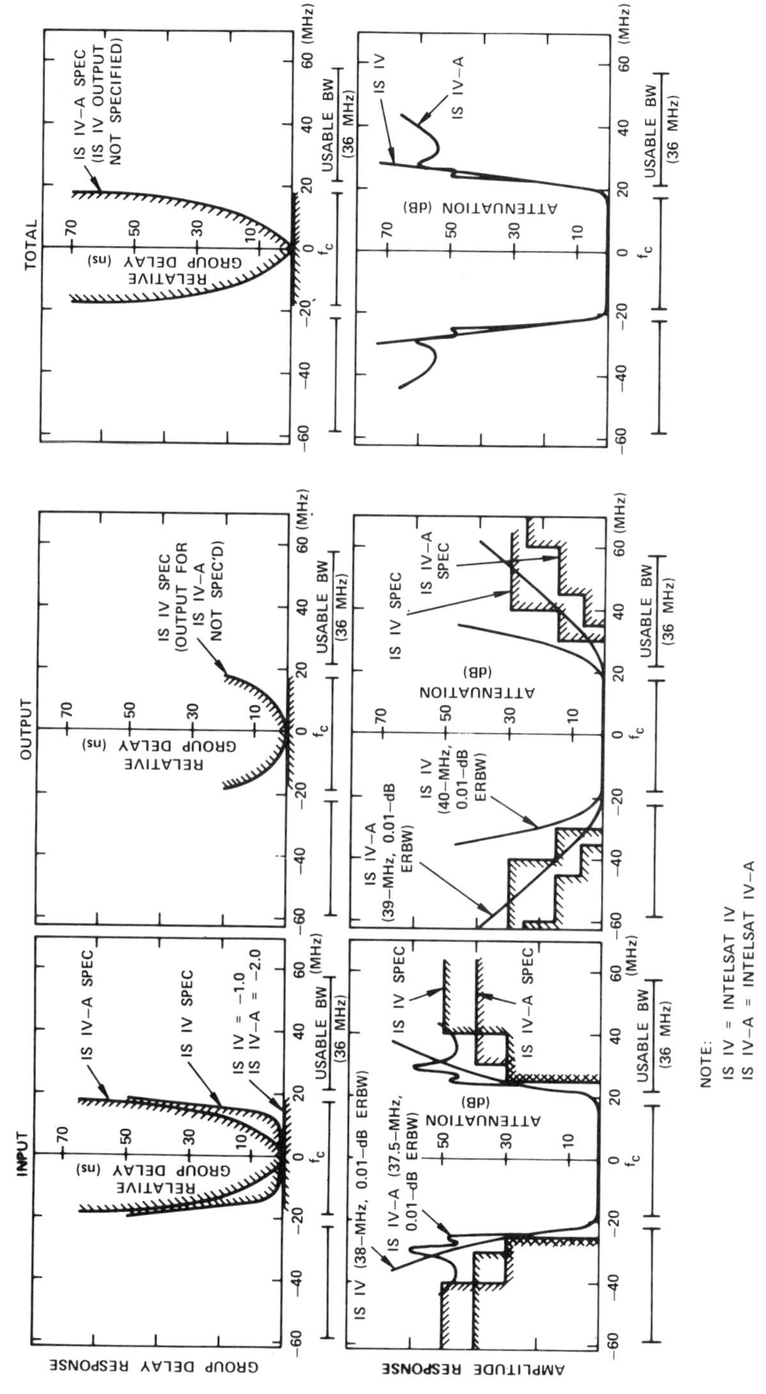

Fig. 8 General comparison of Intelsat IV and Intelsat IV-A input and output filters.

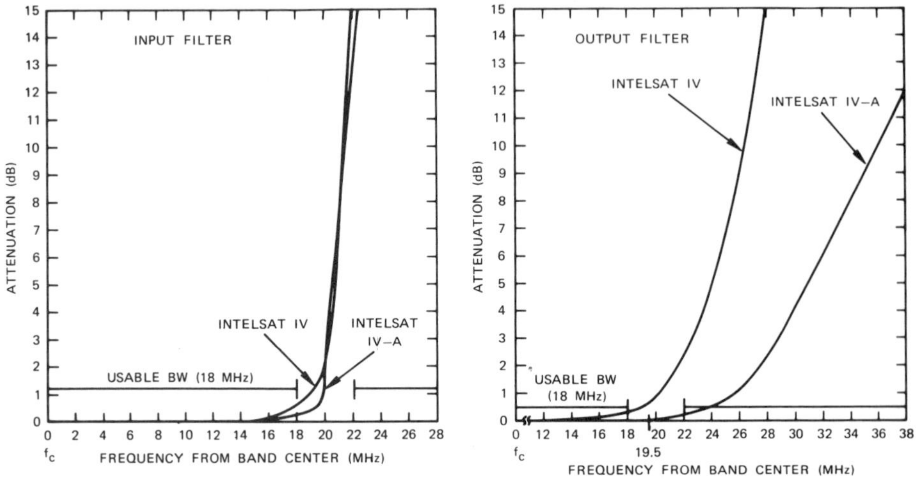

Fig. 9 Detailed comparison of Intelsat IV and Intelsat IV-A input and output filters.

The opportunity for a carrier to travel through an indirect path and be recombined at a receiving Earth station is found near the band edges of each transponder, where the skirts of both the input and output filters overlap. A dual path is thus available to a carrier where it can be transmitted in part through the adjacent channel. This impairment source will introduce phase nonlinearities in the wanted signal and can be treated as group delay noise, which is principally a function of the following conditions: 1) gain difference between paths due to input satellite flux density and transmit antenna gain; 2) phase difference between paths; and 3) filter rejection characteristics. Phase differences are difficult to analyze because they depend upon the geographical location of the receiving Earth station and are hard to determine uniquely.

Dual path distortion is generated primarily by the linear components of the combined group delay response and gives rise to second-order intermodulation products. When measured, the combined group delay response is due to both the direct (intrinsic) and the indirect paths that the signal would follow. The dual path distortion under discussion is coherent, that is, it is the vector sum of the voltage of each path. At the frequency where transmission gain through both paths is equal, the resulting distortion may be zero if the paths are out of phase. Conversely, if the paths are in phase, the maximum distortion will be 6 db. As the relative gain between the paths increases, distortion can be greater than 6 db.

Distortion due to intrinsic group delay and dual path is expected to be held to a 500-pwOp allocation in the Intelsat IV-A system noise budget. Two typical examples are shown in Figs. 10 and 11. (These figures represent the results of a computer system model developed by Comsat under Intelsat sponsorship to simulate the effects of distortion due to dual path. The program has been compared with measured results under various study contracts and is believed to be a reliable model. A discussion can be found in Ref. 6.) The gain difference between these transponders can be seen readily in Fig. 12, which shows the relative gain for each of the 20 transponders in a typical Intelsat IV-A frequency plan. In each example, values have been determined for the total distortion with International Radio Consultative Committee (CCIR) emphasis used in the transmission chain and for a variety of phase possibilities.

As has been mentioned, one reason the group delay specification in the Intelsat IV-A system could be reduced was because only a few large carrier sizes would be required. In

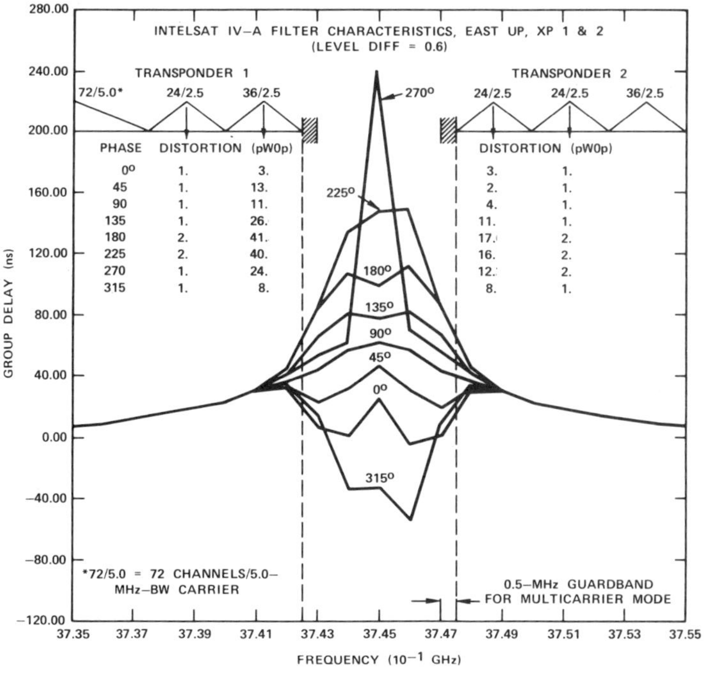

Fig. 10 Typical small-carrier group delay and dual path distortion (Intelsat IV-A).

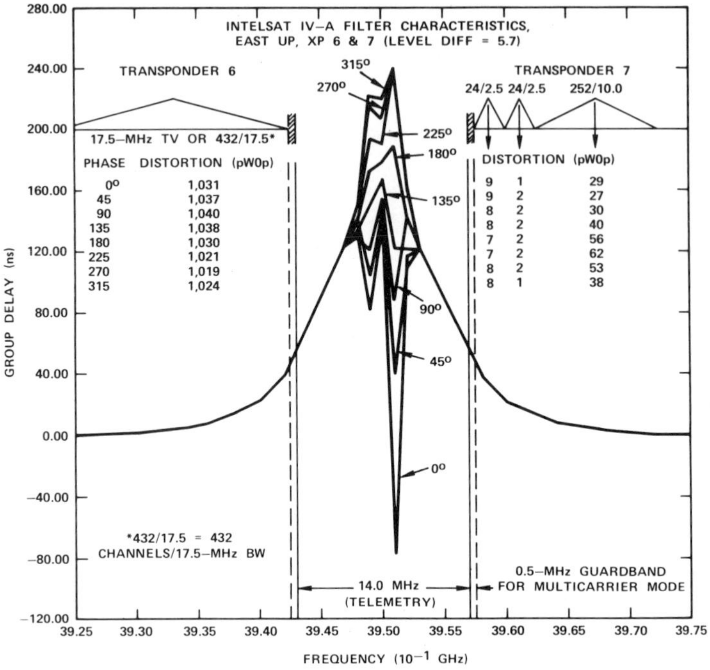

Fig. 11 Typical large-carrier group delay and dual path distortion (Intelsat IV-A).

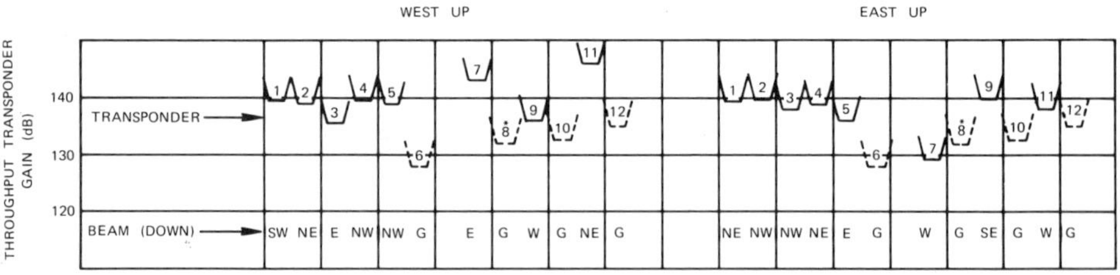

NOTE:

1. 8* USED FOR SPADE AND POSSIBLY TDMA FIELD TRIALS. ONLY THIS TRANSPONDER HAS SPECIAL GROUP–DELAY EQUALIZATION.
2. G = GLOBAL BEAM.
3. SW, NE & NW = DOWN–LINK SPOT BEAM.
4. E & W = DOWN–LINK HEMISPHERIC BEAM.
5. = GLOBAL BEAM (CAN BE ACCESSED ON EITHER E OR W UP–LINK)

** REFERENCE INTELSAT IVA FREQUENCY PLAN IN FIGURE 14.

Fig. 12 Intelsat IV-A transponder gain differences (dual path considerations).

general, carrier sizes with bandwidth units of 2.5, 5.0, and 10.0 MHz will be in the majority. Group delay and dual path distortion for carrier sizes in this category (at band edge) are shown in Fig. 10. There are some cases, however, where larger carrier sizes will be placed into operation. For example, the effect of locating a 432-channel/17.5-MHz carrier at band edge is shown in Fig. 11. The deviation of a large carrier such as this is expected to produce a distortion level greater than the budgeted 500 pw0p. Alternatives to this problem are simply to relocate the carrier toward the center of the transponder or, if relocation is not feasible, to use equalization at the Earth station. Since there are relatively few of these situations, the Intelsat IV-A system is expected to be able to meet the system noise budget of 500 pw0p for group delay and dual path distortion for all cases, even with a filter specification less restrictive than the Intelsat IV.

Adjacent Transponder Interference (Caused by Intermodulation)

When satellite transponders are operated in the multicarrier mode (accessed by many carriers), intermodulation products are generated in the output TWT and are transmitted together with the desired signals. Although the out-of-band products are attenuated by the output filter, they can, in some cases, be of sufficient magnitude to cause considerable interference to signals transmitted from adjacent transponders. For example, when the two television signals in Fig. 13 are transmitted through global beam transponder 12, the TWT is operated very close to saturation. Of the two third-order intermodulation products generated (type 2A-B and 2B-A), the center frequency of the lower one falls 5.75 MHz inside of transponder 11. The output filter attenuation at this point is approximately 1.8 db. The other product falls out of the satellite frequency band. Since the intermodulation from transponder 12 is from a global beam transponder, it is received at all Earth stations. The effect of this interference is twofold:

1) Stations located in the Northeast spot beam which are receiving the large carrier transmitted from the Western hemisphere will experience interference to their reception.

2) Stations located in the Western hemisphere which are receiving the large carrier transmitted from the Eastern hemisphere will experience interference to their reception.

The magnitude of the interference depends on the configurations of both the interfering and interfered with transponder. In this case, where it is required for satisfactory TV performance that transponder 12 be operated at or near its saturation point, the placing of large carriers in the adjacent transponder helps to minimize the amount of interference.

It should be noted that intermodulation products generated by global beam transponders always cause interference to some carriers in adjacent spot beam transponders. However, since the global transponders have the lowest EIRP, whereas carriers emanating from spot beam transponders generally have the highest power levels, this interference can be kept to a minimum. On the other hand, intermodulation products generated by spot or hemispheric beam transponders adjacent to global beam transponders do not always interfere with the global beam carriers. This will depend on which spot or hemispheric beam is activated and the destination of the global carriers.

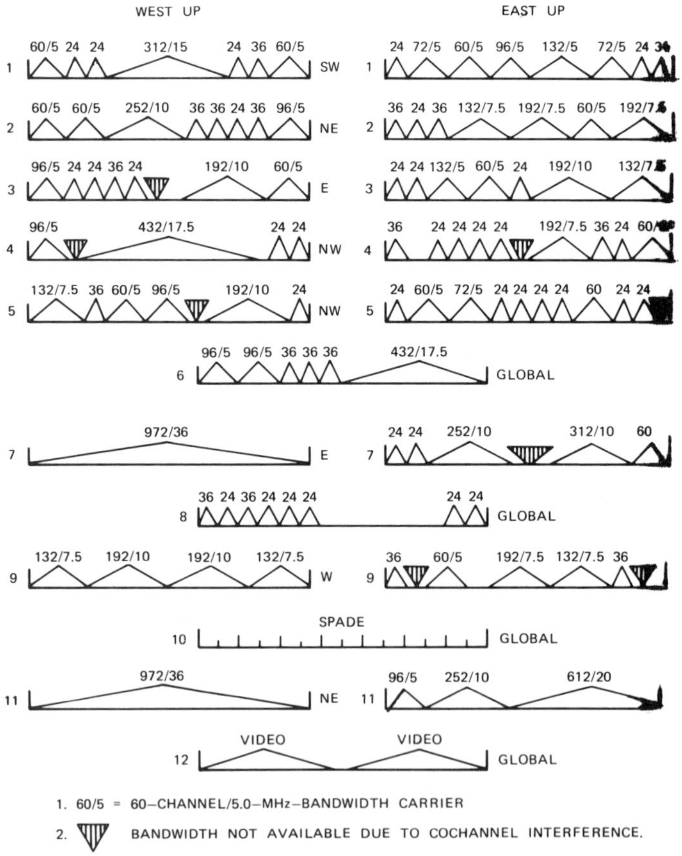

Fig. 13 Typical Intelsat IV-A frequency plan.

Although the actual interference values will depend on the individual transponder frequency plans, the range of interference values expected is summarized in Table 2. In comparison, adjacent transponder intermodulation was not detectable in the Intelsat IV system.

Crosstalk

When FDM-FM carriers are applied to a multicarrier satellite repeater, intelligible crosstalk is produced by a transmission vs frequency deviation, followed by an amplitude-to-phase conversion (AM/PM conversion), that is, as the carrier passes through the TWTA, its phase is changed as a function of any amplitude modulation that may be present. Another carrier passing through the TWTA also will vary in phase at the same rate as the AM modulation and thereby pick up intelligible information from its neighboring carrier.

The crosstalk objective in the Intelsat system conforms with the Geneva CCITT Recommendation[7] for intelligible crosstalk to be -58 db or better. In preparation for the Intelsat IV satellite system, extensive laboratory testing and modeling was conducted to insure that this objective could be met through appropriate specifications.[6] Field tests on the actual system later were conducted with the indication that even worse-case situations did not produce detectable crosstalk levels. There was, therefore, reasonable assurance that the crosstalk specifications for the Intelsat IV-A could be modified because of the different usage envisioned for this satellite. Through supportive modeling and calculations, an Intelsat IV-A contract specification for intelligible crosstalk was established 6 db lower than the Intelsat IV. This, in

Table 2 Intelsat IV-A interference due to adjacent transponder intermodulation

Example No.	Case	Interference In Adjacent Carriers, pWOp
1	2 TV signals (global beam) adjacent to large carrier in multicarrier mode (transponder 12 ⟶ transponder 11 W)	...
2	Multicarrier (hemispheric) adjacent to multicarrier global (transponder 7 ⟶ transponder 8)	≅ 500 - 1000
3	Multicarrier (spot) adjacent to multicarrier spot (transponder 2 NW ⟶ transponder 3 NW)	≅ 1000 - 1500

turn, meant that the gain slope tolerance of an eight-section elliptic function input filter would be compatible with the overall system objectives.

There are some reservations, however, concerning whether or not a crosstalk level of -58 db can be met in the few rare instances where two carriers with large bandwidth units are placed side-by-side in the same transponder. It is felt that situations in this category can be overcome by insuring that proper safeguards are applied in the frequency planning process. At the time of this writing, further investigations are proceeding to determine if there still may be some large carrier cases in which corrective measures are needed at the Earth stations in the form of gain slope equalization before the input of the HPA. There are limitations to this approach in that crosstalk also can be produced in the Earth station HPA.

Whatever the end result of this investigation, it is expected that crosstalk can be held within the CCITT intelligible crosstalk recommendation. If necessary, alterations in the Intelsat IV-A frequency plan will be made to overcome the few potential problems due to crosstalk.

Intelsat IV-A Frequency Planning

Frequency planning for the Intelsat IV-A system is more complex than the Intelsat IV system because of the additional transmission constraints resulting from frequency reuse techniques. A rough visual approximation of the magnitude of this difference can be seen by comparing typical frequency plans for both systems in Figs. 13 and 14.

Referring back to Figs. 1 and 2, a further comparison of the Intelsat IV and Intelsat IV-A systems can be presented. For the Intelsat IV-A, there are two possible uplink modes: hemispheric (East and West) and global. The downlink has three modes: hemispheric (East and West), spot (NW, NE, SW, and SE), and global. In the Intelsat IV, only global access is employed on the uplink, and either global or spot can be chosen for the downlink.

In order to provide such services as SPADE, TDMA, TV, and telephony to Earth stations outside the normal hemispheric coverage zones, four transponders were designated as global beam channels in the Intelsat IV-A. (SPADE is an acronym for single-channel-per-carrier, pulse-code-modulation, multiple-access, demand-assignment equipment.) With some constraints, the remaining eight cotransponder pairs (16 transponders) can

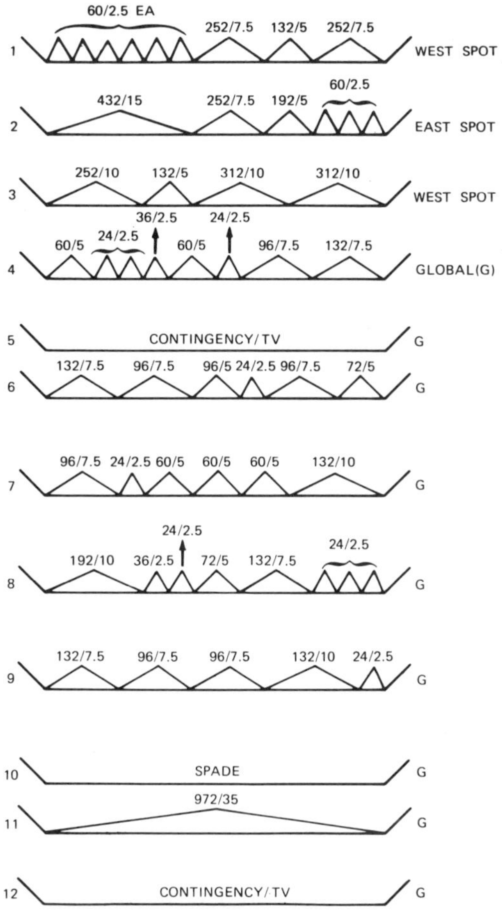

Fig. 14 Typical Intelsat IV frequency plan.

be switched to either a hemispheric or spot beam on the downlink; that is, of the potential 1000-MHz bandwidth made available by frequency reuse techniques, 160 MHz (40 MHz/transponder x 4) will be lost in order to provide unique services on a global scale. Within these bounds, the 20-transponder Intelsat IV-A is expected to be configured to handle 12,000 to 12,500 voice-grade channels (plus one TV and one SPADE transponder) for a typical Atlantic primary satellite frequency plan configuration. By comparison, an Intelsat IV Atlantic primary satellite can be configured for 7500 channels (plus one TV and one SPADE transponder). The net capacity increase of the Intelsat IV-A will be approximately 5000 channels.

Several technical considerations must be taken into account in Intelsat IV-A frequency planning to achieve maximum channel capacity in a frequency reuse environment. A brief explanation of some of these constraints follows.

Cochannel Interference

To minimize interference, the same carrier center frequency should not be assigned to two carriers in the same cochannel set. Spot and hemispheric beams should be paired together only as a last resort. Interference from spot into the hemispheric transponder will be increased 3 db if this configuration is employed.

Adjacent Transponder Interference (Due to Intermodulation)

The output filter skirts on the Intelsat IV-A are much wider than on the Intelsat IV and therefore do not offer as much rejection to adjacent transponder intermodulation. In some cases, judicious frequency planning can resolve the problem. For example, in Fig. 13 the two large intermodulation products produced by the dual video carriers in transponder 12 will create a minimum amount of interference because one of the products will fall out of the satellite band, whereas the other product will fall under the large carriers in transponder 11. Care also must be taken in frequency planning to use different downlink beams for adjacent transponders.

Dual Path and Crosstalk

With the particular multisection input and output filters used in the Intelsat IV-A system, group delay at transponder band edge will be more serious than in the Intelsat IV. As a frequency-planning criterion, carriers with large bandwidth units should not be placed at band edge, where they are more susceptible to distortion due to group delay and its associated dual path impairment. The gain difference between transponders is also greater in the Intelsat IV-A than in the Intelsat IV and will produce greater dual path distortion. To reduce severe crosstalk possibilities, large carrier sizes should not be placed in the same transponder.

Connectivity requirements in the spot and hemispheric beams will mean that each Earth station will transmit several small carrier sizes. For the frequency plan shown in Fig. 13, approximately 3/4 of the total carriers have bandwidth units of 5.0 MHz or less. Having a large number of small carrier sizes in the system has the advantages of 1) less group delay and crosstalk distortion, and 2) a lower overall rf out-of-band emission environment.

Obviously, frequency planning is becoming more complicated with each generation of Intelsat satellite systems. In an effort to expedite technical evaluations of the many carrier arrangements that are proposed before a finalized plan is agreed upon, a computer system model has been developed that will simulate many of the impairments mentioned previously and automatically balance power levels for the best performance. A system model is currently in use for the Intelsat IV satellites and has worked very favorably over the past three years.[1]

Conclusions

Transmission design for the Intelsat IV-A satellite has been developed around a concept that will create a minimum change in present-day Earth station equipment and operation. In addition, the satellite will possess the channel capacity necessary to fulfill international satellite traffic requirements in the Atlantic Ocean region until the early 1980's. To achieve these goals while retaining the use of an Atlas Centaur launch vehicle, a technique of frequency reuse by spatial separation of East and West hemispheric beams will be employed.

Theoretically, frequency reuse should double the available satellite bandwidth allocation and thereby double the number of available channels. In practice, however, a typical Intelsat IV-A Atlantic primary satellite will fall short of this objective by about 20%. Several factors account for this loss:

1) The requirement to provide global beam coverage for special services such as SPADE, TV, secondary coverage zones, and versatility in application with other ocean regions has removed four transponders from the satellite configuration which could have been employed with frequency reuse. This accounts for 160 MHz of bandwidth, or 16% of the total 1000 MHz initially available.

2) With the exception of cochannel interference, the transmission impairments discussed in this paper (adjacent transponder interference, group delay and dual path distortion, crosstalk, and Earth station rf out-of-band emission) can be held within the normal Intelsat system noise budget apportionment by either careful frequency planning or appropriate countermeasures at the Earth stations. On the other hand, cochannel interference, which results from the coupling between spatially separated beams, will add a new contribution to the noise budget in the range of 1000-1500 pwOp. This impairment will cause some carriers intentionally to be left out of certain cotransponder configurations just to minimize its effects.

3) Frequency reuse will require the utilization of many small carrier sizes. Approximately 3/4 of the carriers in a typical Atlantic primary frequency plan will have bandwidth units of 5.0 MHz or less. Maximum channel capacity can be gained when a transponder is operated with a large, single carrier near saturation; this provides the best downlink EIRP and does not produce harmful TWT intermodulation. As more carriers are added, intermodulation increases and the transponder's channel capacity decreases.

4) The size of large, single carriers will be limited by the amount of interference it will create in a cotransponder. To minimize this interference, uplink power must be restricted, which in turn reduces the maximum channel capacity that the carrier might otherwise have been able to attain.

Although frequency planning in the Intelsat IV-A will be more intricate than the Intelsat IV because of the restraints mentioned in the previous paragraphs, automated techniques are presently under development to keep pace with the demands of day-to-day system operation. Although there are limiting factors, the Intelsat IV-A is expected to achieve its goal of handling the Atlantic Ocean region traffic through the early 1980's using the Atlas Centaur launch vehicle and with minimum Earth station disruption. When configured as an Atlantic primary satellite, the Intelsat IV-A should have a capacity of approximately 12,000 to 12,500 voice-grade channels (plus one TV and one SPADE transponder), or a net increase of about 5000 channels over a similarly configured Intelsat IV.

References

[1] Dicks, J. L., Schultze, P. H., and Schmitt, C. H., "Intelsat IV System Planning," Comsat Technical Review, Vol. 2, No. 2, Fall 1972, pp. 439-476.

[2] "Performance Characteristics of Earth Stations in the Intelsat IV System," ICSC-45-13 (Rev. 1), June 23, 1973, Intelsat.

[3] Pontano, B. A., Fuenzalida, J. C., and Chitre, N. K., "Interference into Angle-Modulated Systems Carrying Multichannel Telephony," IEEE Transactions on Communications, June 1973, Institute of Electrical and Electronics Engineers.

[4] Brown, M. P., Jr., "Review of Intelsat Earth Station RF Out-of-Band Emission Criteria," Comsat Technical Review, Vol. 4, No. 1, Spring 1974, pp. 147-168.

[5] Atia, A. E. and Williams, A. E., "New Types of Waveguide Band-pass Filters for Satellite Transponders," Comsat Technical Review, Vol. 1, No. 1, Fall 1971, pp. 21-43.

[6] Berman, A. L., Mahle, C., and Wachs, M. R., "Transmission Modeling;" also Chitre, N. K., "Transmission Simulation;" also Schultze, P. H. and Pontano, B. A., "Field Tests," Comsat Technical Review, Vol. 2, No. 2, Fall 1972, pp. 489-561.

[7] Geneva CCITT Recommendation, G.151, Green Book, Vol. III-1, Dec. 1972, ITU Publication.

ALGERIAN DOMESTIC SYSTEM

Abdelkader Bairi*

Algerian Ministry of Posts and Telecommunications,
Algiers, Algeria

Abstract

In order to develop expeditiously, both economically and culturally, Algeria has endeavored to implement a modern communications network throughout the country, which has more than 2×10^6 km^2 (900,000 miles^2). During the planning stage, it became apparent that the conventional communications means such as cables or radio relays were not a practical economical solution to cover the desert portion of the country, which is 4/5 of the total of the country. Algeria also wishes to provide educational television to sizable but widely separated communities in the heart of the Sahara. A solution that seemed particularly attractive was the use of satellite communications. One alternative was to use a completely independent satellite system for Algeria. This was abandoned temporarily because of the extremely high costs associated with it. The other alternative was the utilization of Intelsat space segment capacity for both international and domestic services. A special Intelsat tariff and a specially designed Earth station network appear to offer the solution.

I. Introduction

Algeria, which is formally known as the Democratic and Popular Republic of Algeria, has a scattered population consisting predominantly of Arab peoples, but with also a sizable population of Berbers. Algeria's population is almost exclusively Moslem in its religious faith, and Arabic is spoken uniformly as the national language, although French and Berber also are spoken. Because of the widely dispersed population, with many people living in desert towns and communities, and nomadic living patterns, it is not

---

Presented as Paper 74-493 at the AIAA 5th Communications Satellite Systems Conference, Los Angeles, Calif., April 22-24, 1974.

*Director of Telecommunication Equipment.

surprising that the Algerian telecommunications network is still in its development phase. There are approximately 1.2 telephones/100 inhabitants, 40 radios/1000 inhabitants, and 15 televisions/1000 inhabitants.

Despite the limitations of geography and population distribution patterns, Algeria has been pressing forward with plans to provide modern telecommunications throughout the country. The principal question has been how to provide the communities living in the heart of the Sahara, many of them important trading, mining, or tourist centers with a population of 10,000 or more, with reliable telecommunications, including urgently needed educational television. These communities, which include Ghardaia, El-Golea, El-Oued, Timimoun, Tindouf, Adrar, In-Salah, Bechar, Taman-Rasset, Ouargla, In Amehas, and Djanet, are separated from each other by distances of greater than 500 km (312.5 miles). These separations across the desert in the past have meant that the only possible means of telecommunications has been high-frequency radio. The possible alternatives for providing service, including microwave, troposcatter, or satellite communications, seemed too expensive, too difficult to install, or too difficult to maintain. The possibility of satellite communications, which could provide service flexibly to as many cities and towns as required, regardless of distance, seemed particularly attractive. Nevertheless, the costs of establishing a separate domestic communications satellite system simply were too expensive. It was at this time that the option of leasing satellite capability from the International Telecommunications Satellite Organization (Intelsat) was given active consideration. Under such an arrangement, sufficient capacity for domestic requirements might be obtained and at much lower costs than deploying an entire domestic satellite system. At the same time, parallel developments within Intelsat made such a planning concept seem reasonable.

II. Intelsat Provision of Domestic Lease Service

The Intelsat Organization, with a current membership of 89 Signatories, was created in August 1964 as an international joint venture, with membership open to all members of the International Telecommunication Union (ITU). This Organization was established "for the design, development, construction, establishment, maintenance and operation of the space segment of the global communications satellite system" (Intelsat Agreement, Article II) and was "to be available on a non-discriminatory basis to all areas of the world" [Intelsat Agreement, Article III(a)].

Under the provisions of the definitive Intelsat Agreements, which replaced the early Interim Agreements in February 1973, the following stipulations were made concerning domestic satellite services:

"The Intelsat space segment established to meet the prime objective shall also be made available for other domestic public telecommunications services on a non-discriminatory basis to the extent that the ability of Intelsat to achieve its prime objective is not impaired" [Intelsat Agreement, Article III(c)].

"The following shall be considered on the same basis as international public telecommunications services:

(i) domestic public telecommunications services between areas separated by areas not under the jurisdiction of the State concerned, or between areas separated by the high seas; and

(ii) domestic public telecommunications services between areas which are not linked by any terrestrial wideband facilities and which are separated by natural barriers of such an exceptional nature that they impede the viable establishment of terrestrial wideband facilities between such areas, provided that the Meeting of Signatories, having regard to advice tendered by the Board of Governors, has given the appropriate approval in advance" [Intelsat Agreement, Article III(b)].

Even before the new definitive arrangements specifically stated how domestic satellite services might be provided on Intelsat facilities, consideration was being given to this matter. In 1971, Intelsat began to give serious consideration to how its space segment capacity might be allotted for certain new types of services on a long-term basis. In particular, inquiries from Brazil for possible domestic services, from Mexico and Spain for full-time television exchange channel, and from Italy on behalf of the European Broadcasting Union for television services indicated that there might be a market for a new type of Intelsat service offering.

At that time, the Interim Communications Satellite Committee (ICSC), which has been replaced by the Board of Governors under the new Intelsat Agreements, requested that appropriate charges for such services be investigated by the Finance Committee of Intelsat. By the Fifty-Fifth Meeting of the ICSC, in September 1971, the Finance

Committee had completed its studies and recommended that a charge equivalent to the charge for 360 units of utilization (or 360 4-kHz voice channels) be established for a full Intelsat IV transponder, which is 1/12 the capacity of an Intelsat IV satellite.

At the same time, the ICSC agreed that Intelsat should charge 180 units for use of one-half of an Intelsat IV transponder, recognizing that a television channel of acceptable quality could be derived in this manner. Subsequently, specific terms and conditions were approved in 1972 for potential requirements that might apply to Brazil with regard to the lease of a full Intelsat IV transponder for domestic telecommunications and to Mexico/Spain with regard to the lease of a half Intelsat IV transponder for television service.

The detailed terms and conditions for these lease arrangements and the definition of the services involved need not be discussed here. What was significant, however, was the precedent of approving terms and conditions for a leased transponder service that would utilize small-aperture, non-standard Earth stations and that equal priority of service for purposes of restoration was guaranteed.

III. Intelsat and the Algerian Lease Arrangement

Even with these landmark precedents established with Intelsat, the solution for Algeria was still far from clear. The 360-unit rate for an Intelsat IV global beam transponder at the 1974 charge of $9000/unit results in a total charge of $3,240,000/yr. Since Algeria was planning on the basis of service to perhaps 10 to 12 locations and an Earth station network costing in the region of $\$5\text{-}7 \times 10^6$, the payment of such a high space segment charge was sufficiently unattractive that the Intelsat approach seemed doubtful.

The idea that emerged to solve this problem was a plan to spend slightly more on the ground segment so that Algeria could operate flexibly and quickly with either the Atlantic major path satellite, the Atlantic spare satellite, the Indian Ocean primary, or the Indian Ocean spare satellite. This design was intended to let Algeria utilize "spare" capacity on the Intelsat system on a pre-emptible basis in exchange for a reduced rate. Based upon the projected rates for a 5-yr service period from 1975 through 1979, an average rate of $\$10^6$/yr was derived on the basis of approximately one-half of the 360-unit transponder rate amortized over the 5-yr period. The service as proposed was guaranteed not to advance the saturation dates for the Intelsat system because the critical satellite is the Atlantic primary satellite, which would be utilized in no way for the Algerian domestic service. This concept works to the benefit of all Signatories by providing revenues not otherwise available and not requiring any additional satellite facilities. On the preceding basis, the Intelsat Board of Governors at its Fifth Meeting, held in Tenerife, Canary Islands, October 1973, approved the Algerian request in principle.

After further definition work concerning the ground segment and preparation of suitable contractual terms and conditions, the Board of Governors, at its Eighth Meeting, in March 1974, approved the specific arrangements for this service and recommended to the Meeting of Signatories that this traffic be considered on the same basis as international traffic, subject to the provisions of Article III(b) (ii) of the Intelsat Agreement for the purpose of determining voting quotas for the Board of Governors.

IV. Designing the Algerian Network

Because the Algerian domestic system will operate on the spare satellite in either the Atlantic or Indian Ocean regions, it was not possible to utilize directly the Lakhdaria standard Earth station, which will operate exclusively with the regular public international service on the Atlantic Primary Intelsat IV satellite. The need for direct interconnection of the domestic system with the international system therefore resulted in a decision to collocate the primary domestic Earth station at the same site.

The domestic system, which in a sense originates from the Lakhdaria domestic Earth station, 70 km (43 miles) from Algiers, will handle telephone, telex, and television traffic between the Sahara and between towns in the Sahara and Algiers. The remote Earth stations in the desert regions at the locations shown in Fig. 1 are not equipped with a television transmit capability.

This condition is, for practical purposes, overcome by use of transportable television equipment. The cost savings thereby achieved seemed warranted in light of the relatively little television programing expected to originate from the desert cities.

The significant features of the Earth stations of this system are the utilization of an 11-m antenna with an

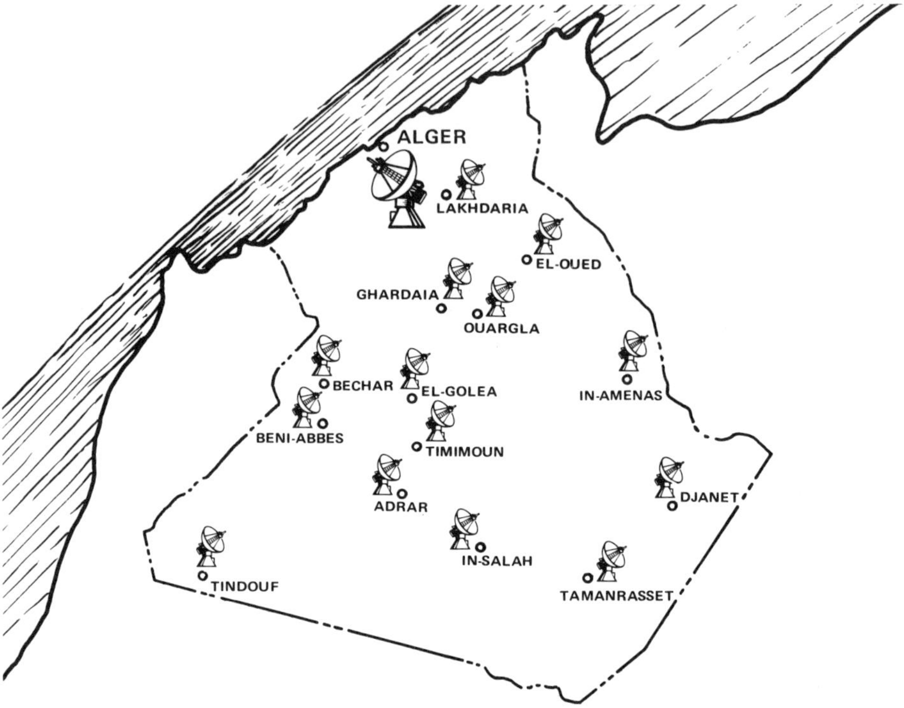

Fig. 1 Algerian domestic satellite system - configuration.

uncooled, low-noise receiver to achieve a G/T of 31.7 db/^{o}K and the use of single channel per carrier (SCPC) voice-processing equipment with a simplified form of demand assignment to achieve maximum telephone capability at the lowest possible cost.

The Earth stations will transmit and receive signals in the 6- and 4-GHz bands, respectively, through a transponder of an Intelsat satellite. The received signals are processed through uncooled parametric amplifiers and SCPC or television receiving equipment. The SCPC voice and telegraph channels are transported via cable to the local telephone exchange and connected through switchboards for local distribution.

Voice and telegraph signals from the local switchboards are processed by the SCPC equipment and modulated on separate FM carriers. The carriers are amplified by a pair of redundant 400-w TWT amplifiers to provide the required radiated power levels from the antenna.

Each channel modulator and demodulator of the SCPC equipment is tunable across the 36-MHz bandwidth of a satellite transponder. Channel spacing is 45 kHz. A frequency synthesizer determines the carrier frequency of each modulator and demodulator under the control of the demand assignment processor.

Voice frequency compandors are used to achieve toll-quality performance at relatively low carrier-to-noise density ratios. The equipment also uses emphasis to optimize noise performance and a peak limiter in the channel modulator to prevent overdeviation and reduce adjacent channel interference. The rf carriers are voice-actuated to minimize satellite loading. Echo suppressors are included as part of the SCPC equipment.

Transmitting carrier frequencies are controlled by a high-stability reference oscillator integral to the system. A pilot frequency is transmitted by the master station to permit automatic frequency control in each receiving terminal to correct the errors in the satellite translation oscillator.

Alarm circuits continuously monitor the system and will switch from one redundant path to the next upon detecting an alarm condition.

The demand assignment multiple access equipment interfaces with the SCPC equipment and continuously monitors the call status of the voice channels at remote stations over a common polling channel and updates the status record at a master station control processor. When a request for service occurs, the master station control assigns a transmit/receive frequency pair (if available) to both stations, monitors the call status, and disconnects the circuit when the call is completed. Initially, manual operation of the exchanges is required, but the system is designed to be used through automatic exchanges planned for the future.

Television video and sound programs for domestic broadcasting are received at the Lakhdaria domestic Earth station via the microwave link from Algiers. The TV transmitting facility, consisting of video modulator, up-converter, video power amplifier, and program channel modulator is normally used for the transmission of domestic TV to the remote stations. In the event that a TV program is to originate at a remote station, the transmit facility is transported from Lakhdaria to the remote station. If it is necessary to transmit a program from the Lakhdaria domestic station while this is being done, the TV transmitting facility of the international Intelsat station can be coupled to the domestic antenna and used for this purpose.

The TV signals are carried over video quality cable to the TV transmission equipments for the local area. Wherever possible, the local TV transmission equipments are collocated with each Earth station. This minimizes the complexity of interconnect and power equipment required between the TV and satellite stations.

A block diagram of the master station is shown in Fig. 2. The remote stations differ only in that they do not include the demand assigned multiple access control processor and the TV transmit facility is not normally installed.

The first stations of the system were placed in operation with Intelsat III in the Indian Ocean on Feb. 17, 1975, providing a temporary television transmitting capability at Bouzarea and receiving capability at Ouargla and Bechar pending the availability of the single channel per carrier telephony equipment. Under these conditions the entire transponder could be used for television transmission. Color and monochrome video transmissions have been of good Intelsat quality. Telephony service is scheduled to commence by mid-1975. The complete system of 14 stations will be in full operation late in 1975. Figure 3 indicates the system configuration.

V. Summary

The Algerian Domestic Satellite System is designed to operate with any of the Intelsat Atlantic Ocean and Indian Ocean satellites; thus it can make the fullest use of any spare satellite capacity available. The FM single channel per carrier radio equipment uses compandors and threshold extension detection devices in order to minimize both the satellite bandwidth and the satellite power required for a telephone channel. Voice activation of the rf carriers is provided in order further to conserve satellite power and thus provide nearly double the capacity which a system without voice activation could provide. As a further means of reducing the bandwidth required from the satellite, a limited number of satellite channels are shared among all of the stations through the use of demand assigned multiple access techniques.

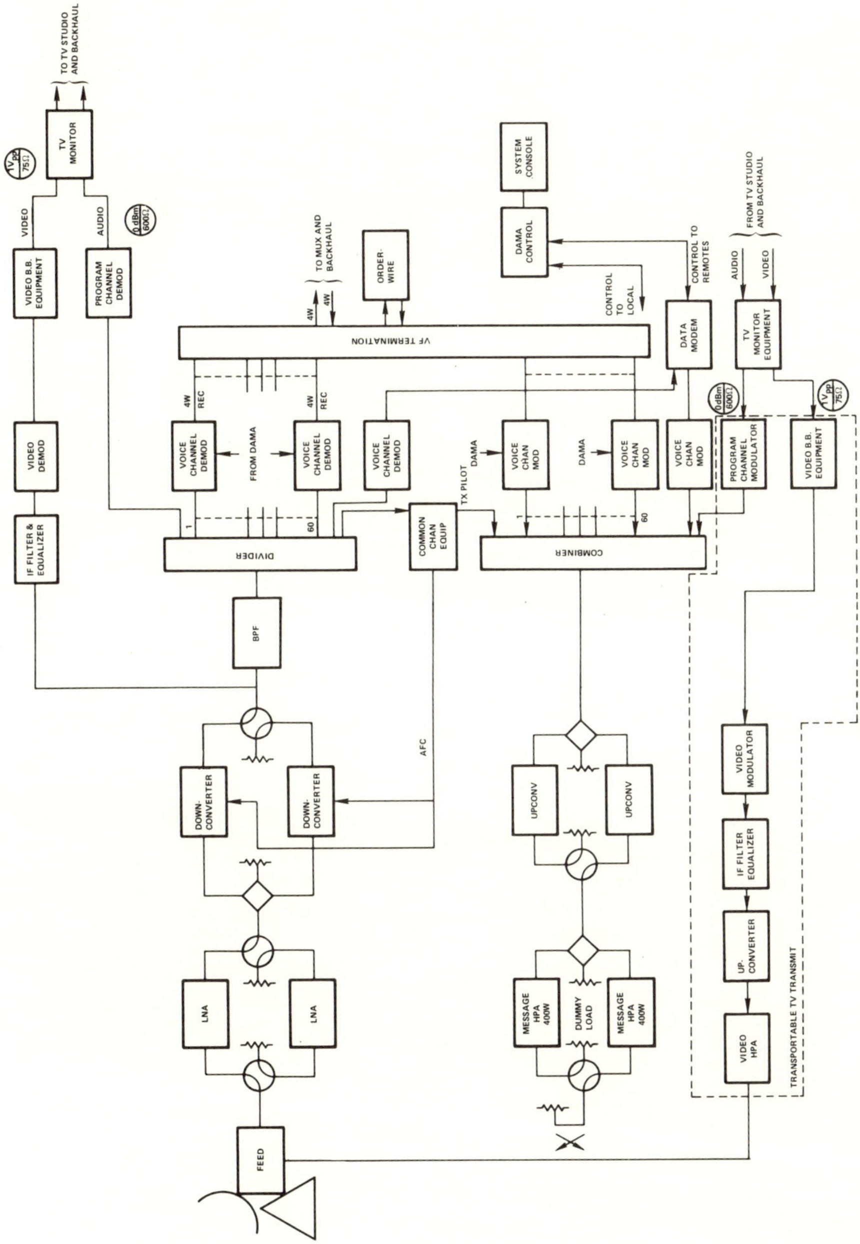

Fig. 2 Master station block diagram.

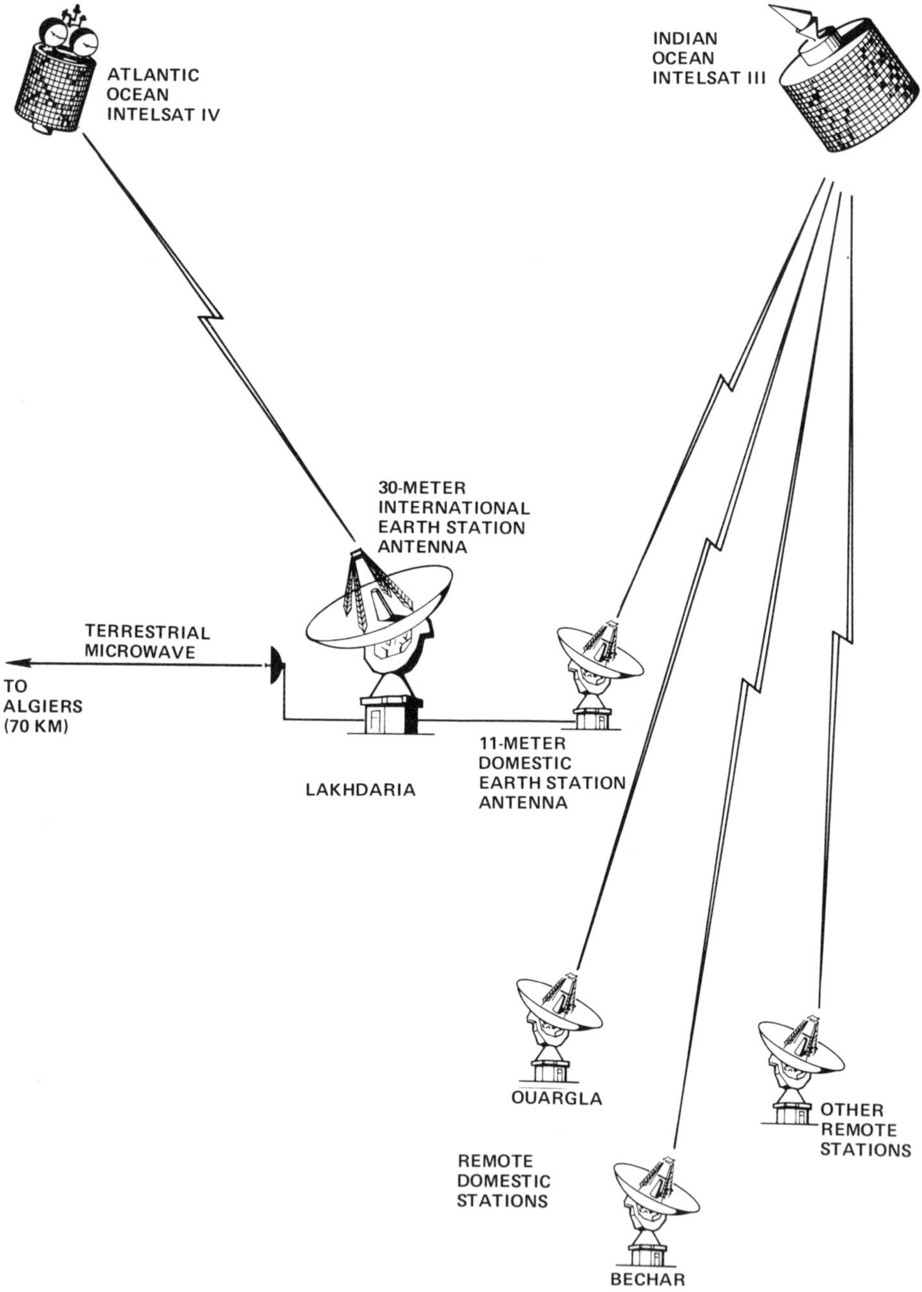

Fig. 3 Algerian domestic system configuration.

The system design results in an extremely cost-effective satellite telecommunications system with a high potential for future expansion at minimum cost. A number of factors contribute to this result. First, the ability to provide initial telephone service and television distribution with a single satellite transponder results in a tremendous saving during the first few years of operation. Several technical design features, resulting in more efficient use of satellite capacity, make this possible. While the necessity to be able to operate with any Intelsat satellite available does result in increased system cost due to the need for slightly larger antennas with the ability for quick repositioning, it results in savings of nearly $11 million over the 5-yr lease period. It is estimated that any terrestrial system capable of providing the same services would have required an annualized expenditure many times that of the combined space and ground segments of the present system.

The satellite telecommunications system will be of immense benefit to the economic development of Algeria. Good quality, extremely reliable telephone, telex,and data communications linking the Sahara communities with the capital and with each other will improve the conditions of trade, spur the growth of tourism,and accelerate the development of natural resources, such as petroleum and minerals. The lives of the citizens of these communities will be improved through access to better health care and educational facilities, and enriched through access to greater cultural opportunities. The entire nation will benefit from the increased political cohesiveness which develops from improved communication between all elements of the society.

OTS PROJECT:
BASIS OF FUTURE EUROPEAN SPACE COMMUNICATION SYSTEMS

R.C. Collette* and B. Stockwell†

European Space Agency,
Noordwijk, The Netherlands

Abstract

The Orbital Test Satellite (OTS), approved in 1972 as part of the European Communication Satellites Programme and presently under final development, is a three-axis stabilized vehicle of the Delta class intended to carry experimental communication payloads. The OTS conceptual design is such that the satellite consists in fact of two separate modules: a service module grouping most support subsystems, and a communications module including the repeaters and antennas. Throughout the design, the accent was on adaptability of the basic platform, in particular the power and control systems, to the requirements of various communication missions and payloads. The first satellite (OTS 1) will be launched in June 1977 and will carry an 11/14-GHz payload, satisfying the needs of an experimental system for public communications and television exchange in Europe. The second satellite (OTS 2 or MAROTS), foreseen at the end of 1977, will carry an L-band payload for maritime purposes. Thereafter, the operational satellite envisaged for the European PTT regional system will be a direct derivative of OTS 1, with the first unit of the series delivered at the end of 1979. In parallel, several proposals and studies are underway for the reuse of the basic OTS for other missions: TV broadcast, small terminal systems such as data transmissions and conference links, mobile services, etc.

Presented as Paper 74-495 at the AIAA 5th Communications Satellite Systems Conference, Los Angeles, Calif., April 22-24, 1974.

*Head, Communication Satellite Program.

†Satellite Manager, Communication Satellite Program.

I. European Communication Satellites Program

The basic objectives and broad lines of this program were already presented in Ref. 1. The program, foreseen as a three-stage one (Fig. 1), is presently in its second phase, i.e., the technological and experimental one, which will last until 1977. The first phase, the program definition, was completed at the end of 1971. The third and last phase, concerned with the final development and implementation of the CEPT (Conférence Européenne des Postes et Télécommunications) operational regional system, will start in 1976. The over-all cost of this 10-yr program is estimated at $ 400 . 10^6 (U.S. dollars, 1975 conversion rates).

It will be recalled briefly that the initial mission envisaged for this intra-European system is to route via satellite, as of 1980, part of the long-distance public communications traffic as well as special transmissions such as TV program exchanges between the various broadcasting agencies. The final configuration of the first-generation operational satellite, known as ECS, will be selected in 1976. To maximize the communications capacity of the system, which will use the 500 MHz available at 11 GHz (downlink) and 14 GHz (uplink), it is intended to reuse the available frequency bands on two orthogonal polarizations.

The present phase of the program has mainly technological and experimental objectives. To this end, as from 1971, advanced development activities were initiated, and a comprehensive $ 40 . 10^6 supporting technology program was implemented; this covered most of the critical onboard hardware in the areas of the communication payload and of the spacecraft systems, such as 11/14-GHz repeater units, antennas, traveling wave tubes, low-noise preamplifiers, flywheels, infrared sensors, thrusters, lightweight arrays, power regulation and battery electronics, drive mechanisms, lightweight structures, etc. In parallel the definition of the over-all communication system proceeded together with ground experimentation of propagation phenomena above 10 GHz and of transmission problems encountered in TDMA systems (Time Division Multiple Access systems).

In-orbit experimentation had always been considered necessary. However, the precise characteristics of such an experimental satellite were not agreed upon until late 1972. It finally was concluded that the most valuable and cost-effective project would be a test spacecraft with a configuration similar to that of the future operational satellite. For most subsystems, emphasis would be put on modular design to permit

PHASE 1 1970-1971

- PROGRAM DEFINITION AND SYSTEM STUDIES
- INITIAL TECHNOLOGICAL DEVELOPMENT

PHASE 2 1972-1977

- COMMUNICATION SYSTEM DEFINITION
- SUPPORTING TECHNOLOGY PROGRAMME
- FLIGHT OF EXPERIMENTAL O T S
- DEFINITION OF OPERATIONAL E.C.S. SATELLITE

PHASE 3 1976-1980

- FINAL DEVELOPMENT OF SPACE SEGMENT
- PROCUREMENT OF OPERATIONAL SATELLITES

Fig. 1 Three-stage communication satellites program.

later use on ECS of essentially the same building elements. These considerations thus led to the inclusion of the Orbital Test Satellite (OTS) in Phase 2 of the program.

The prime motivation for this development work by ESA is to give to European industry a solid technical basis from which to proceed with the commercial utilization of space communication systems. Hence, long-term planning and system adaptability to various missions were important aspects in the definition of the program and, in particular, in the choice of the basic configuration of the spacecraft platform. A three-axis stabilization with Earth-oriented body and sun-tracking array was selected at the outset, in view of the design flexibility and economic advantages of this approach for the larger and higher-power satellites of the future. Furthermore, at the level of subsystem units and subassemblies, maximum emphasis was put on modularity so as to enable the utilization of such equipments in satellite systems of varying layout and size.

OTS 1, as described in this paper, is the first application of that philosophy, this satellite being itself the precursor of the operational 11/14-GHz vehicle intended for the ECS system. OTS 2, or MAROTS (Maritime OTS), is another example where the OTS platform, comprising all support subsystems, is reused essentially unaltered to carry an L-band maritime payload. This latter project is consequently extremely cost-effective in view of the technical and schedule commonalities with OTS 1. Other OTS derivatives, some of them of larger size, are presently being envisaged for missions such as national TV broadcast at 12 GHz, intra-European data transmissions, and communications with oil platforms in the North Sea. In fact, with suitable adaptation of the stabilization accuracy and solar array power, most Earth-oriented geostationary payloads could be accommodated on this basis.

II. OTS Objectives and General Description

The Orbital Test Satellite is an experimental and pre-operational communications satellite that will be placed in orbit in 1977. It will be located at 10°E long in the geostationary orbit and will be used to carry out a number of tests and experiments to confirm the soundness of its general design. OTS is a forerunner prefiguring the operational European Communications Satellite (ECS), which is scheduled to commence operation in 1980 and which will operate at frequencies in the region of 11 and 14 GHz.

The OTS Project has been defined with the following objectives in mind: a) promote applications satellite technology in Europe; b) demonstrate the performance and reliability of all on-board equipment to be used later in the ECS system; c) fulfill, from the communication system point of view, the experimental objectives required by the mission of the future ECS; d) enable comprehensive propagation experiments at 11/14 GHz; e) provide adequate capacity for pre-operational transmissions; and f) demonstrate the platform's adaptability to other missions.

From the point of view of communication experiments more specifically, the aims are 1) to validate the assumptions made with regard to atmospheric attenuation and depolarization by means of extensive measurements in the area to be covered; 2) to evaluate experimentally the transmission techniques envisaged for the operational ECS (e.g., time division multiple access techniques, utilization of orthogonal polarizations for frequency reuse, speech interpolation); and 3) to experiment with the transmission of narrow-band data signals between small Earth terminals.

OTS is to be launched by a Delta 3914 vehicle. It is a three-axis stabilized spacecraft designed for a 5-yr life in geosynchronous orbit, with the individual satellite equipments designed for 7-yr life capability. A modular approach has been adopted in the satellite design, and, in addition, equipments are identical to those that will be used for the subsequent ECS mission. The OTS consists of a service module capable of providing all of the basic service functions and a communications module that carries the payload. A significant advantage of this modular concept is that it allows different payloads to be carried without major redesign of the service module. Also, the structural design of the communications module is such as to allow easy adoption of several alternative payloads, again without major redesign.

The external appearance of the satellite is shown in Fig. 2. The separable communications and service modules are seen in the exploded view of Fig. 3. Two solar paddles (520 w, end-of-life), each consisting of two hinged panels, are attached to the bearing and power transfer assemblies mounted on the north and south sides of the satellite and are axially aligned along the satellite pitch axis. During launch and transfer orbit, the arrays are folded against the satellite north and south sides so that half of their surface is exposed

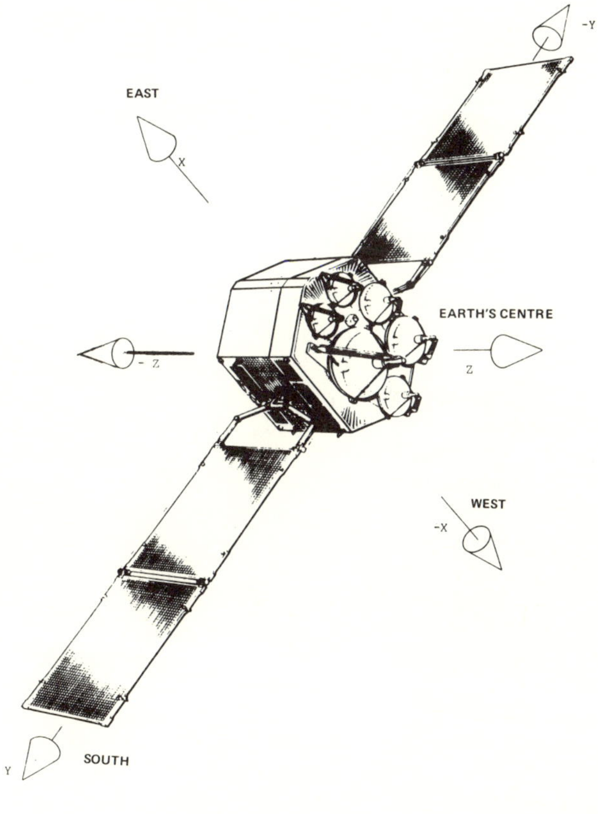

Fig. 2 OTS on station.

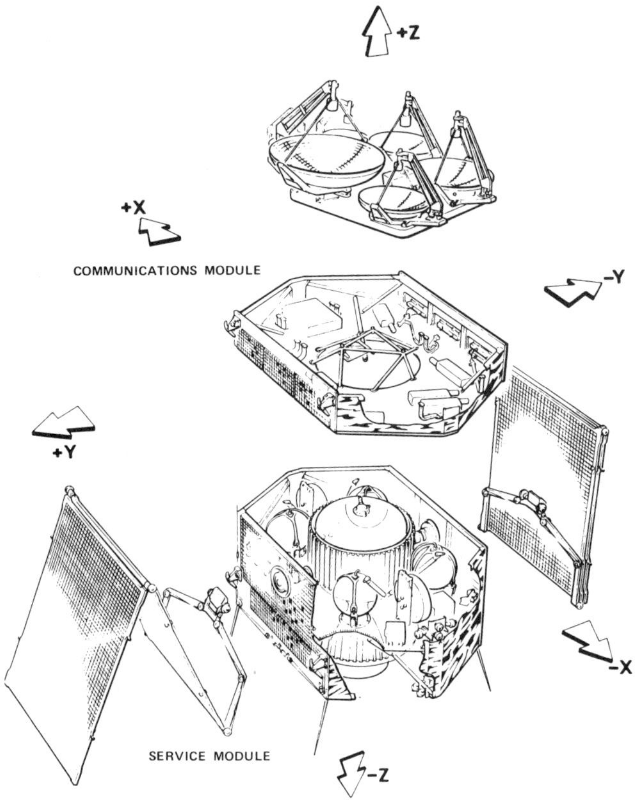

Fig. 3 OTS modular design.

to provide power during this phase of the mission. In normal operation, the arrays are deployed, and the paddles are steered independently to track the sun, revolving around the pitch axis. The six SHF antennas are body-mounted on the face pointing toward the Earth. The satellite subsystems are described in some detail in Secs. IV and V.

The Delta 3914 launch vehicle from Eastern Test Range is capable of placing a 907-kg payload (including the launch vehicle/satellite adapter) into a 28.8° inclined geosynchronous transfer orbit. Optimization of the transfer orbit parameters in order to maximize synchronous-orbit mass led to the choice of a 27.25° inclined orbit, which reduces the apogee boost motor (ABM) mass, and at the same time reduces the launch vehicle capability of 902 kg, which leads to a satellite useful mass (beginning-of-life) of 432 kg, excluding ABM, adapter, and hydrazine for transfer orbit.

Reliability techniques have been used to optimize the allocation of spacecraft mass and cost to the provision of hardware redundancy, in order to minimize the probability of failure. As well as providing redundancy, the design is oriented toward enabling partial, but worthwhile, use of facilities after a failure. At the end of the 3 years, for instance, the probability of survival of the service module alone is 0.89, and for the over-all system, with 4 out of 6 repeaters in use, it is 0.77.

An all-year-round launch window is provided. A diagrammatic representation of the mission event sequence after launch is given in Fig. 4. The launch vehicle will inject the satellite, in a spin-stabilized configuration, into an elliptical transfer orbit whose parameters are chosen to maximize the synchronous orbit payload capability, and whose apogee is near synchronous altitude. A solid-propellant ABM then will inject the satellite into a near-synchronous orbit. After injection into near-synchronous orbit, the satellite will be despun, the solar array paddles deployed, and the three-axis Earth-pointing stabilized attitude acquired. Orbit maneuvers then will be performed to bring the satellite to its on-station position of 10°E long within 21 days. During the subsequent 3-yr life of the satellite, the station will be maintained at $10^{o}E \pm 0.1^{o}$, and the orbital inclination within $\pm 0.1^{o}$. Ground control of the satellite during transfer, near-synchronous, and synchronous orbit phases will be provided by the European Space Operations Centre (ESOC), Darmstadt, Germany. Throughout the life of OTS, a number of tests will be performed to provide in-orbit performance evaluation of the satellite

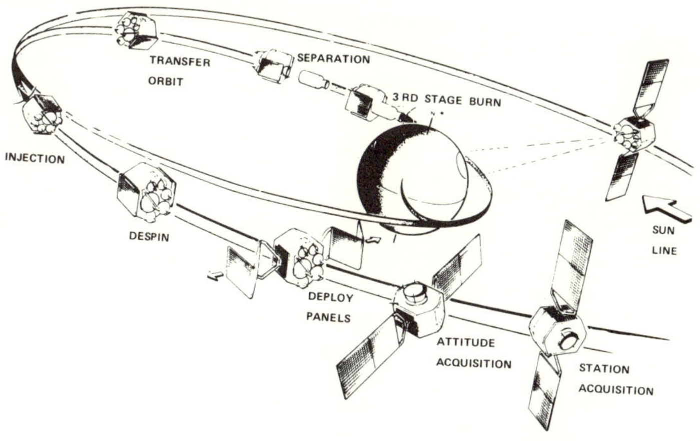

Fig. 4 OTS mission event sequence.

subsystems and payload, and to provide experimental validation of communications concepts and propagation assumptions. The program of experiments to be carried out with OTS is described in Secs. II.A and II.B hereafter.

A. Orbital Test Program (OTP)

The first part, the OTP, is aimed at measuring the satellite behavior in all respects to verify the validity of its design and to assess its performance in orbit, particularly with regard to the communication payload. This part will be carried out by ESA, with the assistance of the P & T Administrations and possibly other organizations. It will cover 1) the attitude and orbit control subsystem in all modes of operation; 2) the power subsystem including generation, conditioning, and storage of power; 3) the thermal control subsystem; 4) the telemetry, tracking, and command subsystem; 5) classical repeater tests such as amplitude response, group delay response, output power, noise figure of receivers, frequency stabilities, phase shift, dynamic range of repeaters, and gain range; 6) in-band dynamic tests: verification of the over-all compatibility between the repeater design characteristics and the transmission of adequate digital four-phase PSK or analog TV signals; 7) antenna tests: verifying the electrical characteristics of antennas (gain beam shape and influence of thermal effects for various sun illumination angles and during sunlight-eclipse transition), verifying the pointing capacity of the spotbeam antenna pointing mechanism, and verifying the performance of the antennas in terms of polarization purity; 8) transmission measurements: tests planned in order to verify in orbit the cochannel and adjacent-channel interferences in the communications equipment and to highlight any strong influence of the propagation media; and 9) propagation experiments: tests planned in order to collect extensive data from long-term measurement of atmospheric attenuation at up- and downlink frequencies, and atmospheric depolarizing effects on circularly and linearly polarized waves.

B. Communication System Tests

The second part will be aimed at demonstrating the feasibility of a satellite communication system operating at frequencies of 11 and 14 GHz using such new techniques as digital transmissions ranging from 60 to 180 Mbit/sec, time-division multiple access (TDMA), and frequency reuse by means of orthogonal polarizations. This part of the program will be performed essentially by the P & T Administrations.

III. Industrial Organization

The preliminary definition of the satellite was initiated at the end of 1972. Three contractors performed Phase A studies: British Aircraft Corporation (BAC), Hawker Siddeley Dynamics (HSD), and Marconi Space and Defence Systems (MSDS). In April 1973, two industrial groups led by BAC and HSD were awarded Phase B contractors for the detailed design of OTS. In November 1973, Hawker Siddeley Dynamics (United Kingdom) finally was selected as prime contractor for the final development, manufacturing, and qualification of the spacecraft, with delivery of the flight unit in April 1977.

In view of the long lead-time and relatively high technology content of the onboard communication hardware, development work in this field had been started previously: in early 1972, AEG-Telefunken (Germany) had won the contract for the 11/14-GHz repeater, and, in early 1973, Selenia (Italy) was awarded the SHF antenna task. Consequently, AEG-Telefunken and Selenia now are responsible, under the over-all prime contractorship of HSD, for the repeater and antenna systems, respectively, which together constitute the essential part of the communications module.

In fact, within most of the OTS subsystems at the level of subassemblies and units, quite a few items are procured from subcontractors who had been awarded development contracts as part of the supporting technology program that started in 1971, thus preceding the satellite system work by about two years on the average.

The basic industrial team led by HSD is shown in Fig. 5. It consists essentially of the firms grouped within the MESH consortium, to which the communication package contractors have been added. The distribution of tasks and the most important procurement sources are listed here:

1)Prime contractorship: HSD (United Kingdom) carries the entire responsibility for all tasks and must deliver the satellites under a strict system of incentives covering the development as well as the performance in orbit. A central project team headed by the HSD project manager and located in Stevenage, United Kingdom, covers the management, system engineering, and subsystem monitoring functions. The assembly and functional testing of the engineering, qualification, and flight units will take place at Matra, Toulouse, France, with a common HSD-Matra integration team in charge. Electrical

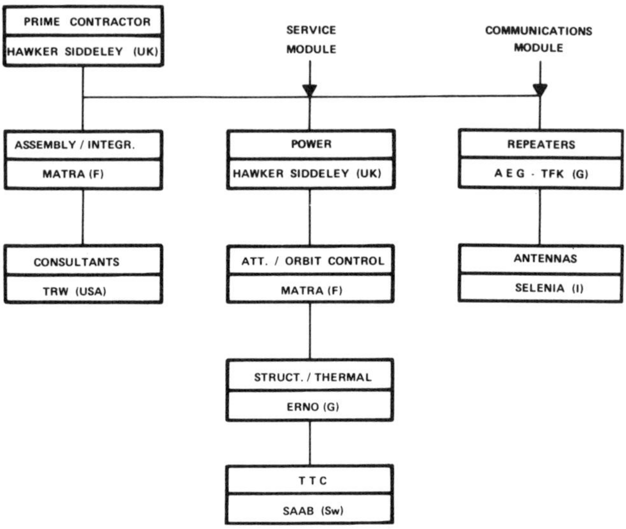

Fig. 5 OTS 1 industrial organization.

ground support equipment is made by Matra (France), CIR (Sweden), and Aeritalia (Italy), and mechanical ground support equipment by HSD, Contraves (Switzerland), and INTA (Spain). Environmental and acceptance testing will be performed using ESA and CNES facilities. TRW, associated with the MESH consortium since its creation about 10 years ago, are technical consultants to HSD as well as U.S. agent for the procurement of high-reliability parts.

2)Structure, thermal control, and propulsion: ERNO (Germany) responsible for all of these subsystems, with manufacturing of the structure subcontracted to Aeritalia (Italy).

3)Attitude measurement and control: Matra (France). The main Earth infrared sensors are procured from both Galileo (Italy) and Sodern (France) and the flywheels from Hollandse Signaal (Netherlands) and Teldix (Germany).

4)Telemetry, tracking, and command: SAAB (Sweden) main contractor, with subcontracting of vhf transponder and shf transmitter to LCT (France) and uhf receiver and test equipment to BTM (Belgium).

5)Power generation and conditioning: HSD (United Kingdom). Solar array from MBB (Germany) using AEG cells, batteries from

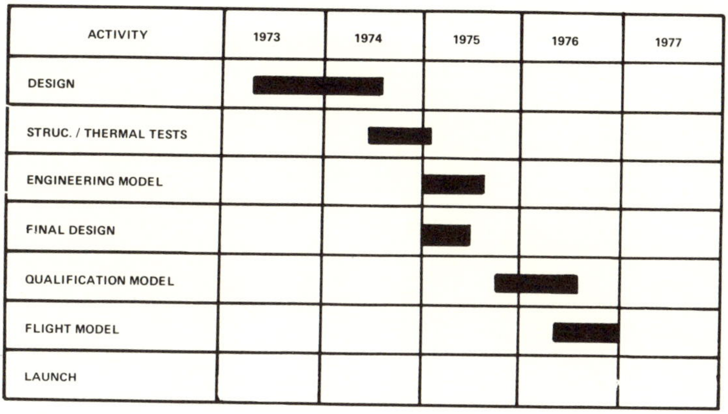

Fig. 6 OTS 1 development schedule.

SAFT (France), with associated electronics made by Hollandse Signaal (Netherlands).

6) SHF antennas: Selenia (Italy). Receive and transmit antennas at 11/14 GHz with dual orthogonal polarization.

7) Repeaters: AEG-Telefunken (Germany). TWT's from Thomson-CSF (France) and AEG (Germany). Subcontracting to Thomson-CSF (France), Fiar (Italy), Selenia (Italy), GTE (Italy), Marconi (United Kingdom), LM Ericsson (Sweden), and ETCA (Belgium).

8) Apogee boost motor: Aerojet (U.S.).

The project development schedule is shown in Fig.6. In the case of MAROTS (OTS 2), the platform again will be provided by Hawker Siddeley Dynamics, with Marconi Space and Defence Systems (United Kingdom) being in this case responsible for the development and procurement of the maritime L-band payload.

IV. OTS Communications Payload

The communications payload is split into two distinct modules, as shown in Fig. 7.

A. Module A

The first module, called Module A, is a preoperational version of the communication payload that will be placed onboard the future CEPT operational satellites. Most repeater parameters are basically the same as those of the projected operational system. Access to this part of the communication payload will be by Earth stations having over-all characteristics similar to the operational ones. This module will provide retransmission of signals such as telephony in PSK/ TDMA and television. Module A contains two chains of 40-MHz nominal bandwidth and two of 120 MHz. Each pair is arranged in frequency reuse configuration employing orthogonal linear polarizations. The telecommand (TC) receivers and the telemetry (TM) transmitters also are part of Module A. The receive antenna system consists of two redundant identical dishes giving full European coverage with a nominal half-power beamwidth of $7.5^{\circ} \times 4.25^{\circ}$ (Eurobeam A). The transmit antenna associated with the 40-MHz channels likewise produced an elliptical beam for European coverage.

The transmit antenna used to transmit the two 120-MHz channels is a spotbeam antenna with circular halfpower beamwidth of nominally 2.5°.

There are two basic sections in the module: the wide-band section and the channelized section. The wide-band section consists of two receiving chains, one active and one standby, for each polarization, thus four in total. The inputs are connected by pairs to the dual-polarization feedhorns of two Eurobeam reflector antennas via orthomode transducers (OMT).

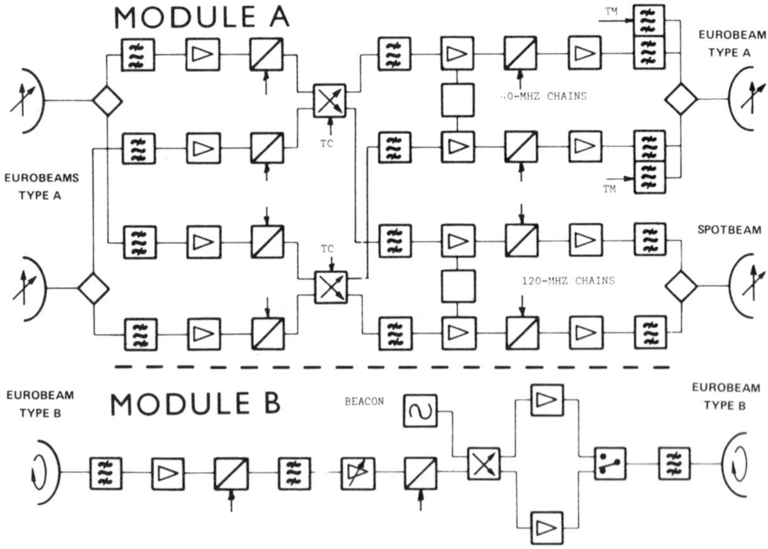

Fig. 7 OTS communications payload.

Each receiving chain includes an input filter, a low-noise parametric amplifier, a down-converter with associated IF pre-amplifier, and a power-dividing network. This section covers the full bandwidth of 500 MHz, i.e., from 14.0 to 14.5 GHz at the input and from 800 to 1300 MHz at the IF output. The telecommand signal (TC) that can be received on either polarization is extracted at IF frequency (i.e.,uhf). The power-dividing network combines, for each polarization, the output of the active and the redundant receiving chains. The power divider is followed by the IF channel filters, which constitute the input stages of the channelized section. Each channel then is handled separately up to the transmitting antennas.

The channelized section of the repeater includes channel filters, IF main amplifiers, variable attenuators using PIN diodes controlled by telecommand via a control box, upconverters to translate the IF frequency up to the appropriate 11-GHz channel, the 20-w traveling wave tube amplifiers (TWTA's), and finally the output filters. Except for the IF filters and the output filters, which are 40 or 120 MHz wide, all other elements are at least 250 MHz wide. The output TWTA's cover the full range from 10.95 ot 11.7 GHz. The outputs of the two 40-MHz channels are fed to a single Eurobeam antenna radiating with two linear orthogonal polarizations. Similarly, the outputs of the two 120-MHz channels are fed to a single spotbeam antenna. The telemetry signal (TM) is diplexed into the Eurobeam antenna and sent on either polarization.

B. Module B

The second module, called Module B, includes repeater chains of 5 MHz nominal bandwidth with a gain substantially higher than the chains of Module A. As such, it will permit access by small and relatively inexpensive Earth terminals for propagation and narrow-band transmission experiments. This module also contains an onboard beacon transmitter. Reception and transmission are via separate antennas with an elliptical beamwidth of nominally 5° x 3.5^{o} (Eurobeam B), and circular polarization is employed. The selection of a definite type of polarization, linear or circular, for the operational system has not been made yet, since there are some uncertainties with regard to both. The communication payload of OTS therefore utilizes linear polarization for Module A and circular for Module B. This will enable the advantages and disadvantages of both types of frequency reuse techniques to be evaluated and compared under realistic conditions.

The wide-band section is similar to that of Module A and the down-converter uses the same local oscillator source. The 5-MHz channel filter is followed by an IF amplifier with variable attenuation using PIN diodes, an up-converter, and a combining network to which the output of a beacon oscillator also is connected. The combining network feeds two TWT amplifiers, one of which is active and the other standby. The output of the active TWTA's is connected to the transmitting antenna via a switch and an output filter.

The total gain of the chain can be varied by telecommand over a range twice as wide as for the Module A chains, and the maximum value to which it can be set exceeds the gain of those chains by more than 20 db. This, together with the relatively narrow bandwidth, makes it possible for experimenters to work with OTS with relatively small Earth terminals.

Both receiving and transmitting antennas are equipped with a circular polarizer and an orthomode transducer. However, only one polarization is used, and the two antennas operate in opposite polarization modes, one left-hand and the other right-hand.

C. Frequencies

Among the bands available for the fixed satellite service in Region 1 between 10 and 15 GHz, the European Conference of Post and Telecommunication Administrations (CEPT) have selected the following for the future ECS system: 1) downlinks: band A (10.95-11.20 GHz) and band B (11.45-11.70 GHz); and 2) uplinks: band D (14.00-14.50 GHz).

The repeater of OTS will operate for the main part in bands B and D only. The detailed frequency plan is as follows.

1. Module A. The uplink frequencies are all contained in band D:

TC receiver	:	14,125 MHz
40-MHz chains	:	14,152.5 to 14,192.5 MHz
120-MHz chains	:	14,242.5 to 14,362.5 MHz

The downlink frequencies are all contained in band B:

40-MHz chains	:	11,490.0 to 11,530.0 MHz
TM transmitter	:	11,575.0 MHz
120-MHz chains	:	11,580.0 to 11,700.0 MHz

2. Module B. The uplink frequencies are contained in band D:

5-MHz chain	:	14,455.0 to 14,460.0 MHz

The downlink frequencies are situated above band B in a band allocated on an equal basis to different services, including the Satellite Broadcasting Service, as follows:

Beacon transmitter	:	11,786.0 MHz
5-MHz chain	:	11,792.5 to 11,797.5 MHz

D. Antennas

All five antennas are made of circular reflector with frontal feedhorn supported by four struts. The spotbeam antenna has a beam of circular cross section, whereas the Eurobeam antennas of both types A and B are designed to produce a beam of elliptical cross section by distorting the profile of the reflector so as to widen the beam in one of the planes of symmetry. The Eurobeam A antennas differ from the Eurobeam B in the width of their beam, which is slightly greater for the former, and in their polarization, which is linear for A and circular for B. Transmitting and receiving antennas of a given type are identical except that their size is scaled to the wavelength.

Since maximum gain and maximum coverage area are conflicting requirements, the design of the antennas has been based on the result of an optimization process in which such parameters as antenna gain, pointing accuracy, distance from satellite to individual Earth stations, and atmospheric attenuation are taken into account.

Peak gains of Eurobeam A, Eurobeam B, and Spotbeam are, respectively, 26.5, 29.1, and 35.5 db. These values include transmission losses to and from the repeater interface points, viz. the input of the receive filters and the output of the transmit filters. Coverages are shown in Fig. 8.

The three Eurobeam A antennas use linear polarization along two orthogonal planes, which coincide with the planes of symmetry of the elliptical beam when related to the propagation along the boresight. The isolation between the two modes is expected to be 30 db at least within the 4-db contour and 26 db in most of the area between this contour and the beam edge. Both Eurobeam B antennas are equipped with a polarizer and an orthomode transducer, permitting operation with circular polarization with either left- or right-hand sense. The sense that will eventually be chosen will depend on which is found to offer the highest degree of polarization purity. The isolation between the two modes is expected to be of the order of 28 db or more within the 4-db contour and 26 db in most of the area between this contour and the edge of the beam. The antenna radiates with two orthogonal planes of polarization. Although the beam cross section is nominally circular, the antenna has two physical planes of symmetry which affect its radiation pattern, especially with regard to cross-polarization level.

The two planes of polarization relating to the propagation in the direction of the beam axis coincide with the planes of symmetry of the antenna and are aligned with those of the Eurobeam A antennas. The isolation between the two modes of transmission is expected to be 30 db or more within the 4-db contour and 26 db in most of the area between this contour and the edge of the beam.

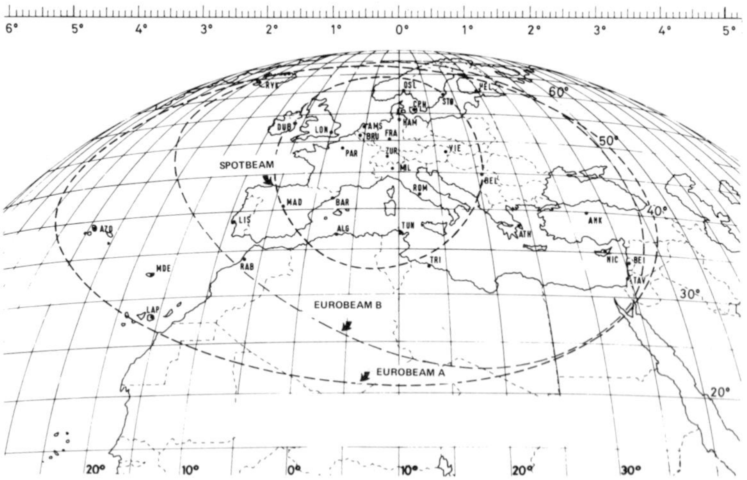

Fig. 8 OTS antenna coverages.

V. Service Module Subsystems

A. Structure

The structure subsystem consists of two basic elements: 1) the communications module, on which are mounted the antennas and repeater units, including the necessary thermal radiators, as well as some of the service module equipment; and 2) the service module, on which are mounted the subsystem equipments providing functional services to the payload.

The OTS structure forms a hexagonally shaped body, on the north and south faces of which there is adequate volume between the launch vehicle shroud to stow the solar arrays in the launch configuration and adequate surface area to allow sufficient power to be generated by the stowed array in the transfer orbit and to achieve a positive energy balance. The subsystem employs a central tube, together with two equipment floors to attach the ABM to the satellite, to interface with the launch vehicle adapter, and to provide adequate mounting surface for all equipment, while maintaining adequate strength and stiffness for the launch as well as the required tolerances for alignment of critical equipment and the communication payload structure.

B. Thermal Control

The function of the thermal control subsystem is to provide a satisfactory thermal environment for all onboard equipment during all mission phases. It incorporates aluminum radiator plates for thermal control of the traveling wave tubes (TWT) and an appropriate combination of thermal finishes, coatings, and superinsulation for over-all satellite thermal regulation. Heaters are provided where necessary to maintain the required thermal conditions during transfer orbit and eclipse, and to establish satisfactory temperatures immediately after eclipse periods.

C. Power Generation and Electrical Distribution

The power requirement of the OTS at end of life is approximately 520 w. This power is provided by two, independently steerable, rigid solar arrays, each consisting of two flat panels. The solar arrays are aligned continuously to the sun using two bearing and power transfer assemblies (BAPTA's) that are rotated at orbit rate. Power and signal transfer from the solar array to the spacecraft are achieved via an adequately redundant slip-ring assembly within the BAPTA. A digital shunt regulator system consisting of several switchable sections, together with a proportional shunt section, regulates the main power bus voltage during sunlight. During eclipse, power is provided from a nickel-cadmium battery via a PWM (pulse width modulation) boost regulator that controls the main bus voltage. The main bus voltage during both sunlight and eclipse is 50 v d.c. $\pm$ 1%. Separate regulators are used for the AOCS, TTC, and instrumentation subsystems, and to control power to the repeater subsystem of the communications payload. The EPC's (electronic power conditioners) of the TWTA's (traveling wave tube amplifiers) are supplied directly from the 50 $\pm$ 1% main bus.

The battery package is sized to provide 122 w at end of life of the OTS for housekeeping functions. The battery energy is, however, sufficient to allow operation of up to two repeaters during short eclipses for experimental evaluation of the sun/eclipse transient behavior of the communication subsystem parameters. The battery is charged via redundant series chargers; battery charge and discharge control is achieved by using an ampere-hour meter and battery cell monitoring.

The electrical distribution subsystem provides for electrical interconnection of the various equipment and for protection of the main power bus against failures; it is designed, through proper wiring, shielding, and grounding techniques, to minimize electromagnetic interference problems.

D. Telemetry, Tracking, and Telecommand

Prior to normal-mode operation, and also in backup modes, the TTC subsystem operates at vhf. In normal-mode operation it operates at shf, 14.125 GHz for the uplink and 11.575 GHz for the downlink, within the communications payload frequency bands. In order to maintain the basic modular concept, the TTC subsystem is divided into the service module part, consisting of vhf antenna and branching unit, vhf transponder, priority select and interface unit, and the service module decoder and encoder, and the communication module part, consisting of uhf receiver, which is fed by signals extracted from the repeater subsystem following down-conversion, the shf transmitter, which feeds the TM signals and carrier via a diplexer into one of the Eurobeam transmit antennas, and the payload decoder and encoder.

E. Attitude and Orbit Control Subsystem (AOCS)

The AOCS consists of the various attitude sensors, measurement and signal processing electronics, and actuators to perform the required functions of attitude measurement, attitude control and orbit control during transfer orbit, drift to station, and on-station operation.

During transfer orbit, the AOCS uses a V-beam sun sensor, infrared pencil-beam Earth sensors, a passive nutation damper, control electronics, and hydrazine thrusters to provide attitude sensing, attitude determination, and correction (with the aid of ground station control).

Following ABM burn, in order to acquire the operational three-axis controlled attitude and station, additional sun sensors are used, together with the monopropellant hydrazine reaction control system (RCS). During the Earth-pointing, three-axis-stabilized operational mode, the AOCS uses a fixed momentum wheel in conjunction with the RCS as the primary control actuator and a two-axis infrared sensor for the attitude reference in pitch and roll. Wheel momentum dumping and on-station orbit control are performed using the RCS, in conjunction with a rate integrating gyro (RIG) for yaw attitude error control.

The performance of the AOCS is such that, taking all sources of error into account, the electrical boresight of the antennas is maintained to within a half-cone angle of 0.2° ($3\ \sigma$) and the satellite yaw error to within $\pm\ 0.5^{\circ}$ ($3\ \sigma$) under all orbital conditions. Following acquisition of geostationary position, periodic stationkeeping maneuvers are performed using the hydrazine thrusters in order to maintain that position within $\pm\ 0.1^{\circ}$ E-W and the inclination within $\pm\ 0.1^{\circ}$, for the operational lifetime of the satellite.

F. Apogee Boost Motor

A solid-propellant apogee boost motor is mounted within the central tube of the spacecraft structure. The motor is sized to impart sufficient velocity increment to the satellite at the apogee of the transfer orbit to inject it into a near-synchronous circular orbit with approximately zero inclination, subsequent fine corrections to acquire operational orbit and station being provided by the hydrazine system.

VI. MAROTS Satellite

This satellite (see Fig. 9) is part of the ESA maritime program. The project is presently in the final definition stage (Phase B). Launch is foreseen for the fall of 1977, with positioning at 12.5°W long.

Briefly,[2] the ESA maritime mission, in accordance with IMCO guidelines, is intended to provide an experimental and preoperational capability to evaluate and demonstrate the following:

1) In the area of general communications: a) various types of ship terminals; b) ship/satellite L-band links in telephony and telegraphy with various signal qualities and modulation signal/noise processing techniques;

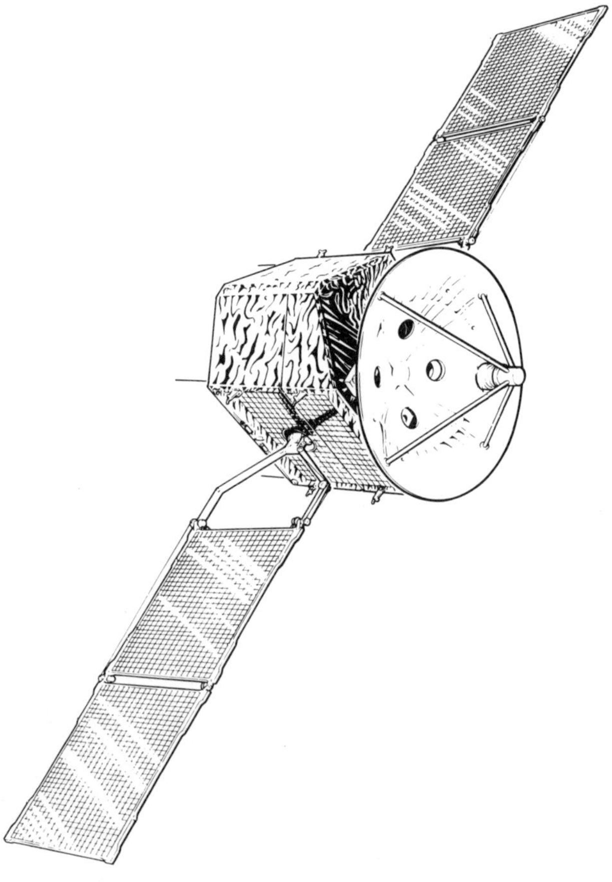

Fig. 9 MAROTS on station.

c) direct connection to public telephony and telegraphy networks; and d) access to satellite communication by multiple ship and shore stations.

2) In the area of distress, search and rescue, and safety: a) immediate priority access techniques for distress communications; b) special distress equipment, including emergency position indicating beacons (IPERB); and c) "all ships" information broadcasts and individual mobile weather routine via satellite.

3) In the area of radio determination: ranging techniques for line of position determination.

MAROTS, from the service module point of view, is essentially the same spacecraft as OTS. The payload consists of a solid-state L-band repeater with its associated antenna, which is 2 m diam. Communications with Earth terminals are at 11/14 GHz. Figure 10 shows that the principle of modular design also is preserved, a fact that will facilitate later adaptation of the vehicle to an operational mission.

VII. Future Derivatives of OTS

In the design of OTS, as already shown in the previous sections, maximum attention has been paid to the possible reuse of the basic vehicle for other missions. In addition to the necessity of such flexibility within the framework of the communications satellite program itself, where OTS prefigures the later operational satellites, the reuse of the basic platform has been a major goal since the inception of the project. The potential of the basic OTS for growth and flexibility is in two main areas: 1) the potential for satellite mass increase if other launchers are used; and 2) the relative independence of the spacecraft platform from the payload package.

The basic OTS presently is designed for launch on the Thor Delta 3914. Growth for adaptation to even bigger launchers is relatively easy because of the modularity of most subassemblies within the spacecraft.

Because of the relative independence of the spacecraft platform from the payload package, mass increase can be reflected into considerable increases in payload mass capability or into payload power provisions by augmenting the platform solar arrays. The latter is of particular importance in the case of high-power missions, e.g., television broadcast missions, where normally the payload constraint is in terms of power rather than mass.

However, even without the increase available from a launcher extension, the basic OTS offers the flexibility of exchanging payload power for payload mass and of augmenting payload capability by trading off against other operational characteristics, for example, if the mission permits a relaxation of orbital control constraints (e.g., free N/S inclination drift). The relaxation of N/S station-keeping requirements gives, for example, a payload increase capability of approximately 25 kg or a power increase of approximately 450 w. Two specific examples of probable reuse of OTS, or growth versions of it, for mission other than the ECS or Maritime systems are 1) TV broadcast, and 2) small terminal communications systems.

Western Europe has a well-developed infrastructure of networks for broadcasting television. In fact, these

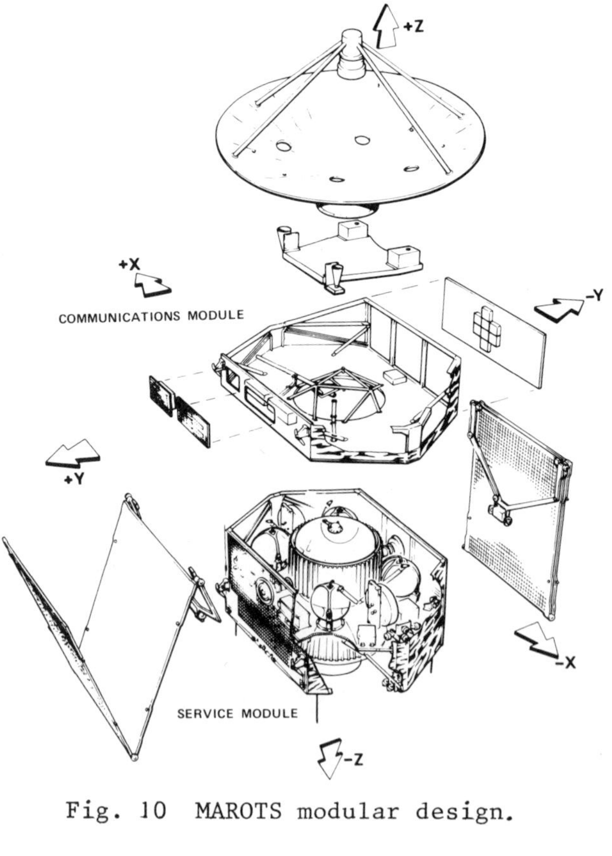

Fig. 10 MAROTS modular design.

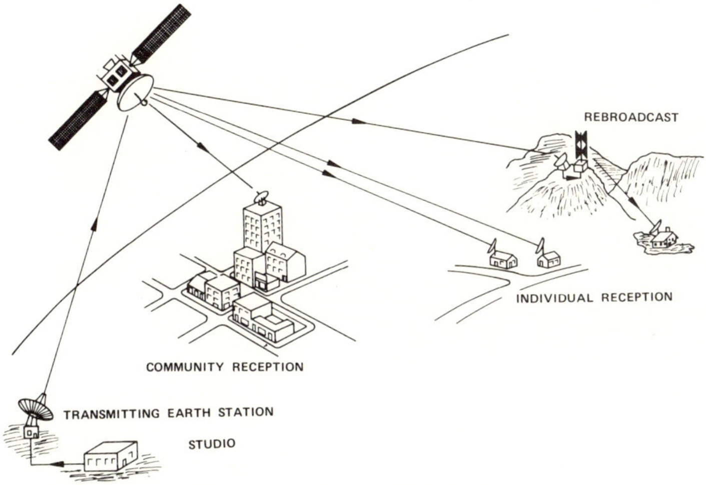

Fig. 11 TV broadcast system at 12 GHz.

networks utilize most of the vhf and uhf frequency bands allocated to the broadcast service in Region I. The frequency resources for over-the-air broadcast of television programs may be exhausted before the end of this decade. Further extension of the number of TV programs will require that new broadcast media become available. The frequency band 11.7 to 12.5 GHz recently was allocated to both satellite and terrestrial broadcast services. On the ground, the design of a 12-GHz broadcast network is complicated by the shadowing effects of buildings in urban and suburban areas and the topography of rural areas. The use of satellites to relay broadcast transmissions offers the unique possibility of covering the whole of a country with individual direct reception of the TV programs or via community cable networks (see Fig. 11).

A growth version of OTS for TV broadcast, launched by Delta 3914, would lead to useful payload capacities, ranging from 120 kg and 1000 w to 70 kg and 2000 w, which certainly are quite adequate for preoperational systems.

In the field of small Earth terminal systems, there is a whole array of future applications in sight: thin route traffic, facsimile, data exchanges, conference networks, electronic mail, etc. A particular mission presently is the subject of intensive work in Europe: communications with North Sea oil and gas platforms (see Fig. 12). Off-shore radio service (hf), line of sight (shf), troposcatter links (uhf), and submarine cables all have significant limitations. In view of the large number of installations foreseen, automation of

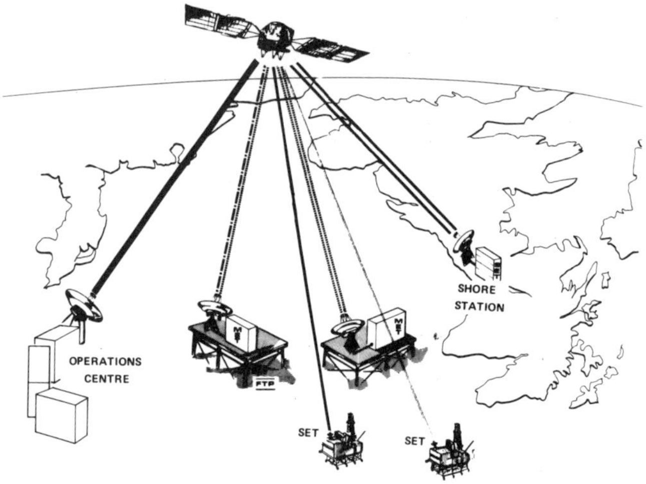

Fig. 12 North Sea oil platform data satellite system.

production is becoming very desirable, and only satellite systems could insure the flexibility and continuity of operations necessary.

VIII. Conclusion

In view of the sizeable development work now undertaken in Europe in the field of communication satellites such as the ESA technology program and the OTS and MAROTS experimental systems, European industry soon should reach the proper level of technical experience needed to embark into commercial space ventures. Indeed, repeated flights of OTS and its derivatives should insure that very high performances and reliability will be achieved for the operational vehicles.

Furthermore, as is being proposed, if several geostationary systems make use of the same basic vehicle in future, sufficiently high production rates will be reached, with consequent decrease of costs, which should then become extremely competitive. Unsurpassed technical excellence and reasonable rate of manufacturing are indeed two very basic ingredients for successful commercial space applications.

References

[1]Collette, R.C., "The European Satellite Program," AIAA Progress in Astronautics and Aeronautics: Communications Satellite Systems, Vol. 32, edited by P.L. Bargellini, The MIT Press, Cambridge, Mass., 1974, pp. 213-230.

[2]Vandenkerckhove, J.A., "The ESRO MAROTS Programme," Conference on Maritime Satellites, British Interplanetary Society, April 18, 1974, London.

DEFENSE SATELLITE COMMUNICATIONS SYSTEM

Gilbert E. LaVean*

Defense Communications Engineering Center, Reston, Va.

Abstract

The Phase II Defense Satellite Communications System (DSCS), which employs synchronous communications satellites, is capable of providing flexible and readily adaptable wide-band service to any location on the surface of Earth between 70° N and S lat. A major goal of the Defense Communications Agency (DCA) is to capitalize on the unique characteristics of communications satellites to furnish high-quality (primarily digital) communications service to a large variety of users. The Phase II DSCS will have several distinct periods or stages, each providing different communications capabilities. In the first period, Stage 1a, Phase II provides a point-to-point operational capability. In Stage 1b, the Phase II system will operate in the FDMA mode to provide a multipoint communications capability and the code division multiple access (CDMA) mode to provide point-to-point protected communications. Stage 1c will begin when PCM, TDM, PSK, and later coding equipment is introduced into the system and most of the analog system is converted to a digital system. FDMA will be used primarily to access the satellite until TDMA is introduced in approximately 1980.

I. Introduction

The Defense Communications System (DCS) must continue to use a mix of transmission media in order to provide the reliability, availability, and survivability needed to satisfy the critical U. S. command and control communications requirements. Satellite communications in general and the Defense Satellite Communications System (DSCS) in particular represent only one

---

Presented as Paper 74-457 at the AIAA 5th Communications Satellite Systems Conference, Los Angeles, Calif., April 22-24, 1974.

*Deputy Chief, Satellite Communications Engineering Division, Defense Communications Agency.

of these media. Thus, the portion of the overall requirements to be satisfied by the DSCS must be chosen carefully so that the unique characteristics of the DSCS are used optimally in fulfilling the overall mission. The DSCS may then assume a principal, supporting, or backup role appropriate to the particular communications requirement under consideration. This paper describes the DSCS, the satellite communications system DCA has developed to satisfy critical Department of Defense (DoD) requirements.

II. Role of the DSCS

Evaluation criteria have been developed to filter user-to-user requirements to determine those best satisfied by satellites. These evaluation criteria have been utilized to separate potential satellite requirements into two categories: those which can capitalize on the unique capabilities of satellites (e.g., flexible service to military operations), and those of a secondary nature for which satellites must qualify competitively in terms of cost, performance, and versatility advantages (e.g., DCS trunking). The DCS applications that capitalize on unique satellite capabilities (category 1) are as follows:

1) Provision of adaptable and flexible communications to support operational military needs. Included are the applications for extending the DCS rapidly into remote areas during emergency situations, expanding channel capacities, rapidly restoring communications between locations whose facilities have been damaged or destroyed, and servicing mobile and rapidly transportable terminals (particularly those beyond line-of-sight).

2) Provision of access to headquarters isolated by distance, lack of commercial communications, or political actions or pressures. Effective single-hop, long-haul communications for direct point-to-point service over distances up to 7000 miles can be provided by satellites.

3) Provision of wideband channels that can support high-data-rate signals. This application satisfies military intelligence missions and secure voice service that will range from a few kilobits to tens of megabits per second.

Category 2 applications of the DSCS require detailed tradeoff analysis considering existing terrestrial facilities, the satellite terminal locations established under category 1, and the cost of entirely new facilities. This analysis should therefore be made after the category 1 analysis is completed.

A detailed requirements analysis was performed and has resulted in a categorization of requirements and the definition of the desired characteristics and attributes of the DSCS. These results have been translated into design goals and have established the criteria for designing the future DSCS. They are also used as standards to evaluate the existing and soon-to-be-implemented DSCS.

III. System Development

The Secretary of Defense established the Defense Satellite Communications Program (DSCP) in 1962. The purpose of the DSCP was to provide long-haul links between fixed, transportable, or shipboard terminals. The Defense Communications Agency (DCA) was assigned overall responsibility for the DSCP. The DSCP has consisted of two phases. DSCP Phase I was an outgrowth of Army participation in NASA's Project Syncom and an initial system concept calling for medium-altitude satellite orbits. By participating in the SYNCOM Program, the Army gained a nucleus of ground terminals that were originally developed for Advent and operating experience applicable in the DSCP Phase I. In June 1966, the first seven Defense Satellite Communications System (DSCS) Phase I satellites were orbited successfully by a Titan IIIC booster.

The period immediately after the first launch was used for system testing between deployed terminals. In particular, a two-channel full duplex link capability between two AN/MSC-46 terminals was demonstrated. In July 1967, the Pacific network was placed in operational status, and integration of DSCS into the DCS was begun. Thus designed as an R&D system, the DSCS almost immediately became an operational system.

Four launches placed 26 operational DSCS Phase I satellites into near equatorial orbits at a near synchronous altitude of approximately 18,300 naut miles. The satellites used a conservative design and did not include a spacecraft command capability. They were placed in a drifting orbit that did not require commands and were thus immune to accidental or intentional capture. The technical support of the launching phase was provided by the Air Force Satellite Control Facility (AFSCF). Since the completion of the launches, the SCF has provided orbital tracking data and telemetry monitoring to determine satellite health.

Eleven of the initial 26 satellites are still operational. The satellites, as viewed from Earth, drift from west to east at about 30° long/day. A single satellite stays within view of a particular Earth terminal about 4 1/2 days. A varying distribution of satellites encircling Earth exists, since each satellite was released from the dispenser at a slightly different orbital velocity. The differential velocities are arranged to reduce "bunching" of satellites, thus enhancing satellite availability. Originally designed for a mean time to failure (MTTF) of 1 1/2 yr (with a goal of 3 yr), the satellites have exceeded this goal. The satellites are simple spin-stabilized spacecraft and weigh approximately 100 lb. Figure 1 is a picture of the Philco-developed satellite. The transponder of the DSCS Phase I satellite is a double-frequency-conversion, hard-limiting repeater operating in the 7- to 8-GHz range. The EIRP has a design minimum of 7 dbw.

Experience with the Phase I DSCS demonstrated that satellite communications can satisfy worldwide communications needs of DoD. Therefore, in June 1968, DoD announced its decision to acquire six satellites of a new design and additional Earth terminals as the second phase of the DSCS. The Phase II satellites (Fig. 2) are also spin-stabilized and have two steerable narrow-beam (NB) (2.5°) and one Earth-coverage (EC) (17°) antenna on a despun platform. The effective design-radiated powers per antenna are 40 dbw (NB) and 28 dbw (EC), respectively. The antennas can be cross-strapped to permit any antenna to receive and transmit via any other. The various transponders and cross-straps have different bandwidths selected on the basis of expected needs. The satellites have tunnel diode inputs and TWT output amplifiers.

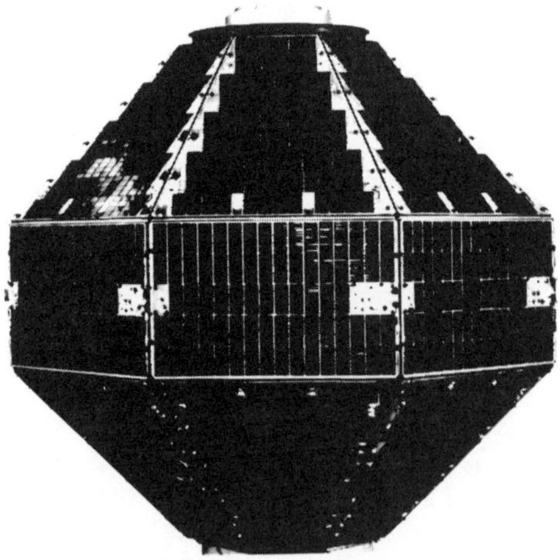

Fig. 1 Phase I satellite.

The first launch of two satellites from the initial procurement of six occurred on Nov. 2, 1971. The Titan IIIC was used as the launch vehicle for lifting the two satellites into synchronous orbit. Both spacecraft

Fig. 2 Phase II satellite.

experienced anomalies shortly after separation from the launch vehicle. The no. 2 spacecraft incurred power and stability problems that resulted in a failure after a short operational period. The no. 1 spacecraft, experiencing stability and telemetry problems, provided limited operational capability for approximately 1 1/2 yr but failed due to power distribution problems. The initial problems were corrected in subsequent spacecraft, and two additional spacecraft were launched on Dec. 13, 1973. These satellites have undergone extensive testing and are now operational. At least three of these satellites are needed to supply a global coverage capability that provides connectivity among the geographical locations dictated by military requirements. The concept of sparing for this three-operational satellite system is still under consideration.

The Phase II DSCS will have several distinct periods or stages, each providing different communications capabilities. In the first period, now in process, Stage 1a, Phase II operates in the FDMA and code division multiple access (CDMA) modes and provides a point-to-point operational capability.

In Stage 1a, the point-to-point terminal linking arrangement results in an operational system similar to that of Phase I. However, in this case, many links will be handled simultaneously by each satellite.

In Stage 1b, to be initiated this year, the Phase II system will operate in the FDMA mode to provide a multipoint network satellite communications capability and the code division multiple access (CDMA) mode to provide point-to-point protected (i.e., jam-resistant) communications. New FM and

PSK modems and FDM multiplexers will be purchased to provide the capacity required to fulfill the Stage 1b requirement. Stage 1b will use multiple receive and transmit carriers at selected locations (nodes) to support links in a multipoint network operating via a single satellite. Traffic will range from 12 to 48 voice channels on the AN/MSC-46 pairs to three voice channels between the AN/MSC-46 and AN/TSC-54 terminals. A few links will be used to provide wideband digital traffic between selected areas in support of wideband data and digitized voice requirements.

Stage 1c will begin when the newly developed digital equipment (PCM, TDM, PSK, and, later, coding equipment) is introduced into the system. FDMA will be used to access the satellite until TDMA is introduced in approximately 1980.

The introduction of digital operation will result in increased system capacity through a more efficient combination of different types of traffic (voice, TTY, secure voice, data, and wideband data) than is possible in an analog system, and through the use of coding, which allows more efficient use of satellite power. As the time division multiplex (TDM) pulse code modulation (PCM) and other analog-to-digital conversion equipment becomes available, the system will phase from an analog system to a hybrid, and finally into a digital system. Protected traffic will continue to be provided using pseudo-noise (SSMA) equipment. However, the CDMA (SSMA) modems employed will be of an advanced design.

The introduction of TDMA equipment into the DSCS will signal the beginning of Stage 2, Phase II DSCS. With the TDMA complementing the other digital equipment, more effective use of the Phase II satellites and significantly greater system capability are expected to be achieved. All links in a TDMA network will be established on a time base (rather than frequency), allowing each terminal to function as a multiple link or nodal terminal. A full complement of modified Phase I and newly procured terminals will be available for use during this stage. The requirements analysis has indicated the DSCS in this time frame will serve a broad spectrum of terminals, ranging from a few feet in diameter to large 60-ft terminals. Therefore, several access modes will be necessary, including a demand access capability for low-duty-cycle users, FDMA for certain applications, and TDMA for high-capacity, well-disciplined networks. The requirements analysis also shows the need for effective utilization of these resources to satisfy the dynamic demands of a military communication satellite system and the resulting requirement for an automatic adaptive control system.

IV. Phase II Subsystem Characteristics

The Phase II DSCS is well defined and will be the operational DSCS for the remainder of this decade; therefore, it is pertinent to discuss in some detail the subsystems of the Phase II DSCS. The DCS normally interfaces with the DSCS at a DCS technical control facility and provides all of the signal conditioning to transmit information over the DSCS. Figure 3 is a block diagram of the Phase II DSCS and shows the relationship between the three DSCS subsystems: 1) space subsystem, 2) Earth subsystem, and 3) control subsystem.

Space Subsystem

The Phase II satellite transponder consists of a multichannel repeater with crosslinked channels, a receive and a transmit EC antenna, and two steerable NB antennas, each capable of receiving and transmitting simultaneously.

This arrangement will provide four different modes of operation: 1) Earth coverage to Earth coverage (EC-EC),

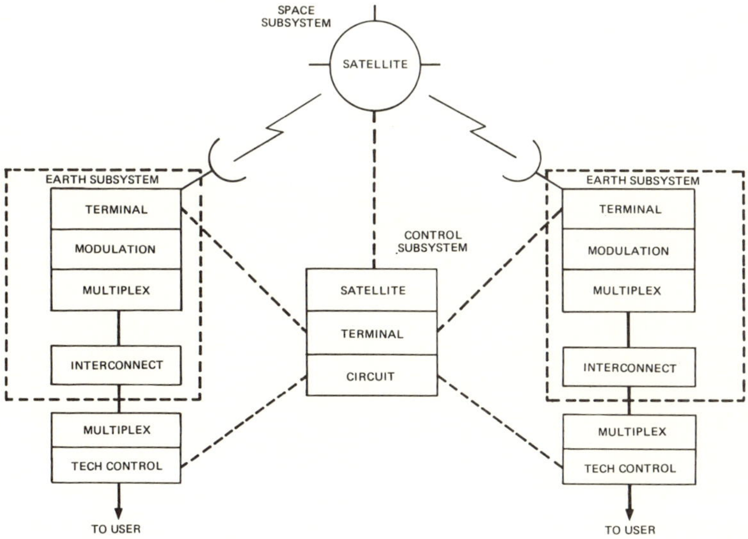

Fig. 3 Phase II block diagram.

2) Earth coverage to narrow beam (EC-NB), 3) narrow beam to Earth coverage (NB-EC), and 4) narrow beam to narrow beam (NB-NB). The basic interconnection for these four modes is shown in Fig. 4.

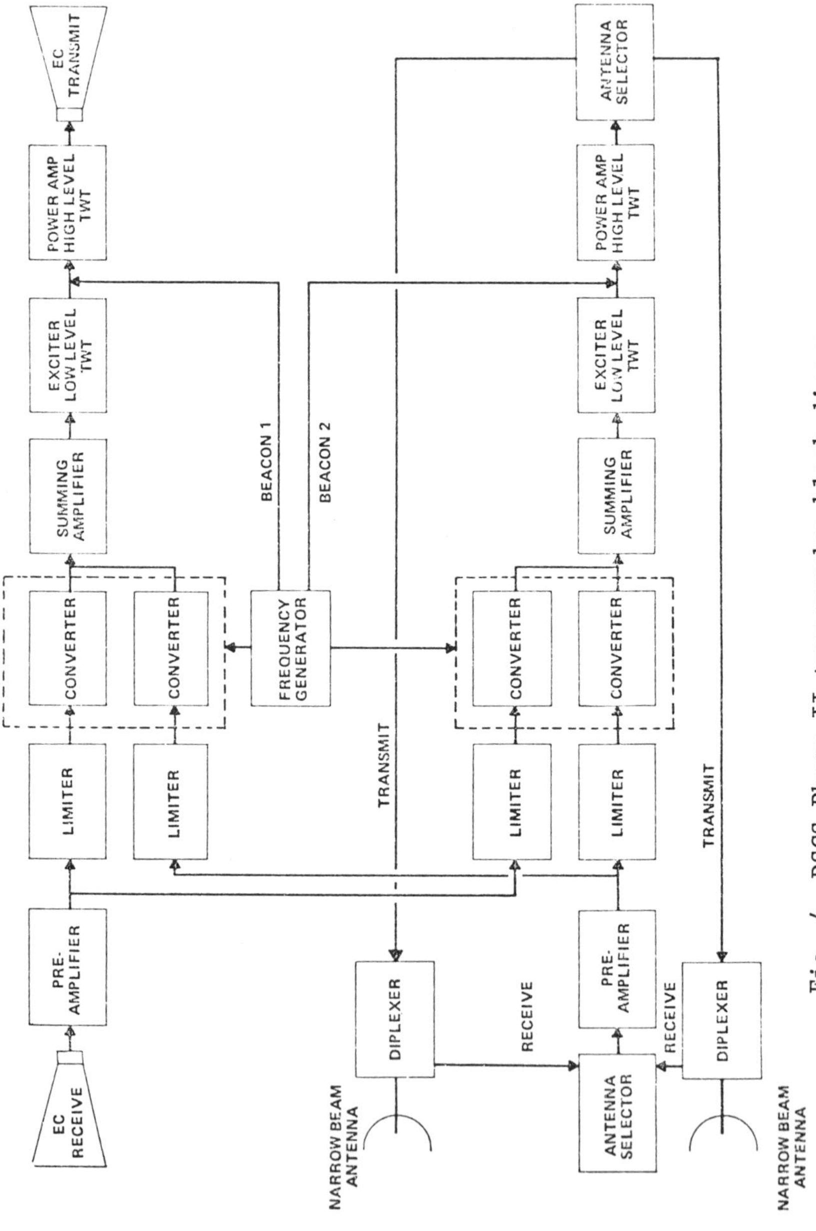

Fig. 4 DSCS Phase II transponder block diagram.

The uplink frequency transmitted by a Phase II Earth terminal determines which of the four modes will be used. (This assumes that an Earth terminal planning to use either the NB-NB or NB-EC channels is within the geographical area covered by one of the NB antenna patterns.) The received signal is amplified and retransmitted in the 7250- to 7750-MHz band. A simplified frequency diagram presented in Fig. 5 relates the various channel modes, together with their related frequency translations and satellite antennas. Both the EC-EC and NB-EC channels share the output power of a 20-w TWT amplifier, using the EC antenna. Likewise, the NB-NB and EC-NB channels are combined and transmitted via a second 20-w TWT, using either one or both of the narrow coverage antennas. The bandwidth of each channel is presented in Fig. 5 and represents 410 MHz of usable bandwidth. The satellite EIRP is 28 dbw for the EC mode and 43 dbw for a NB antenna (or 40 dbw each if two are used). This allows a marked increase in capacity and flexibility compared with the 7 dbw of the Phase I satellite. The Phase II synchronous satellites will be maintained within $\pm 3^{\circ}$ of their designated orbital positions; however, they can be repositioned at least once during their operational life to any other equatorial point at a rate of 15°/day.

All active components within the transponder are redundant. The selection of the active components and NB antenna switching and steering is done by ground command. To achieve maximum in-orbit usage of the Phase II satellite, all onboard systems have been sized to provide a minimum 5-yr operational lifetime.

All transmitting antennas are left-hand circularly polarized, whereas the receiving antennas are right-hand circularly polarized. The two NB antennas are capable of being steered independently $\pm 10^{\circ}$ from each of two orthogonal directions, and the half-power beamwidth covers

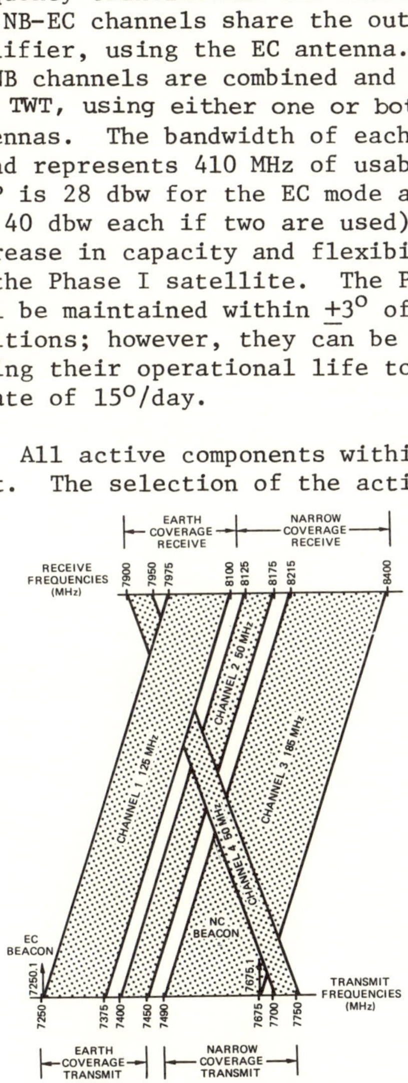

Fig. 5 DSCS Phase II satellite frequency translation plan.

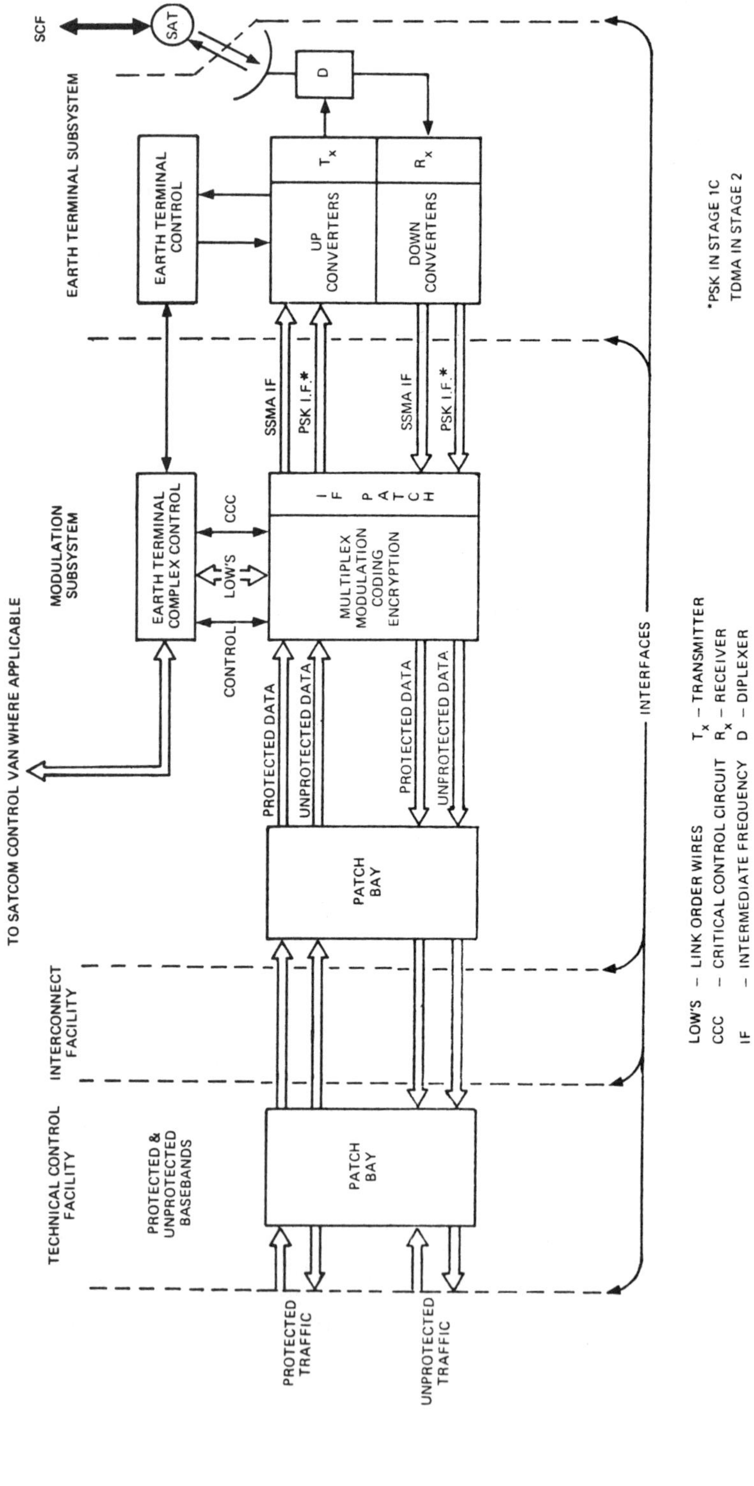

Fig. 6 Functional block diagram of the Stage 1c, Phase II DSCS.

an area approximately 750 miles in diameter near the subsatellite point. The EC antennas (transmit and receiver horns) provide coverage to approximately one-third of Earth's surface. Electrical power is supplied to the satellite for eclipse operation.

Earth Subsystem

The DSCS consists of many different types of satellite Earth terminals (Table 1). Most of these terminals fall into the rf tunable class of terminal (frequency selection accomplished at rf) with a standard 70-MHz IF frequency. This basic configuration is shown in Fig. 6. The older terminals such as the AN/FSC-9, AN/MSC-46, AN/TSC-54 have been modified to include new up and down converters with frequency synthesizers and reliability improvements. The newer DSCS terminals such as AN/MSC-60 (HT) (see Fig. 7) and AN/MSC-61 (MT) include many special reliability and control features. The contingency and Army tactical terminal (i.e., AN/TSC-86, see Fig. 8, AN/TSC-85, and AN/MSC-59) are small, rapidly deployable terminals. The AN/SSC-6 and AN/WSC-2 are shipborne terminals, and the AN/ASC-18 is planned to be used on the AABNCP. The special terminals, including the Diplomatic telecommunications terminals (DTS) and NSA, are for specific applications. The modems and baseband equipment are generally conventional in nature, but with added flexibility to achieve the best performance to utilize the capacity of the satellite effectively. Nonconventional modems (SSMA) for ECCM applications are being improved to give increased protected performance in the presence of intentional or unintentional interference.

Fig. 7 AN/MSC-60 (HT).

Control Subsystems

The current control subsystem for the DSCS is primarily manual in nature, with the resulting inherent slow responsiveness. The control subsystem is in the process of being improved with the goal of a fully automatic adaptive control system by the mid-1980 time frame. The current control system and the planned improvements are discussed in detail in Ref. 1.

Table 1 Characteristics of existing and planned Earth terminals

Terminal type	Antenna size, ft	G/T, db	EIRP, dbm	Power amplifier output, kw	Instanteous bandwidth, MHz		IF interface frequency bandwidth	
					P.A.	LNA	TX	ICX
AN/MSC-60(HT)	60	39	124	TWT(5)	500	500	70/40	70/40
			127(AJ)	TWTs(10)(AJ)				
AN/MSC-61(MT)	40	34	120	TWT(5)	500	500	70/40	70/40
			123(AJ)	TWTs(10)(AJ)				
AN/FSC-9	60	37	128	KLY(12.5)	125	500	70/40	70/40
AN/MSC-46	40	34	117	TWT(3)	500	500	70/40	70/40
			123(AJ)	KLY(12.5)(AJ)	125			
SC-2(DTS)	35	31	118	KLY(2)	50	50	70/40	70/40
NSA	33	31	117	KLY(3)	50	500	70/40	70/40
AN/TSC-54	18 (Eo.)	26.5	117	KLY(5)	50	500	70/40	70/40
AN/TSC-86	8	18	103	KLY(1)	50	500	70/40	70/40
AN/TSC-86	20 (Nom.)	27	111	KLY(1)	50	500	70/40	70/40
AN/TSC-85(V)	8	17	100	KLY(0.5)	50	500	70/10	70/10
AN/MSC-59	8	17	93.5	TWT(0.1)	500	500	70/10	70/10
AN/SSC-6	6	15	110	KLY(12.5)	125	500	70/40	70/40
AN/WSC-2	8	18	101	KLY(3)	500	500	70/40	70/40
			107(AJ)		50(AJ)			
AN/ASC-18	2.75	7	104	KLY(12.5)	125	500	70/40	70/40
SC-1A(DTS)	18	26	104	KLY(0.5)	20	50	70/20	70/20
S-1A(DTS)	6	17.5	94	KLY(0.5)	20	50	70/20	70/20
Special DSCS transportable	6.3	16	99	KLY(0.5)	50	50	70/40	70/40

V. Summary

Fig. 8 AN/TSC-86 (LT) SHF satellite communications ground terminal.

The DSCS is part of a continuing effort to satisfy the unique requirements of the U. S. Government for control of military forces on a worldwide basis. The DSCS has evolved from an R&D phase in the 1960's to provide a significant operational capability for the 1970's. During the latter years of the 1970's, system improvements will be implemented to improve the responsiveness and performance of the DSCS.

References

[1] Drummond, R. L., "Network Control and Coordination within the U. S. Defense Satellite Communications System (DSCS)," AIAA Paper 74-477, April 1974, Los Angeles, Calif.

U. S. NAVY FLEET SATELLITE COMMUNICATIONS

N. L. Wardle*

Naval Research Laboratory, Washington, D. C.

Abstract

To reduce the impact of the vagaries of high-frequency radio and make fleet communications equal to the demands of modern Naval tactics, an extensive satellite communications system is being built. Operation is largely in the military uhf band (225-400 MHz) and uses channelized translating repeaters. Key among the reasons for use of uhf are timeliness and terminal cost and size. The channelized repeater was selected to simplify equipment and operation. A constellation of four geosynchronous satellites will accommodate Navy ships, submarines, and aircraft and Air Force aircraft. Launch of the satellites presently is scheduled to begin in late 1976.

Background

The history of Navy fleet communications began on Jan. 5, 1776 when the Naval Committee of the Continental Congress instructed Commodore Esek Hopkins, the first Navy Commander-in-Chief, "to give out to the Commander of every ship such signals and other marks and distinctions as may be necessary

---

Presented as Paper 74-458 at the AIAA 5th Communications Satellite Systems Conference, Los Angeles, Calif., April 22-24, 1974.

*Lieutenant Commander, U. S. Navy.

for their direction." These directions consisted of a few tactical maneuvers and battle orders. Manipulation of sails or the positions from which pennants, the ensign, or other national flags were flown provided combat direction in 1776. Lights and flares served a similar purpose at night.

Fleet communications, or, the in-vogue term, telecommunications, meaning communications to, from, and among users in Naval ships on the high seas, have been marked by repeated instances of unsatisfactory performance; communications problems have hindered tactics. The first board to investigate a fleet communications problem was convened by President Theodore Roosevelt in 1906. Partially as a result of that board, development of wireless telegraphy for the U. S. Naval Fleet was pushed energetically. This was the greatest leap forward, before or since, for fleet communications, and it was said justly at the time that in few fields of national endeavor were public funds invested more wisely.

In spite of the steady progress made in the use of radio and other media, communication has persisted as a limitation as operational tactics have evolved and become ever more rapid-paced. Many unsatisfactory performances have been documented in very clear terms by the victimized Naval Commanders. For example, in 1923 Admiral Hilliary P. Jones, Commander-in-Chief, U. S. Fleet, after a major fleet exercise, stated "The Commander-in-Chief considers that rapid communications within the fleet, between the fleet and its bases, and between the fleet and the Navy Department is neither satisfactory nor reliable. The maneuvers of the United States Fleet off Panama during both 1922 and 1923 have demonstrated this fact in the most profound manner; and although there has been some improvement during the past two years, the Commander-in-Chief believes that the subject of communications still warrants the serious and immediate consideration of the Department."

Admiral Jones' report resulted in the establishment of another board of investigation, which found, among other things, that "too much was being attempted with too little by poorly qualified people." Since 1923, the deficiencies have been escalated. Instead of the three transmitters and eight receivers carried by the USS California, a capital ship of 1923, one of today's aircraft carriers has 56 transmitters and 84 receivers, but still our communications "is neither satisfactory nor reliable."

Over the years, many boards and committees have been convened to diagnose and prescribe for the correcting of Naval communications problems. The most recent is the CIACT or CNO Industrial Advisory Committee on Telecommunications. It was no surprise to many that the CIACT initial report in 1972 found essentially the same problems that Admiral Hilliary Jones saw in 1923. There has been, somehow, a monotonous sameness to the major points of the diagnosis down through the years. The question is now, will the next board have to come to the same conclusions? There are grounds to expect that it will not. The most significant single innovation, which the Navy is now in the process of introducing, is the use of satellite relay as the primary mode of over-the-horizon communication with the fleet. The Navy is today engaged in building the uhf Fleet Satellite Communications System, or FLTSATCOM, which will give many improvements, one of the greatest being freedom from the present nearly total dependence on the ionospheric vagary-ridden high-frequency band for over-the-horizon communications.

The Navy was the pioneer exploiting space technology for communications in using passive relays that were more predictable than the ionosphere to redirect radio waves to a receiver beyond the horizon. Following a research program, the Navy established in 1959 a two-way operational teletypewriter link between Honolulu and Washington, D. C. by reflecting 400-MHz waves off the moon. In 1962, the moon-bounce system was operated extensively from a ship at sea to a shore terminal. By 1964, the Navy was participating in the SYNCOM project, which, in 1965, involved the first active satellite relay for ships at sea. Following SYNCOM, which operated at 7 GHz up and 2 GHz down, the Navy participated in both the Initial Defense Communications Satellite Program and in the Joint Tactical Satellite Communications Program (TACSAT), which provided both Atlantic and Pacific fleets with essential operational communications.

FLTSATCOM Conceptual Design

In 1970, with technical feedback and strong operational endorsement from the fleet on the TACSATCOM experiment, the Navy proposed the uhf FLTSATCOM system. The uhf band was selected not for technical reasons as much as for the omnipotent twins, timeliness and economy. The relatively low-cost development of the shipboard and airborne transceivers is permitting the Navy to have nearly the entire fleet equipped by the time the satellites can be developed, built, and orbited. Selection of uhf as the dominant band for FLTSATCOM should not be construed as Navy rejection of the shf band. On the contrary, the two bands are seen as being complementary rather than competitive. Nor is uhf seen as the interim solution, with shf as the exclusive ultimate.

Several of the salient properties of the two frequency bands are shown in Fig. 1. A uhf transceiver is packaged conveniently to occupy a few inches in a standard 19-in. rack, whereas an shf terminal fills several racks. Especially in aircraft, submarines, and smaller ships, size is very important. The cost disparity is from $100,000 plus for an installed uhf terminal to a couple of million for shf. From experiences with uhf terminals at sea in the TACSATCOM experiment and with shf terminals operating in the IDCSP and SYNCOM, the

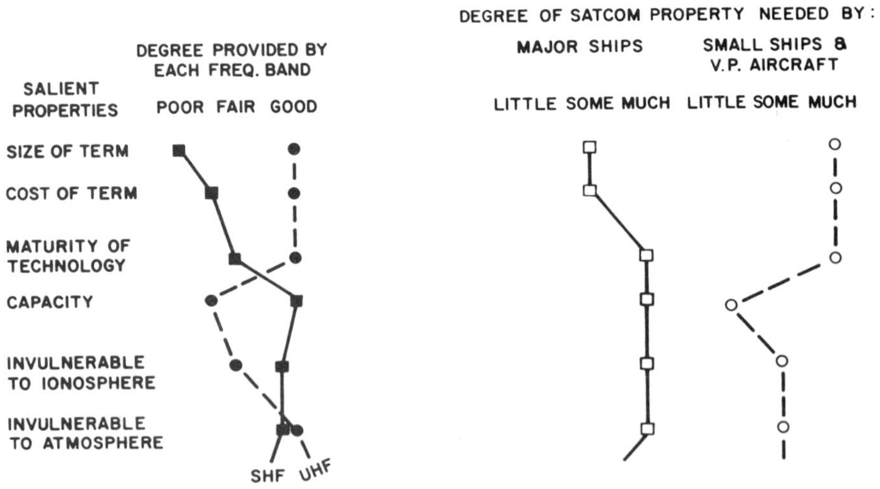

Fig. 1 Comparison of uhf and shf for fleet use.

conclusion has been reached that shf shipboard terminal hardware is much less mature than uhf . Although, in a theoretical sense, system capacity in bits per second can be shown to be essentially independent of frequency, if one designs the two systems each to best exploit its natural advantages, the uhf system will have much lower capacity. The lower capacity is not a limitation for aircraft, nor is it expected to be for small ships, but large ships can be expected to need capacity levels best attained at shf. The vulnerability of uhf SATCOM to ionospheric vagaries, a problem occurring much less than outages of high-frequency sky-wave, was unanticipated by most satellite advocates, although fully predictable by radio astronomers. Fortunately, scintillation is a serious problem only near the geomagnetic equator and poles, near the equinox, and at night. The only sense in which it can be said, as represented by Fig. 1, that small ships and aircraft are more tolerant of ionospheric scintillation outages is that they seldom carry tactical commanders, and so only their own readiness is impaired by interruptions of their communications. The last category, effect of weather, of course shows very little difference between uhf and shf but is the strongest reason for holding back on employment of yet higher frequencies for SATCOM in the fleet.

Although the choice of uhf resulted from a positive lesson learned in the TACSATCOM experiment, the choice of a channelized frequency translating repeater in lieu of the wideband repeaters used in TACSAT and LES 6 was a result of a negative lesson. Control of intermodulation product interference and power robbing proved to be an unmanageable problem in the tactical mobile platform. These problems are avoided in FLTSATCOM by providing separate channels for each access at any instant in time.

The selection of the bandwidth of the individual channels followed from the following five considerations:

1) The users of the uhf spectrum are planning to move eventually to a 25-kHz center-to-center allocation plan, making 25 kHz a convenient upper bound on bandwidth.

2) The futility of using wide instantaneous bandwidth for processing gain with mobile uhf terminals is clear.

3) Future expansions of data rates up to 9.6 kbit/sec with biphase PSK (Phase Shift Keying) require a linear phase bandwidth of about 20 kHz.

4) Capturing the full satellite power with low uplink power demands that the satellite bandwidth be kept narrow. Easily generated ERP's (Effective Radiated Power) on ships and aircraft are such that decreases below 10 or 20 kHz are no longer rewarding.

5) A much narrower bandwidth was expected to require an additional down-conversion stage to permit practical filtering.

The dominant theme of FLTSATCOM communications for the mobile users, ships, and patrol aircraft in the fleet is illustrated in Fig. 2. There will be no fixed point-to-point service through FLTSATCOM; for that we look to DCS. The ships

COMMUNICATION LINKS

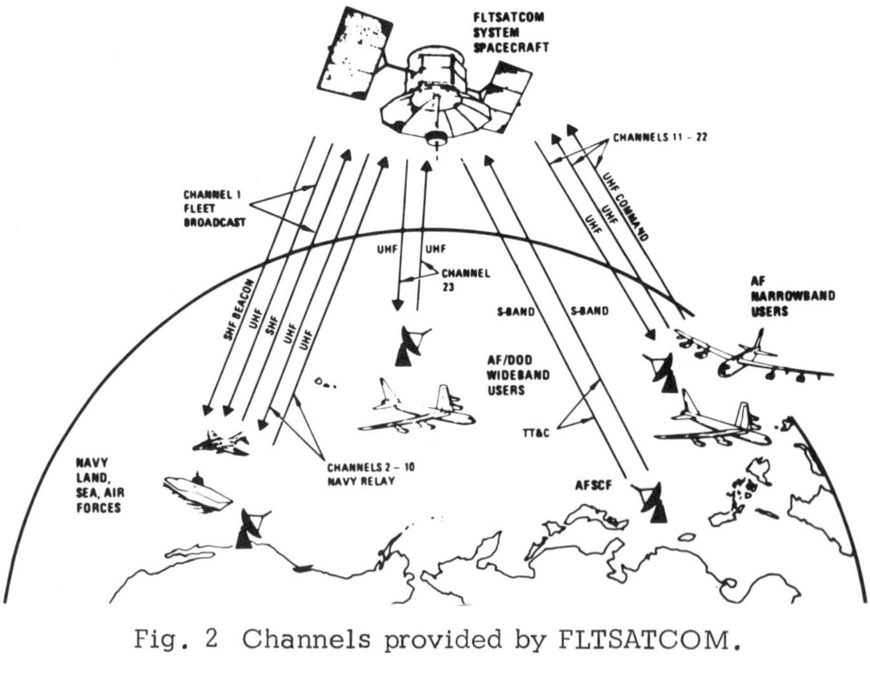

Fig. 2 Channels provided by FLTSATCOM.

and aircraft within the zone of visibility of each satellite will be served by one shore-to-sea fleet broadcast channel using an shf uplink and uhf downlink. The ships and aircraft will be given additional access to one or more of nine two-way or half-duplex channels, each of which will be time-shared by a number of users. The two-way channels are uhf both up and down. Net discipline on most of these channels will be automated to remove human operator delays and assure efficient usage.

At about the time the FLTSATCOM system concepts had progressed to the point now described, the Office of the Secretary of Defense decided to add on each satellite some 13 additional channels of different descriptions for use by the U. S. Air Force in hopes of attaining spacecraft economies of scale and reducing overall Department of Defense costs.

COMMUNICATION SUBSYSTEM

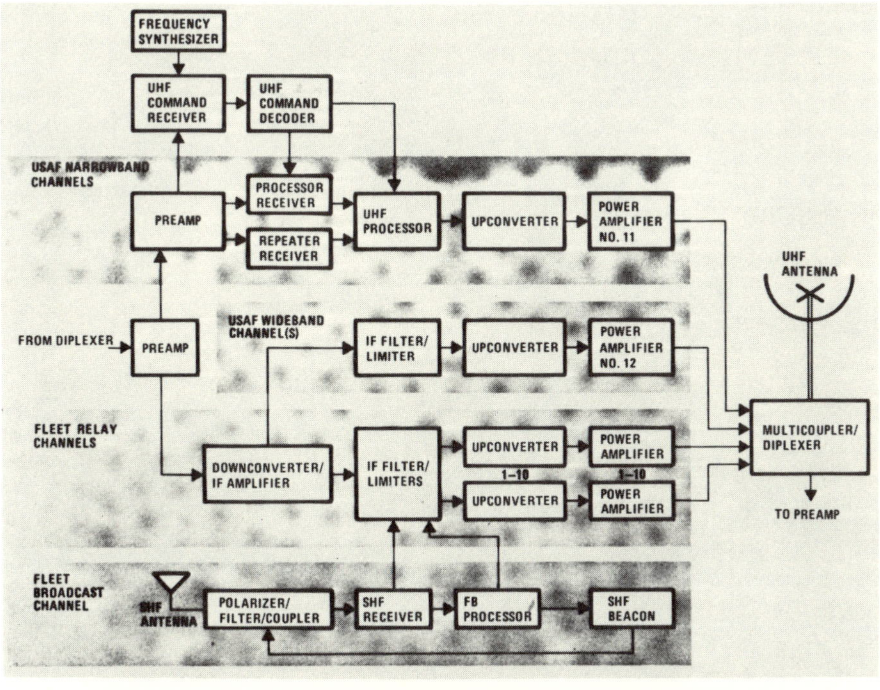

Fig. 3 FLTSATCOM repeater block diagram.

Spacecraft

A top-level block diagram of the communications payload on each satellite is shown in Fig. 3. Note that a common uhf antenna is shown for transmit and receive of all of the uhf channels. The multicoupler with the requirements for low loss, isolation, and suppression of spurious emissions is proving to be one of the most challenging features of the repeater. After proceeding to system-level tests, it was found necessary to employ separate antennas for receive and transmit functions. An alternative of using a smaller number of amplifiers, each of which would amplify several carriers, was rejected early in the design process as being more troublesome and less efficient. A major design modification was the elimination later in the

Fig. 4 FLTSATCOM artist's conception.

design process of many-step level adjustments on the output of the final amplifiers. This was done to save weight and enhance reliability through parts count reduction.

The artist's view of a deployed FLTSATCOM spacecraft, which will be launched singly by Atlas Centaurs beginning in late 1976, is shown in Fig. 4. The spacecraft is being built by TRW Systems Group under a contract let by SAMSO for the Navy. Onboard consumables will be carried for a designed life of five years. Both the 2-x4-m solar panels and the 5-m-diam uhf antenna reflector are folded for launch. The central structure of the three-axis stabilized satellite is, as shown, an octagonal prism 2.5 m across and 1.3 m high. Earth and sun sensors, through the stabilization system, reaction wheels, and solar array drive motors, maintain the antennas pointed at Earth center, the solar array axle parallel with the Earth axis, and the solar cells facing the sun while the satellite maintains geosynchronous station.

Terminals

Now one comes down to the mundane problems of building transceivers that can survive and operate in a warship or tactical aircraft and that are inexpensive enough to be bought for and installed in the entire fleet. The paramount consideration, overshadowing any other performance parameter, is availability: is it working when needed? Allied against availability are the Navy's own set of local shipboard environmental pollutants depicted in Fig. 5. Direct impact of the shipboard environment on the SATCOM transceivers is seen in the two key SATCOM parameters ERP and G/T. The limited available space and power, together with the impracticability of installing high-gain uhf antennas on ships or aircraft, has resulted in a system designed with lower Earth terminal ERP than usual. Furthermore, the same antenna limitation, together with the high equivalent antenna temperature due to RFI, has yielded a lower G/T (antenna gain divided by system noise temperature) than is usual for most SATCOM systems.

The G/T problem is somewhat alleviated in the fleet broadcast receiver system being installed throughout the surface fleet by the use of a predetection antenna combiner in the AN/SSR-1. Instead of incurring the cost and maintenance

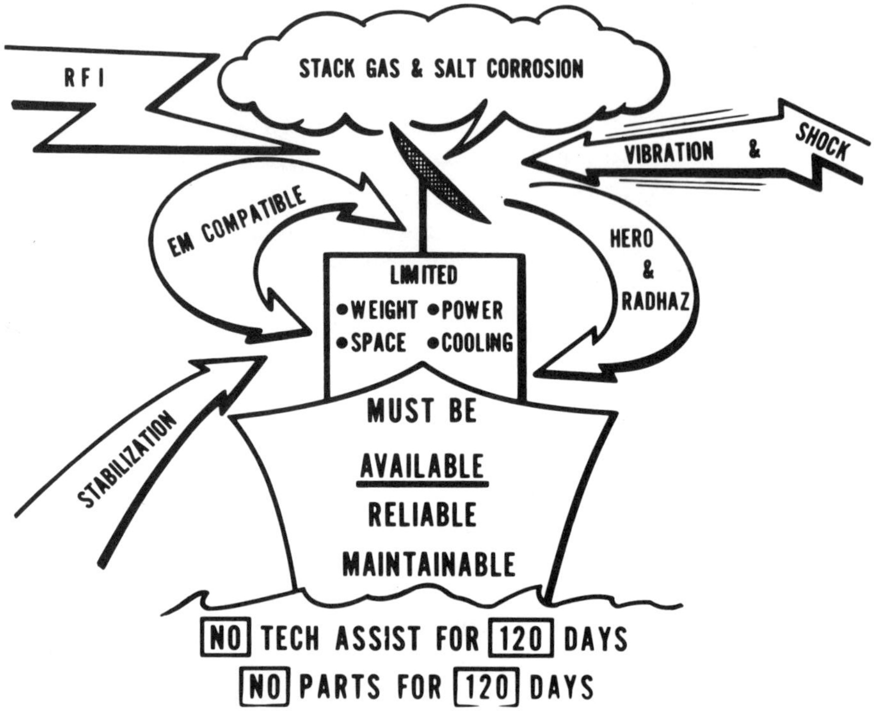

Fig. 5 The shipboard environment.

burden of a mechanically pointed antenna to provide a minimal gain, the IF signals from each of four omnidirectional antennas are fed, in the Motorola-designed and built system, to a combiner. The combiner, in effect, adjusts the phase of the four signals to make them cophasor, weights them according to their respective signal-to-noise ratios, and adds them together to provide several decibels improvement above the gain which any one omnidirectional antenna alone could provide and indeed a few decibels improvement above a switched selection of the best single antenna.

Because we have not yet attained transmit performance on a simple array of omnidirectional antennas equivalent to the receive performance of the AN/SSR-1 receive-only system, we use a more conventional approach for the antenna used aboard ship with the AN/WSC-3 transceiver. The antenna is built by Collins Radio and consists of an array of four crossed dipoles arranged on a mechanically oriented ground plane. The antenna is not stabilized to compensate for roll and pitch but does have servo drive controlled by a gyro input to maintain correct true azimuth and a remote but manual elevation control. The shipboard transceiver that is built by ECI features a fully digital synthesizer for 25-kHz channelization and a built-in modem providing FSK, PSK, FM, and AM modulations.

For Navy aircraft that will be afforded access to the FLTSATCOM system, an existing uhf radio already integrated into the avionics suit will be modified to serve SATCOM as well as line-of-sight uhf functions.

Both the AN/SSR-1 and AN/WSC-3 are currently in production and will be broadly installed throughout the fleet by the time the FLTSATCOM spacecraft are launched. To provide the lesser satellite capacity necessary to permit use of the fleet terminals while their numbers are being built up, uhf satellite access will be leased from COMSAT General, a subsidiary of COMSAT. That spacecraft project is variously known as Marisat, GapFiller, or GapSat. This service is expected to start in the summer of 1975.

In summary, by the end of the decade the Navy expects to see satellite relay become the primary means of over-the-horizon communications for the fleet through the FLTSATCOM system. Furthermore, a new era of more dependable communications to provide control of the fleet will be ushered in, permitting direction of this great nation's fleet through communications that are both satisfactory and reliable.

Index to Contributors to Volume 41